▶ **Applied Communication Research**

▶ **Applied Communication Research**

JUDITH M.
BUDDENBAUM

KATHERINE B.
NOVAK

Iowa State University Press
A Blackwell Science Company

About the Authors

JUDITH M. BUDDENBAUM, Ph.D., is a professor at Colorado State University Department of Journalism and Technical Communication and author of three other Iowa State University Press books.

KATHERINE B. NOVAK, Ph.D., is an assistant professor at Butler University Department of Sociology.

© 2001 Iowa State University Press
A Blackwell Science Company
All rights reserved

Iowa State University Press
2121 South State Avenue, Ames, Iowa 50014

Orders: 1-800-862-6657
Office: 1-515-292-0140
Fax: 1-515-292-3348
Website (secure):   www.isupress.com

Authorization to photocopy items for internal or personal use, or the internal or personal use of specific clients, is granted by Iowa State University Press, provided that the base fee of $.10 per copy is paid directly to the Copyright Clearance Center, 222 Rosewood Drive, Danvers, MA 01923. For those organizations that have been granted a photocopy license by CCC, a separate system of payments has been arranged. The fee code for users of the Transactional Reporting Service is 0-8138-2017-0/2001 $.10.

⊗ Printed on acid-free paper in the United States of America

First edition, 2001

Library of Congress Cataloging-in-Publication Data

Buddenbaum, Judith Mitchell
Applied communication research / by Judith M. Buddenbaum and
Katherine B. Novak.— 1st ed.
p. cm.
Includes bibliographical references and index.
ISBN 0-8138-2017-0 (alk. paper)
1. Communication—Research—Methodology.   I. Novak, Katherine B.
II. Title.
P91.3   .B78 2001
302.2′07′2—dc21   2001003539

The last digit is the print number: 9 8 7 6 5 4 3 2 1

Contents

Preface	xi
Acknowledgments	xiii

Part I. Introduction — 1

  1. Social Science Research — 3
    Ways of Knowing — 5
      Traditional Ways of Knowing — 5
      Science as a Way of Knowing — 7
    The Nature of Social Science Research — 10
    Issues in Scientific Research — 12
      The Role of Theory: Deductive vs. Inductive Reasoning — 12
      The Purpose of Research: Basic vs. Applied — 14
      Data Analysis: Quantitative vs. Qualitative — 14
    Main Points — 15
    Terms to Know — 16
    Questions for Discussion and Review — 16
    Readings for Discussion — 17
    References and Resources — 18

Part II. Getting Started — 21

  2. Fundamental Concerns — 23
    Limiting Factors — 23
      Cost — 23
      Ethics — 24
        (Dealing with Subjects—Working with Sponsors—
          Analyzing Data—Reporting Findings)
    Quality Control — 27
      Reliability — 29
      Validity — 31
    Main Points — 35
    Terms to Know — 36
    Questions for Discussion and Review — 36
    Readings for Discussion — 37
    References and Resources — 37

3. Planning Research Projects — 39
- Preplanning — 39
  - Setting Parameters — 39
    - (Define the Problem—Determine the Purpose—Identify Constraints)
  - Reviewing the Research Literature — 45
  - Refining the Project — 47
  - Deciding How to Proceed — 51
- Planning — 52
  - Choosing the Method — 53
  - Developing the Methodology — 54
    - (Choose the Subjects—Create Measures)
  - Working Out the Logistics — 56
- Doing the Work — 56
- Main Points — 57
- Terms to Know — 58
- Questions for Discussion and Review — 59
- Readings for Discussion — 59
- References and Resources — 60

4. Choosing Subjects — 63
- Probability Sampling — 64
  - Sampling Theory — 65
  - Types of Probability Sampling — 68
    - (Simple Random Sampling—Systematic Random Sampling—Stratified Random Sampling)
- Nonprobability Sampling — 73
  - Types of Nonprobability Sampling — 73
    - (Quota Sampling—Purposive Sampling—Snowball or Network Sampling—Convenience Sampling—Self-selection Sampling)
- Multistage Sampling — 76
- Sample Size — 78
  - Probability Samples — 78
  - Nonprobability Samples — 80
- Validating a Sample — 81
- Main Points — 82
- Terms to Know — 83
- Questions for Discussion and Review — 83
- Readings for Discussion — 85
- References and Resources — 85

5. Creating Measures — 87
- Conceptualization — 87
- Operationalization — 88
  - Manipulation — 88
  - Observation — 89
  - Self-reports — 90
- Levels of Measurement — 91

Nominal Level Measurement	92
Ordinal Level Measurement	92
Interval Level Measurement	93
Ratio Level Measurement	93
Implications	94
Measurement Scales	95
Likert Scales	96
Semantic Differential Scales	97
Thurstone Scales	98
Guttman Scales	98
Feeling Thermometers	99
Ranking Scales	99
Composite Measures	100
Quality Control	104
Reliability	106
(Concurrent Reliability—Alternate Forms Reliability—Split-half Reliability—Test-retest Reliability—Cronbach's Alpha—Intra- and Intercoder Reliability—Test-item Analysis)	
Validity	109
(Content Validity—Criterion Validity—Construct Validity)	
Main Points	111
Terms to Know	112
Questions for Discussion and Review	112
Readings for Discussion	114
References and Resources	114

Part III. Basic Research Methods	**117**
6. *Experiments*	119
Communication Applications	119
The Logic of Experiments	120
Causation	120
Controlling the Manipulation	121
Controlling for Confounds	121
(Elimination—Blocking—Holding Constant)	
The Experimental Setting	123
Ethical Concerns	124
Experimental Designs	124
True Experiments	125
(The Classic Experiment—Posttest-only Control Group Design—Solomon Four-group Design—True Experiment Variants)	
Quasi-experiments	128
(Equivalent Group Pretest-posttest Design—Nonequivalent Control Group Design—Separate-sample Pretest-posttest Design—Time Series—Counterbalanced Designs)	
Pre-experiments	131
Quality Control	132

viii Contents

        Reliability   132
        Internal Validity   133
        External Validity   135
    Data Analysis   135
    Main Points   140
    Terms to Know   143
    Questions for Discussion and Review   143
    Readings for Discussion   145
    References and Resources   146

7. *Surveys*   149
    Communication Applications   150
    Ethical Concerns   151
    True Survey Designs   152
        Cross-cultural Designs   153
        Longitudinal Designs   153
    Quasi-surveys   154
        Nonprobability Sample Surveys   154
        Interview Surveys   155
    Data-collection Techniques   155
        Researcher-administered Techniques   155
            (Telephone—Face to Face)
        Self-administered Techniques   157
            (Group Administration—Mail—Disk by Mail—Internet)
    Preparing the Survey Questionnaire   158
        Questionnaire Length   159
        Types of Questions   159
        Question Wording   160
        Question Order   162
        Helps   163
            (Instructions—Transitions)
        Questionnaire Layout   164
    Quality Control   165
        Reliability   165
        Internal Validity   166
        External Validity   166
    Data Analysis   167
        Quantitative Analysis   168
            (Univariate Statistics—Bivariate Statistics—
              Multivariate Statistics)
        Qualitative Analysis   181
    Main Points   186
    Terms to Know   187
    Questions for Discussion and Review   188
    Readings for Discussion   190
    References and Resources   190

8. *Content Analysis*	193
Communication Applications	194
Ethical Concerns	195
Types of Content Analysis	195
Readability Studies	195
Textual Analysis	196
Choosing Documents	197
Indexes and Databases	198
Periodicity	199
Measurement in Content Analysis	201
Units of Analysis	201
Category Construction	201
Enumeration	202
Code Sheets and Code Books	202
Using Computers for Content Analysis	203
Preparing Documents for Qualitative Analysis	204
Quality Control	205
Reliability	205
Internal Validity	206
External Validity	207
Data Analysis	208
Quantitative Analysis	208
Qualitative Analysis	208
Main Points	209
Terms to Know	211
Questions for Discussion and Review	211
Readings for Discussion	212
References and Resources	214
9. *Focus Groups*	217
Communication Applications	218
Ethical Concerns	218
Planning the Study	219
Participants	219
(Locating Potential Participants—Recruiting Participants—Ensuring Participation)	
Groups	221
(Size—Number and Composition)	
The Moderator	222
(Requisite Skills—Training Moderators—Number of Moderators)	
The Materials	224
(The Moderator's Guide—Supplementary Materials)	
Logistics	225
(Scheduling Sessions—Location—Recording Provisions)	
Alternatives to the True Focus Group	226
The Nominal Group Technique	227
The Delphi Technique	228
Q-methodology	228

Technology-based Techniques	230
Quality Control	230
Reliability	230
Internal Validity	231
External Validity	231
Data Analysis	231
Quantitative Analysis	232
Qualitative Analysis	232
Main Points	234
Terms to Know	236
Questions for Discussion and Review	236
Readings for Discussion	238
References and Resources	239
10. Observational Research	**241**
Communication Applications	242
Ethical Concerns	243
Planning for Observation	245
Choosing Settings and Sites	245
Gaining Access	246
Selecting a Role and Stance	247
(Role—Stance)	
Collecting and Recording Information	249
Types of Information	249
Determining What Information to Collect	250
Techniques for Recording Data	250
(Note Taking—Technological Assists)	
Managing Data	252
(Observation Notes and Field Notes—Supplementary Materials)	
Quality Control	253
Reliability	253
Internal Validity	253
External Validity	254
Data Analysis	255
Quantitative Analysis	255
Qualitative Analysis	255
Main Points	257
Terms to Know	258
Questions for Discussion and Review	258
Readings for Discussion	260
References and Resources	261
Appendix A. American Sociological Association (ASA) Code of Ethics	263
Appendix B. Evaluating Research Reports	271
Appendix C. Politics, Religion, and Media Use Survey 2000	275
Appendix D. Code Book and Code Sheets for Content Analysis	283
Appendix E. Writing Research Reports	293
Glossary	303
Index	321

Preface

BECAUSE COMMUNICATION IS fundamental to all human endeavors, much social science research is explicitly or implicitly communication research. Organizations use, commission, or conduct research to find out what people are thinking and doing. They turn to research to decide how to advertise or promote their products, services, and causes and how to design and implement projects or improve their day-to-day operations, and then they use it again to evaluate their efforts. Journalists report on the latest research findings. Sometimes they even conduct their own research.

In a world where social science research seemingly is everywhere, communication students who know how to evaluate, plan, and carry out research and then use the findings to make good business decisions have an advantage over those without such an understanding. They tend to get better first jobs. Once on the job, they advance more quickly.

But the upper-level undergraduate students and graduate students for whom this book is intended often have a difficult time seeing the practical utility of a course on research methods. Part of the problem stems from the nature of the texts available for classroom use.

Students who come together in communication research courses may be interested in very different kinds of communication. Their careers will very likely take them in very different directions. Yet all must use the same textbook. Many of those books emphasize basic, theoretical research. Most of them concentrate on just one kind of communication. As a result, many students have trouble seeing how material presented in a "media research book" could be used in public relations or advertising or how research in speech communication applies to print.

Research methods are, of course, generally applicable across disciplines. But to help students see that, we have tried to make this book truly interdisciplinary. Because many more students will use or conduct research on the job than will do basic, theoretical research, we have chosen to emphasize applied research—the kind used by and done for businesses and nonprofit agencies in order to solve very real, practical problems. Because we want to help students understand why they should learn about research methods and how they can use research skills regardless of the career path they may follow, we provide sections on communication applications throughout the book. In those sections and in the examples, questions, and lists of readings at the end of the chapters, we have conscientiously tried to incorporate illustrations from each communication subdiscipline: interpersonal and organizational communication, journalism, specialized and technical communication, public relations, advertising, and marketing. In doing so, we have illustrated the relevance both to new media and the traditional print and broadcast media as well as to businesses and nonprofits that are not really a part of the communication industry.

We have also included information on techniques that rarely show up in texts because they are used infrequently by basic researchers even though they are commonly used in applied research. These include the mall intercept survey, readability testing, Q-methodology, and the nominal group and Delphi techniques for decision making. And because organizations today frequently turn first to the Internet to find the information they need, we used that approach. Rather than provide a disk of data for students to use in data-analysis exercises, we send them to the Internet to locate data sets, download data, and analyze them much as they might sometimes do for an employer or client.

As a result of what we have learned from our students and from clients for whom we have conducted applied research, we have also incorporated a number of other features that further distinguish this book from other research texts.

First, we have dispensed with the chapters on statistics that are usually placed at the end of methods texts. Instead, we incorporate information on both quantitative and qualitative data analysis in the methodology chapters. In doing so, we necessarily had to sacrifice some detail. But we think the sacrifice is one worth making. Putting the basics where they are most commonly used while actually doing a research project will, we believe, better serve the many students who find it difficult to figure out what to do when data-analysis techniques are presented separately from the methods themselves. Others who want or need more can find it by consulting works listed in the References and Resources section at the end of each chapter.

Because students often have a hard time understanding that no research is likely to be perfect but that most projects can be improved, we have added a "Quality Control" section to each methodological chapter. We have also spread the discussion of ethical concerns throughout the book. This, we believe, will reduce the tendency of many students to dismiss out-of-hand published studies that are less than perfect. We also believe it will help them choose from among options and make appropriate trade-offs as they work with research firms or plan and conduct their own studies.

To further facilitate learning, we have included several kinds of questions in the list of questions at the end of each chapter. Some of these are "knowledge" questions. They are designed to reinforce points made in a chapter. Others are "application" questions that are designed to help students learn to plan and carry out research. Still others are "discussion" questions. Like the journal articles and short chapters included in the Readings for Discussion list at the end of each chapter, these are intended for classroom use.

In choosing items for the Readings for Discussion list, we looked for ones that would appeal to students. The ones we chose expand the discussion of issues raised in a chapter or illustrate practical applications of methods or techniques. Although each item on the lists was either written by a respected researcher or published in a juried journal, some authors raise points or made methodological decisions that might be questioned. By examining and then discussing these articles, students can learn much about the relationship between methodology and research findings.

Acknowledgments

MANY PEOPLE HAD a hand in helping us write this book. We gratefully acknowledge their contributions.

First, there are our students. From them we learned what they like and don't like about textbooks and what does and doesn't help them learn. In trying to communicate to them our love for and understanding of research, we learned much about research methods and methodology and how to present that information to our students. We thank all of our students, particularly those who read and critiqued parts of this book.

We also acknowledge the contributions of our colleagues and research clients. They told us what they needed in texts and what they wanted students to know and know how to do as the result of taking a research methods class. We are grateful for their help. Because of their advice, this book is much better than it would have been had they not shared so generously of their time and talents.

We thank Russell B. Williams for preparing the illustrations for this book and all of the professionals at Iowa State University Press for their work turning our manuscript into a published work. We especially thank Judi Brown, acquisitions editor at Iowa State University Press, whose patience, humor, and friendship we value greatly. We also thank the American Sociological Association for giving us permission to include portions of the ASA Code of Ethics as Appendix A.

And, of course, we thank our husbands, Warren Buddenbaum and Mark Novak, for their support, love, and understanding even when we were frustrated, grumpy, or preoccupied as we worked on this book.

Part I

Introduction

1

Social Science Research

WHY TAKE A course in communication research methods? The answer is quite simple. It is one of the most practical subjects you can study. Learning how to evaluate and use research findings can help you make wise personal decisions. That knowledge plus some ability to conduct your own research can also make you very attractive to employers and enhance your chances for success, regardless of what career path you choose to pursue.

One can hardly pick up a newspaper or magazine or turn on a television newscast without being bombarded with someone's latest research findings. On just one day, a small daily newspaper reported that research shows children should not watch television before going to bed, while an editorial raised the question of whether using cellular telephones while driving causes or contributes to automobile accidents.

In many cases, the research findings reported by the media seem contradictory:

- A politician releases evidence purporting to show the news media have a liberal bias; another points to findings showing news coverage promotes conservative causes.
- One expert contends that increasing crime rates are linked to violent television programs, movies, and video games. Another contends that, even as violent programming is increasing, crime rates are falling.

In the face of claims and counterclaims, deciding what to believe is difficult at best. Deciding whose evidence to believe and whose to discount virtually demands the ability to perform an independent evaluation of the strengths and weaknesses of the research evidence and of the interpretations and extrapolations made from it. That, in turn, requires some knowledge of basic research methodologies.

The situation is similar in the business world. Every day business people are bombarded with information about how to achieve success for themselves and for the companies they manage. But the advice from the latest instant expert never quite matches that from the previous one. As in everyday life, making informed decisions about the quality of the advice requires the ability to perform an independent analysis of the evidence in order to separate the credible from the incredible, the relevant from that which is unlikely to work in a particular situation.

However, in the business world, evaluating and then putting into practice findings from the best available research is never quite enough. Businesses often need information that directly addresses their own particular situation. Therefore, they often must gather, analyze, and interpret their own information or hire someone to do it for them.

In the communication industry:

- Radio and television stations need to know how many people tune in to their programs in order to set their advertising rates. They want to know whether the on-air personality they are considering hiring or the programs they are considering airing will attract listeners or viewers or send them scurrying to a competing station. They need to know how proposed legislation may affect them. And, of course, they need to know whether the charges made by their critics or the accolades heaped on them by supporters have merit and, if so, how to respond.
- Like their counterparts in broadcasting, publishers need to know what their audience likes and dislikes, what and why they attend to some content and not to others, and what effect, if any, their messages have on individuals and on society. They need evidence to help them decide on editorial policies and to convince businesses to buy advertising.
- Whether they work in print or broadcasting, journalists engaged in in-depth reporting assignments may need to determine the effects of a new policy or the effectiveness of an old one. They may want to investigate alleged instances of discrimination, consistency of sentencing across courts, the cost-effectiveness of services run by government vs. those operated by private businesses, or the reasons students in some schools consistently perform well on standardized tests whereas students in others consistently score below average.
- Public information officers, public relations practitioners, and those who work in advertising need to know whether their messages are reaching their publics and whether their campaigns are effective. They need to understand the image of the organizations they represent and know about the products, services, and causes they promote. In addition, they want to know how to improve that image if need be, how to attract and hold supporters or customers, and how to mobilize people to action when necessary to increase sales or further organization goals.

Even those in fields that at first glance seem far removed from the communication industry need similar information.

- Both businesses and nonprofit agencies need to know how to communicate with their employees, clients, and customers. They may want to know whether monitoring computer use and phone calls is necessary for success or whether it is counterproductive in certain situations. Even those too small to hire an advertising agency may need information to help them decide whether to advertise and, if so, whether money would be better spent on a display ad in the phone book or the local paper or should be used instead for classified ads or spot advertising on radio or cable TV.
- In the manufacturing and high-tech sectors, companies need assurance that the manuals for their products are accurate and that they meet the needs of end users.
- Politicians, government leaders, interest groups, nonprofit agencies, and many businesses need to know which issues people care about, how to increase issue awareness, and how to garner and maintain support.

The list could go on and on, but by now the point should be fairly clear. In all professions, people have some very similar needs for information. Increasingly they are turning to research for answers. Therefore, many businesses look for and value employees who have some understanding of research—what it is, when and why it may be appropriate for answering questions, how to conduct it or contract for useful studies, how to analyze research findings, and how to decide how much faith to put in findings or interpretations of research findings.

This kind of research deals directly or indirectly with people as individuals and as members of small groups that are, in turn, embedded in communities and cultures. Therefore, it relies primarily on basic social science research methods: experiments, surveys, content analysis, focus groups, and observational techniques. These methods have their roots in science and scientific inquiry as a way of learning about and understanding real-world phenomena.

Scientific research represents one way in which people learn about and come to understand the world. By scientific research we mean the systematic, objective collection and analysis of information to uncover facts, patterns, and relationships. It contributes to the creation of knowledge that can enhance understanding of the world and provide answers to practical questions. Kerlinger (1986) defines it as a "systematic, controlled, empirical, and critical investigation of hypothetical propositions about the presumed relations among phenomena."

Ways of Knowing

Over the course of their lives, people come to know many things. They acquire facts and impressions, fit them together into patterns, and from them form opinions and beliefs about the way things are and about what one should do in certain situations. They develop a repertoire of information and skills that helps them make sense of their world and decide what to do in a given situation. In doing so, they use four basic approaches: intuition, tenacity, authority, and science. None is necessarily the "correct" approach; none is always "wrong." Each has strengths and weaknesses that make it appropriate in some situations but not necessarily in others.

Traditional Ways of Knowing

As a way of knowing, **intuition** rests on the assumption that something is true because it is "obvious," "self-evident," or "just plain common sense," or it "stands to reason." The criterion for truth is the reasonableness of the argument.

With **tenacity,** the logic is that something is true because it is commonly known to be true or has always been true. Truth rests on the assumption that what is customary is true; what worked in earlier eras will still work today.

As a way of knowing, **authority** vests truth in the opinions of people who have greater knowledge as a result of their education, experience, or position in society. Truth is what those in positions of power or influence say it is. One variant is the "agreement reality" commonly used to set standards. "Good art," for example, is what artists, critics, or professors agree it is. As another variant, religion points to the supernatural, to god or the gods, or to sacred writings as the source of truth and to clergy or other leaders as its interpreter.

All three methods may be directly or indirectly based on observation. There is wisdom in the old adage that "seeing is believing." Indeed, you could hardly make it through a day without drawing some conclusions from what you observe. However, the observations that undergird both intuition and tenacity may be wrong because they are too few in number or are not appropriate for the problem at hand. With these methods of knowing, there is no real safeguard against error. Moreover, what seems "obvious" and "reasonable" and what is "customary" are often culture-bound. What is reasonable and customary may seem that way because of the voice of authority.

In the method of authority, what experts say is true may be based on their own use of the methods of intuition or tenacity. But what experts say may also differ over time as well as from country to country, from one subculture to another within the same country and even from person to person within a subculture.

It seems obvious, for example, that the sun appears to rise in the east and set in the west. Therefore, it seems reasonable to believe that the sun moves while the earth stands still. Religious and political leaders once said this was true, and most people accepted it. But even before Galileo, some people had doubts, but today very few people would seriously argue for a stationary earth and a moving sun.

At the time of the American Revolution, many people who had grown up living under a monarchy argued that it was obvious that British rule had always provided security and that democracy couldn't work. Pointing to the battle for control between Catholics and Protestants, many believed it seemed obvious that religious freedom could tear a country apart, and the authority of crown and church supported their belief. But to leaders of the American patriot cause, just the opposite seemed obvious and reasonable; they found their own experts to buttress their claims. Today throughout much of the Western world, both intuition and tenacity seem to support the underlying tenets of democratic government and its corollaries: freedom of speech, press, and religion.

Intuition seems closely related to the concept of the "marketplace of ideas." Through the free flow of information, Truth will emerge. With tenacity, what most people in a society believe and what customs they follow are necessarily the way things are, must, and should be. Majority rules. But, of course, majority rule elected Hitler in Germany. In the United States, moreover, it led to prohibition, which didn't work very well.

In some cases, relying on authority makes sense. Very few people would doubt that people with health problems should consult an expert. However, in China authority might suggest that a person with back pain should consult an acupuncturist; in the United States it would argue for consulting a medical doctor or perhaps a chiropractor. In some subcultures, however, people will turn to a religious leader for prayer and faith healing.

In business, one successful business leader will advocate a top-down communication style plus tight controls on employees' personal use of the telephone and computers. Another equally successful one will argue for a more collegial style that gives employees great latitude.

Authority leaves open the question of whose expertise should count; the answer will usually be based on tenacity/custom plus intuition. In many cases, traditional methods indicate that each of the competing claims is correct and also that each is wrong. Some people get well when they turn to medical doctors, chiropractors, acupuncturists, and faith healers; some die. Some companies employing an autocratic management style flourish; others flounder.

The best way to sort out the competing claims is to turn to science as a way of knowing.

Science as a Way of Knowing

As a way of knowing, science is empirical. That is, it is based on observations—the gathering and testing of evidence that people other than the investigator can agree on and accept. In turning to this way of knowing, people make three assumptions:

1. The world is an orderly place.
2. Regularities exist in the physical and natural world, in institutions, and in the way people think and behave.
3. These regularities can be discovered and explained by aggregating observations of real-world phenomena.

Although relatively few people think of themselves as scientists, almost everyone makes some use of **naive science.** People observe and categorize things in an effort to discover regularities in their world and then make decisions on the basis of what their discoveries lead them to believe is most likely to happen.

Intuition and tenacity have roots in personal observation; the authority of expert sources often comes from their personal observations or from the observations of others who explicitly use science as a way of knowing. Children act as naive scientists as they explore their world. They taste and touch everything, observing the reactions of parents and others as they do so. From that process, they categorize their experiences: Some things feel good when touched, but other kinds of things hurt; some things are good to eat, others aren't; some behaviors elicit smiles, others produce frowns, scoldings, and "time-outs."

Often the methods of naive science "work." But like traditional ways of knowing, naive science is prone to the faulty conclusions that stem from inaccurate, selective observation and from the illogical reasoning that can come from overgeneralization or from assuming that something is necessarily the cause because it precedes something else or occurs along with it.

True science involves research. This research has seven characteristics that set it apart from the more naive or primitive form and also from other ways of knowing. These characteristics help guard against the fallacies that are more or less inherent in other ways of knowing. They also make science well suited for discovering patterns among real-world phenomena that everyone can, or should be able to, agree are true.

1. Scientific research is question-oriented. It begins with a question or problem rooted in the real world and seeks answers or a solution based on observations of real-world phenomena. In doing so, it seeks to find new information about relationships that will be generally applicable, that will provide explanations, and that ultimately will have predictive value.

    A researcher trying to explain why employees in a certain company are not producing as much work as those in another company would not simply say the unproductive employees are not working as hard as the productive ones. Instead, the researcher would try to uncover factors—employer-employee communication, for example—that would explain and, ideally, predict varying levels of productivity. If successful, the findings would point to the solution to the problem of low productivity.
2. Science is empirical. Because the questions of science are about real-world phenomena, relationships, and behaviors, it depends on gathering evidence based on things that can be observed directly or indirectly.

> **Box 1.1.** *Common errors (or fallacies) in reasoning*
>
> People often make several kinds of errors when they inquire casually about the way the world works. Science is geared to guard against all of these common errors in reasoning.
>
> **Inaccurate Observations.** People often misperceive things they see because much of our daily observations are casual or at a semiconscious level. We may think one person is chasing another when, in fact, both are running to catch a bus, or we may glance at the clock and think it says 5:30 when it is really 6:25.
> - Science reduces the likelihood of errors by requiring careful, *conscious* observation.
>
> **Overgeneralization.** People often make generalizations or draw conclusions about what is true based on only a few, limited observations. For example, I have had only three female teachers, and they were all really hard, so I conclude that all female teachers are hard.
> - Scientists guard against overgeneralizations by committing themselves to a sufficiently large representative set of observations in advance and through the replication of studies.
>
> **Selective Observation.** When observing the world, people often don't notice all instances or events equally. Once we have an idea about how the world works, we often focus on those observations that support our perceptions and ignore those that don't. Having been ignored or treated rudely by some clerks, I expect that kind of behavior and fail to notice all those instances when I was served promptly and courteously.
> - Scientists guard against this by systematically measuring and sampling phenomena before reaching a conclusion.
>
> **Illogical Reasoning.** People are often illogical in their day-to-day reasoning. They prematurely jump to conclusions or argue on the basis of faulty assumptions. For example, we may presume that a consistent run of good or bad luck foreshadows its opposite.
> - Science avoids the pitfall of illogical reasoning by using logic consciously and explicitly.
>
> **Association Is the Same as Causation.** People often assume that if two events occur together or one precedes the other, one must have caused the other. For example, the mere fact that people who watch a lot of violent television programs are more likely to be aggressive doesn't necessarily mean that watching television causes someone to be aggressive; people who are or have become aggressive for some other reason may choose to watch television.
> - Scientists guard against this fallacy by recognizing that association is just one of the criteria that must be fulfilled in order to show a cause-and-effect relationship. These criteria are discussed in detail in Chapters 3 and 6.

Our scientist would not try to explain the different levels of productivity between the two companies by saying that god favored one company over the other. It might, of course, be "true" that the gods smiled on one company and not the other, but the mind of god cannot be observed, and so the explanation would not be science. For a scientific explanation that believers in the god and nonbelievers alike should be able to recognize, the scientist would look for observable and measurable differences between the two companies such as the amount and kind of feedback employees receive from their supervisors.

3. Science is objective. By this we mean that researchers try to guard against personal biases by carefully specifying what will and won't count as evidence and how that evidence will be collected and analyzed. They also look for evidence that could contradict their preferences and presuppositions. Because science deals with facts, in writing up findings, researchers make sure that interpretations, conclusions, and recommendations are appropriately labeled and flow logically from the evidence.

In conducting the research on productivity, the investigator might expect and hope that differences in supervisors' communication styles will explain levels of productivity: Encouragement increases productivity, criticism reduces productivity. But in investigating that possibility, the researcher would need to specify what counts as encouragement and criticism as well as consider other possibilities that might negate the assumption. Productivity might depend on other factors such as job experience or available resources.

4. Science is systematic. Making appropriate observations and analyzing results requires careful planning. True scientific research is conducted in an orderly manner according to procedures that are specified in advance and then carefully followed. Although there is no one, single scientific method, certain steps in the process are generally followed although they often do not follow one another in a linear fashion. Relationships among them are often reciprocal. The plan may develop in an iterative fashion, undergoing a series of revisions or partial revisions.

These steps are described more fully in Chapter 3. However, as an example, our researcher might plan to investigate kinds of feedback by observing worker-supervisor interactions, but an examination of the literature could suggest the need to include other kinds of observations. That, in turn, could affect decisions about whom to study and what additional information to collect. However, decisions about whom to study have implications for observation and measurement. Each possibility for observation and measurement has implications for choosing the people to study. Those possibilities also impinge upon methods for data analysis that, in turn, may force a reexamination of procedures that would send the researcher back to the literature in a search for more information and ideas about how to proceed.

5. Science is public. Scientific research serves a purpose beyond simply providing answers to practical questions faced by individuals and organizations. At its core, its purpose is to advance understanding. Therefore, researchers cannot claim to possess personal skills or insights; they must be able to share their methods, observations, evidence, and findings in ways that increase the information available to everyone. To do this, they must document their procedures and observations, saving both so they can be shared with and examined by others. They must also report their findings fully and accurately, including those that are not what

the researchers or their sponsors might have hoped they would be. Whereas those researchers who work for universities or government laboratories usually freely share their procedures and data with others, those employed in the private sector may, by the terms of their employment, be forced to treat certain details as proprietary. However, they would still be expected to disclose their procedures and report all of their findings to those who commissioned the study.

Therefore, a researcher investigating productivity for a corporate client could not tell that client the methods used in the study are secret or selectively release findings to make the conclusions more palatable. The demands of science would require sharing the methodology and all of the findings with the client. Once information from the study would no longer put the client at a competitive disadvantage, ideally the investigator would be freed to present information about the research at conferences, publish it in trade magazines and research journals, and perhaps even share data with other interested researchers.

6. Science is replicable. As a way of knowing, science derives its advantage from the principle that its findings should be reproducible by other investigators using the same procedures. However, this can happen only if both the findings and the methods used to produce them can be shared with other investigators in sufficient detail to allow other researchers to verify the results.

By drawing on and perhaps partially replicating previous work, our researcher is in a position to compare new findings to previously published results. This ability to compare research methods and findings across studies provides the client with some assurance of the quality and trustworthiness of the work.

7. Science is cumulative, yet tentative. Science deals with facts, but to qualify as a "fact," the possibility that something is not true must always remain. A single study, or even multiple studies, can never fully prove something is true. Research can show only that something has not yet been shown to be false.

Researchers use previous work as building blocks to refine, improve, and correct their procedures and their findings. In this way, science adds to the body of knowledge.

Just as our researcher on productivity used other studies to help plan the investigation commissioned by the client organization, this new study will produce information that is useful to the client but that can also serve to correct or strengthen conclusions based on previous investigations. That, in turn, will advance public knowledge and understanding of the role of employee-supervisor communication in fostering worker productivity.

The Nature of Social Science Research

The physical sciences—physics and chemistry—provide the model for science. These sciences are often referred to as "hard sciences" because, through replication, their findings can ultimately produce "truths" that are "hard." That is, they are firm and universally applicable. Ultimately the findings may achieve the status of scientific "law"—the law of gravity, for example. Laws no longer require further testing; they can be presumed to be true.

But **social science** isn't quite like that. Social scientists study people instead of inanimate things. Therefore, social science is more like biology than like chemistry or physics. Chemical compounds, electricity, or light do not have minds of their own; their characteristics are fixed. They cannot change from place to place or over time. Plants and animals, however, do differ, and they can and do change. Two dogs of the same age and breed may have quite different temperaments; bacteria and viruses change, with the result that a medicine that was once effective against them may no longer be effective.

People, in particular, have the ability to think and decide what to believe and how to behave. They have free will, and they use it in ways that can confound. Every medical doctor has seen patients who, by all logic, should recover from an injury or illness simply give up and die. Others who had virtually no chance of survival sometimes recover.

Findings from the life sciences and especially from the social sciences are not nearly as "hard" as those from the physical sciences. At the same time, these sciences, and particularly the social sciences, are much "harder," if by "hard" we mean "difficult."

People differ from one another in intelligence, physical strength, and life experiences. They also think, decide, change, and create. As part of that creative process, they produce organizations, institutions, and cultures that differ from place to place and over time; in turn, those organizations, institutions, and cultures have their impact on people. Therefore, social scientists must take many more things into account than must their counterparts working in physics or chemistry laboratories; they also face the almost impossible task of exactly replicating each other's work.

A chemist who creates a new compound in a lab in Rome, New York, would have every reason to believe that a chemist working in Rome, Italy, could create the same compound if that chemist followed exactly the same procedures as the New York researcher did even if the replication were attempted 5, 10, 50, or 100 years later. Our scientist studying the relationship between employer-employee communication and worker productivity in Rome, New York, however, could not expect, with the same degree of confidence, that a colleague replicating the study in Rome, Italy, would get the same findings even if the two studies were conducted at the same time. In fact, the New York scientist might not even be able to get the same findings if the study were conducted in two different companies or again in the same company just a few weeks later. Moreover, our chemists would expect—and find—that all of the elements behaved the same way; they combined to produce the new compound. But the social scientists would almost certainly find at least a few anomalous cases—people who do not behave the way the majority behaves.

As a result, findings from social science research are always somewhat "soft." That "softness" occurs because of the need to consider many factors, some of which can be observed, measured, or accounted for imperfectly at best. Nevertheless, social science is real science because its research shares the characteristics of the scientific method.

In social science, the goal is to find patterns and regularities. The emphasis is on aggregates rather than on individuals. That some people do not fit the pattern does not negate the usefulness of science as a way of learning about people and their behavior. In naive science, it is easy to mistake those instances of idiosyncratic behavior for the majority. But for social scientists, anomalous cases simply point to the need to conduct more studies that will help uncover the reasons for apparently idiosyncratic behavior.

In their everyday life, people do not always have the time or the resources to turn to true science. In many situations, traditional ways of knowing provide adequate answers. However, like naive science, they provide few of the safeguards against errors resulting from overgeneralizations based on selective, inadequate, or inaccurate observations that are inherent even in the social sciences where findings are necessarily softer.

But for all its advantages, social science cannot answer every question. Social science can show, for example, that "media content x is most likely related to behavior y, except when condition z occurs." However, it cannot answer the question of whether content x or behavior y is good or bad. To label x and y as good or bad, social scientists would have to turn to standards that come from traditional ways of knowing. But even in those cases where authority, backed by tenacity and intuition, could agree that "bad x" leads to "bad y" except when "good z" is present, science cannot serve as the sole authority for decisions about whether "bad x" should be prevented or punished in an effort to decrease the instances of "bad y" or whether the problem should be addressed by taking steps to increase "good z."

Because science depends on observation and measurement of real-world phenomena, neither the natural sciences nor social science can determine whether there is a god or, if so, what that god demands. Neither can science serve as the sole source of knowledge for creating standards or for deciding what to do. Science is, however, the method best suited for discovering patterns and relationships among real-world phenomena, showing the conditions under which those patterns and relationships exist and predicting what is most likely to happen in the future if those patterns and relationships remain in place or if they are in some way interrupted.

Issues in Scientific Research

Although social scientists generally agree on the nature of scientific research, members of the research community hold different opinions on a number of issues. These include the appropriate role of theory, the purpose of research, and the question of how data should be analyzed.

The Role of Theory: Deductive vs. Inductive Reasoning

Many of the benefits of science as a way of knowing stem from the role theory plays in the research process. But where scientists are likely to say, "There is nothing as useful as a good theory," laypersons are more likely to retort, "It's just a theory." The difference in perception stems from the very different meanings people attach to the word "theory."

In everyday speech, people use "theory" to mean an assumption, idea, or guess. In scientific research, however, a theory is much more than that. A **theory** is a logically interrelated set of propositions about empirical reality. Beginning with theory, researchers plan their investigations around those logically interrelated propositions. They develop **hypotheses**—statements about relationships among phenomena that can be tested by empirical means.

In science, as in everyday life, a theory is always tentative. It is subject to correction or refutation as new evidence becomes available that supports, partially supports, or fails to support the propositions. However, so long as the theory has not yet been shown to fail, it is accepted as a reasonable explanation for those phenomena included within its scope. Through the testing of hypotheses derived from theory, science achieves its self-correcting, tentative yet cumulative nature. Still, there is controversy about the appropriate role of theory in research.

To the purist, research is a **deductive** process. Hypotheses are derived from theory and then tested using empirical methods. The results from those tests are then examined in light of the theory. Based on that examination, theories may be accepted, but they can never be proven true. Further work could turn up evidence that may cause

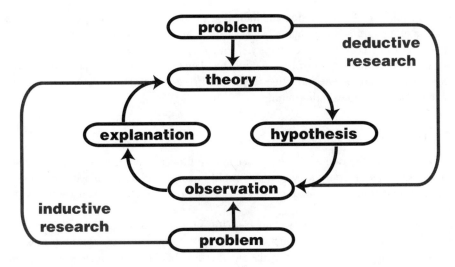

Figure 1.1. The research cycle

them to be rejected. More often they may be expanded or modified to incorporate new classes of phenomena or to include contingency conditions that then become the basis for further hypothesizing and testing. Through this process, knowledge accumulates and grows; theories themselves become more all-encompassing. Some propositions within them may take on the characteristic of laws.

This is the classic process of deductive reasoning. But expecting theory to always serve as the guide and inspiration for research is unrealistic. Unless one were to accept the unscientific proposition that theories arrive full blown from the gods, theories must come from somewhere in the real world. Often they develop through **inductive** reasoning. Inductive research begins with observation; those observations become the raw materials for developing propositions that, in time may be combined to create theories, which then are subject to further testing. Whether theory should come before or after observation is subject to endless debate.

Although deductive reasoning is the classic model, many scientists contend that science is better served by following the inductive method. Those who believe theory should precede observation contend that theory serves as a guide by helping researchers ask appropriate questions, make appropriate observations, and test their hypotheses. Using it saves time by guarding against the dangers of collecting inadequate or irrelevant data. It can also offer protection against illogical reasoning.

However, proponents of the inductive approach, such as Barry Glaser and Anselm Straus (1967), make persuasive cases that theory can be a trap, especially in the social sciences. Because no theory is all-encompassing enough to explain all of human behavior, beginning with theory can cause researchers to overlook, ignore, or downplay factors that lie outside their chosen framework. Therefore, they argue, research should be "grounded"; the methodology should be that of "naturalistic inquiry." Instead of beginning with presuppositions rooted in theories, researchers should gather as much information as possible by observing and talking to people and continue doing so until there is no new information to collect. As patterns emerge, they should be checked against the understandings of the people who are the subjects for the research before those regularities become the stuff of theory.

Although those who prefer the deductive method often consider the inductive model atheoretical, in reality deduction and induction coexist. The relationship between them is more of a reciprocal one than a one-way street leading either from theory to observation or from observation to theory. Neither approach need be atheoretical. The deductive researcher uses theory as a guide but then tests and refines that theory on the basis of observation. The inductive researcher makes observations first, then uses or develops a theory to explain findings.

The Purpose of Research: Basic vs. Applied

Debates over the role of theory in research spill over into questions about whether basic research is the only kind that can qualify as truly scientific or whether applied studies also count. But again, the two kinds are more often complementary than truly at odds with each other.

Basic research is sometimes referred to as theoretical, academic, or scholarly research. It is most often conducted in college or university settings by professors and/or graduate students. Much of the funding for this kind of research comes from government agencies, foundations, or the schools themselves either directly through grants to cover their costs or indirectly by providing salaries and work space. Because its methods, data, and findings are generally public, basic research contributes immeasurably to the advancement of knowledge.

The goal of basic research is to create, test, and improve theory. However, it would be a mistake to think that all, or perhaps even most, social scientists devote their research time to the creation and testing of theory. Some do research simply to document "how things are," secure in the knowledge that what seems obvious today will be less obvious in 10, 20, or 100 years. Their work may truly be atheoretical—devoid of any reference to theory and without any interest in contributing to the creation of theory. Others simply collect data because they work in new areas where no theory exists to serve as guide and not yet enough data to begin to create one.

Still others researchers conduct **applied** research. The goal for this kind of research is to provide solutions to real-world problems. Theory building and testing is not its goal, but findings from purely applied research often provide the raw stuff from which researchers can develop theories. Although pragmatic concerns are rarely the driving force behind basic research, basic research has practical applications. Much applied research is "theory using." Applied researchers often design their studies to incorporate questions, hypotheses, or measurements derived from more theoretical work. They also turn to theory to help them explain their findings and make them more meaningful by fitting them into a larger body of work.

Therefore, applied research is neither totally atheoretical nor bereft of value to the basic researcher. Basic research is not simply an exercise in "ivory tower" theorizing bereft of real-world benefits. The two kinds of research complement each other, working together to enhance understanding.

Data Analysis: Quantitative vs. Qualitative

Also closely related to the deductive/inductive issue is the debate over whether research should be quantitative or qualitative.

In quantitative research, numbers are attached to observations in such a way that statistics can be used to analyze results and present findings. For qualitative studies, observations are not reduced to numbers. Instead, findings are presented in narrative form.

In general, those who prefer the deductive approach argue for quantitative research because the use of statistics can provide better tests for hypotheses. However, proponents of the inductive approach, and especially those who argue for naturalistic inquiry, more often argue that qualitative research is better because evidence presented in narrative form lets people "see" and "hear" the evidence and draw their own conclusions about its quality and relevance.

Much of the concern about qualitative research stems from an erroneous understanding of empiricism and science. Because science as a way of knowing has its roots in the hard sciences, where research is almost always quantitative, people tend to see empiricism and quantification, science and numbers, as synonymous. However, empiricism really means "based on observation." Observations—findings—can be conveyed by attaching numbers to them and using statistics, but they can also be presented in words and pictures.

Science is an approach to gathering and analyzing data that has certain built-in safeguards against errors in observation and logic. Therefore, "doing science" does not depend on quantification. Rather, it depends on following procedures that are faithful to the nature of science.

The decision to conduct quantitative or qualitative research depends on the researcher's interests and skills. Those who follow the deductive model test hypotheses. Therefore, they prefer and become skilled in using data-collection methods such as experiments, surveys, and perhaps content analysis, which lend themselves to collecting the kind of quantitative data that can be analyzed statistically to test their hypotheses. Researchers who follow the inductive model may also conduct quantitative research. However, those inclined toward grounded or naturalistic inquiry gravitate toward methods for which quantitative research is less feasible or appropriate: focus groups, in-depth interviews, observation, and some forms of content analysis.

Neither quantitative nor qualitative research is inherently better. Reliance on numbers and statistics, even when the audience for research findings understands them, can squeeze the "life" out of findings. Word pictures that let the audience see and hear the evidence and evaluate it for themselves, without the crutch of statistical tests of significance, can make the findings seem more real and more understandable. At the same time, the evidence may seem softer and the findings more subjective. Quantitative research better lends itself to the rigorous testing of hypotheses. However, the decisions that go into measurement and the data manipulations associated with quantitative research can bury information in subtle ways that can belie the more objective image that quantitative research enjoys. They can also make its findings less useful than they could be.

Because basic researchers more often present their findings to academic audiences, they can generally assume that those who read the journals in which they publish results can understand and interpret statistics. However, many applied researchers prefer to rely in whole or in part on qualitative research because many of their clients can understand and evaluate evidence presented in words more readily than they can understand numbers and interpret statistics properly.

Main Points

- A knowledge of research methods is necessary to effectively evaluate research presented in the media, to make informed decisions in the business world, and to produce it oneself.

- Scientific research represents one way in which people learn about and come to understand the world around them. Other ways of knowing include intuition, tenacity, and authority. They are more often subject to errors in reasoning.
- Scientific research is the systematic, objective collection and analysis of information undertaken to uncover facts, patterns, and relationships to increase our understanding of the world.
- The main assumptions underlying science as a way of knowing are that the world is an orderly place, there are regularities in social life, and these regularities can be understood and explained by aggregated observations of real-world phenomena.
- The seven key characteristics of science are that it is question-oriented, empirical, objective, systematic, public, replicable, and cumulative yet tentative.
- Because social scientists study people instead of inanimate things, findings are "soft" rather than "hard." The focus is on finding patterns and regularities rather than universal laws.
- Theories are a set of interrelated propositions that explain phenomena. Theories are always tentative and are subject to correction or refutation as new evidence becomes available that supports, partially supports, or fails to support the propositions.
- Theories provide two main functions: In deductive research they guide the research, and in inductive research they integrate observations.
- The main goal of basic research is to create, test, and improve theory in order to advance knowledge, whereas the main goal of applied research is to provide solutions to problems faced in the world.
- Quantitative research uses numbers and statistics to analyze and present data. It is most often deductive in nature. Qualitative research is a method that emphasizes the description of how people experience the world. It is most often inductive in nature. Data take the form of words, descriptions, and narratives.

Terms to Know

scientific research
intuition
tenacity
authority
science
hard science
soft science
hypothesis

deductive research
inductive research
basic research
applied research
qualitative research
quantitative research
theory

Questions for Discussion and Review

1. From the standpoint of a consumer, why is it important to study research methods?
2. Find some examples of scientific research presented in the news media. Which of the methods discussed in this chapter do you normally use to decide whether to believe the report?

3. Why is social science research usually better than the alternative ways of knowing? In what situations would other ways of knowing be more appropriate?
4. Review letters to the editor and opinion pieces in your local newspaper. Identify any of the common errors in reasoning: inaccurate observations, overgeneralization, selective observation, and illogical reasoning. In what ways does science help guard against these common errors?
5. What does it mean to say that science is empirical? Give several examples of questions that scientific research cannot answer.
6. For the relationship suggested on page 12, think up some possible examples of x, y, and z. Discuss the implications of various ways of knowing for examining the relationship and deciding what to do about it.
7. What is the key difference between hard and soft science? What is responsible for the difference?
8. Distinguish between deductive and inductive reasoning. Which do you use more often in your daily life? Why? When might a communication researcher use each kind of reasoning? Justify your answer.
9. Explore the websites for the American Evaluation Association (http://www.eval.org) and the Center for Applied Behavioral and Evaluation Research (http://198.69.134.119/index.html). What is evaluation research? Give at least two examples of such a study. Is this a type of basic or applied research? Justify your answer.
10. Go to one of the following websites, and find an example of qualitative research and of quantitative research:
    AEJMC Convention Paper Abstracts
    http://www.aejmc.org/convention/index.html
    U.S. Centers for Disease Control and Prevention
    http://www.cdc.gov
11. For each study you located in question 10, determine who funded the research and who conducted it. Explain whether the researchers used deductive or inductive reasoning and whether each study is an example of basic or applied research. Discuss how the findings might contribute to basic research and how people working in the communication industry or in other fields might use the research findings.

Readings for Discussion

Berger, C. R. and Chaffee, S. H. (1987). What communication scientists do. In C. R. Berger and S. H. Chaffee, eds., *Handbook of communication science.* Newbury Park, CA: Sage, pp. 99–122. The authors take stock of research practices and trends.

Campbell, D. T. (1988). Can we be scientific in social science research? In D. T. Campbell, *Methodology and epistemology for social science: Selected papers.* Chicago: University of Chicago Press, pp. 315–334. In this article, a master researcher comments on social science as a scientific discipline.

Cuzzort, R. P. and King, E. W. (1980). Can science save us? In R. P. Cuzzort and E. W. King, eds., *20th century social thought,* 3rd ed. New York: Holt, Rinehart, and Winston, pp. 114–130. The authors raise questions about the value and limitations of science as a way of knowing.

Halloran, J. D. (1998). Mass communication research: Asking the right questions. In A. Hansen, S. Cottle, R. Negrine, and C. Newbold, eds., *Mass communication research methods.* New York: New York University Press, pp. 9–34. This chapter raises thought-provoking questions about the impact of culture on findings in communication research.

Merton, R. K. (1968). The bearing of sociological theory on empirical research; The bearing of empirical research on sociological theory. In M. Brodbeck, ed., *Readings in the philosophy of the social sciences.* New York: The Macmillan Company, pp. 465–495. A pioneer in the field comments on the reciprocal relationship between theory and observation.

Miller, G. R. (1983). Taking stock of a discipline. *Journal of Communication, 33*(3):31–41. The author explores the state of mass communication research.

Schoenbach, K. and Bergen, L. (1998). Commentary: Readership research—challenges and chances. *Newspaper Research Journal, 19*(2):88–102. The authors explore the importance, strengths, and weaknesses of research as it applies to the newspaper industry.

Stavitsky, A. G. (1995). "Guys in suits with charts": Audience research in U.S. public radio, *Journal of Broadcasting and Electronic Media, 39*(2):177–189. This article traces the development of research in the field.

Tankard, J. W. Jr. (1976). Reporting and scientific method. In M. McCombs and D. L. Shaw, eds., *Handbook of reporting methods.* New York: Houghton Mifflin, pp. 42–77. The author explores similarities and differences between the methods used by reporters and those employed by social scientists. Other chapters in this and in subsequent editions edited by M. McCombs, D. L. Shaw, and D. Grey treat the use of basic research methods for in-depth reporting.

Wimmer, R. D. and Reid, L. N. (1982). Researchers' response to replication requests. *Journalism Quarterly, 59*:317–319. The authors examine the implications of current practice as it relates to the requirement that science be public.

References and Resources

Anderson, J. A. (1987). *Communication research: Issues and methods.* New York: McGraw-Hill. This is one of the classic methodology texts for quantitative research. A special feature is the commentaries by experts in the field.

Denzin, N. K. and Lincoln, Y. C. (2000). *Handbook of qualitative research,* 2nd. ed. Thousand Oaks, CA: Sage. This comprehensive work is a classic.

Eastman, S. T. (2000). *Research in media promotion.* Mahway, NJ: Lawrence Erlbaum Associates. This is the first book devoted to promoting and marketing traditional and online broadcast programming.

Ferguson, S. D. (1999). *Researching the public opinion environment.* Thousand Oaks, CA: Sage. Public relations practitioners will find this a useful source of information on the purposes, theories, and methodologies for studying publics.

Glaser, B. G. and Straus, A. L. (1967). *The discovery of grounded theory: Strategies for qualitative research.* New York: Aldine Publishing. This classic provides the theoretical and methodological rationale for using the inductive method.

Guba, E. and Lincoln, Y. S. (1981). *Effective evaluation.* San Francisco: Jossey-Bass. This is a standard text on inductive, qualitative, and naturalistic research.

Jones, J. P., ed. (1998). *How advertising works: The role of research.* Thousand Oaks, CA: Sage. This basic text provides detailed information about standard research methods and their application in marketing and advertising.

Jones, S., ed. (1999). *Doing internet research: Critical issues and methods for examining the net.* Thousand Oaks, CA: Sage. Chapters by experts provide a good overview of problems and strategies for overcoming them.

Kaplan, A. (1964). *The conduct of inquiry.* San Francisco: Chandler Publishing. This is one of the classic works on scientific method.

Kerlinger, F. N. (1986). *Foundations of behavioral research.* New York: Holt, Rinehart, and Winston. Definitions and explanations from this classic show up in many methodology texts.

Lowery, S. and DeFleur, M. L. (1995). *Milestones in mass communication research,* 3rd ed. White Plains, NY: Longman. This standard text examines theories developed through basic communication research.

Sharp, N. W. (1988). *Communication research: The challenge of the information age.* Syracuse, NY: Syracuse University Press. This is one of the earliest works to examine the research implications of emerging information technologies.

Tichenor, P. J. and McLeod, D. N. (1989). The logic of social and behavioral science. In G. H. Stempel III and B. H. Westley, *Research methods in mass communication,* 2nd ed. Englewood Cliffs, NJ: Prentice-Hall, pp. 10–29. This chapter explores the nature of social science research as it applies to the field of mass communication; the book is a classic in the field.

Williams, F. (1988). *Research methods and the new media.* New York: Free Press. The author discusses ways in which basic methodology can be applied to problems presented by the newer communication technologies.

Zimmerman, D. E. and Muraski, M. L. (1995). *The elements of information gathering: A guide for technical communicators, scientists and engineers.* Phoenix: Oryx Press. The authors provide an overview of social science research methods and their application in technical communication.

Part II

Getting Started

2

Fundamental Concerns

AS A WAY OF knowing, science gains strength and credibility from the care that goes into planning and executing each individual research project.

Although the terms are often used interchangeably with little resulting confusion, "methodology" usually refers to the overall strategy for conducting the research. "Methods" are the individual techniques that are used in a given study.

Those unfamiliar with scientific research often believe that a single correct methodology works best for each study. But that is rarely the case. In most instances, some methodologies may be better than others; within each methodology, some methods may be more or less appropriate. But there is almost always more than one way to conduct a study. The trick is to find and then use the most appropriate methodology for a given study.

But to come up with the most appropriate methodology, researchers must first take into account certain limitations they face every time they set out to answer their questions by the scientific method. At the same time, they should do whatever they can to ensure the quality of their work.

Limiting Factors

In research, a wide gap often exists between the ideal and the possible. But even if one could plan and execute the ideal study, that study would not be perfect. Cost and concern for the humans who are both the ultimate subjects and consumers of research impose limits on what is possible in any given situation. Moreover, the results of even the best study will be less than perfect because some error always occurs.

Cost

In research, costs are of two kinds: money and time. Every study has a budget and a timeline. The budget defines how much money is available to cover the cost of planning and executing the project. The timeline tells how long the study can take. Although the two are separate, they are also intimately connected.

The budget must cover out-of-pocket costs. The most common of these costs are materials, travel, payments to subjects, and salaries of those conducting the study. In some cases, **overhead** must be added. Overhead is an amount tacked on to out-of-pocket expenses to help defray the costs of building or maintaining facilities such as office and laboratory space or providing utilities such as heat, water, electricity, and telephone service.

Keeping both out-of-pocket and overhead costs in mind, some kinds of research are likely to be more expensive than others. Research that requires sending investigators to distant sites where they will engage in observations over long periods of time are likely to be the most expensive. In most cases focus groups and small-scale laboratory experiments will be relatively cheap. Surveys and content analyses may be expensive or inexpensive depending on how they are done.

However, actual costs may vary greatly among studies using the same basic techniques. Studies that require creating or using special materials or equipment for data collection will be more expensive than those requiring just paper and pencil to record information. In general, the more data one collects, the more expensive the study—at least partly because of the additional time required to collect and analyze the data.

The longer a study takes, the more expensive it will be. The more researchers who can work on the project and the more time each can devote to it, the faster the work can be accomplished. However, additional researchers cost money, so any savings related to accomplishing the study more quickly may be offset by higher labor costs.

Even when labor costs are fixed, as may be the case when the researcher is already on the payroll, or nonexistent, as when students conduct research as a class project, **opportunity** costs will very likely crop up. That is, spending time on one project forecloses the opportunity to work on something else that may, in fact, be a better use of researcher time. Therefore, it is important, before committing to a project or to a particular design for it, to make sure that the project is feasible and worth doing and that the methodology will indeed produce useful and credible results.

Ethics

Part of the credibility of science as a way of knowing stems from the public's confidence that researchers are trustworthy. However, a little bit of that trust erodes each time news surfaces that scientists behaved unethically or illegally.

Law and ethics are not the same thing, but they are intimately related. Law deals with what one must or must not do in order to avoid legal consequences—jail, fines, or damages paid to those who have been wronged. Ethics involves decisions about what is right, moral, and just in a given situation. At their best, law reinforces ethics and ethics compels people to obey the law.

Occasionally the law may require or prevent people from doing what they should or should not do in a particular situation. More often, the law allows people to do things that they should neither ethically nor morally do. Both situations can come into play in the process of conducting social science research. Both are important in social science research because, at its core, social science is about people.

Together, law and ethics put some constraints on research in order to protect people. In some cases law and/or ethics may force researchers to use what they consider a second-best strategy to investigate a particular project. Ideally one might want to investigate the effects of talking to infants on cognitive development, but in doing so researchers could not isolate some babies so they never hear a human voice. Similarly, those who investigate the effects of pornography usually work with "normal," "average" people. Studying those who may actually use it would be more likely to put both subjects and society at risk.

Since the mid-1960s researchers who work for institutions, including colleges and universities that receive federal funds, must follow federal regulations. These regulations require receiving consent for their work from an Institutional Research Board, which must also follow federal guidelines for approving and keeping track of studies involving human subjects. Those whose work is funded by the government must follow similar regulations even if they do not work for organizations that receive public funding.

Failure to follow the rules can have severe consequences. For several months in fall 1999, for example, all human research—medical research and social science research—at the University of Colorado was suspended while irregularities in institutional oversight of research projects were remedied.

For those who conduct research on behalf of and funded by private organizations, foundations, and other nonprofit agencies, government oversight is minimal. The only guidelines that they must follow are those imposed by the funding agency and/or the investigators' own sense of right and wrong. For some studies, however, professional codes of ethics may be controlling. Medical doctors and psychologists who fail to adhere to certain standards could face professional sanctions.

Although some researchers will not be legally bound to adhere to ethical standards, all should act as if they are legally bound to them. Failure to do so can create a public relations nightmare for the sponsoring agency. It can also result in civil, and occasionally criminal, lawsuits in cases where people suffer emotional distress or physical harm as a result of participating in research studies.

Codes of ethics differ in detail. A typical one can be found in Appendix A. Each research method presents its own ethical challenges. These are treated in subsequent chapters. However, ethics in research can be summed up by the Golden Rule—Do unto others as you would have them do unto you—and by the maxim, Minimize risk; maximize benefits.

First and foremost, those principles apply to the treatment of the people who participate as the research subjects. They also apply to researchers' dealings with their sponsoring agencies and to their data analysis and presentation of findings.

Dealing with Subjects. The ethical treatment of subjects begins with their recruitment. Participation must be based on informed consent. That is, before people agree to participate, they must be told what the study is about, what risks they may face, how serious those risks are, what the researcher can do for them if those risks occur, and what benefits will accrue to them or to society if they choose to participate.

Participation should also be voluntary. There must be no sanctions or repercussions for those who decline to be part of a study. Bribery is never OK, but small gifts or payments for subjects' time are usually permissible. Any benefits that accrue simply from participating in a study are also OK unless those benefits would work to the detriment of nonparticipants. The distinctions are particularly important when subjects are recruited from among employees of a company for a study in which the company has a vested interest. They are also important when academic researchers use their students as research subjects.

It would be unethical to give employee-participants priority for promotions or reward them with new, upgraded computers while forcing nonparticipants to use outdated machines that cannot handle modern software. All students may benefit from participating in a research project in order to see how data are collected and analyzed, but giving those who participate extra credit for their work is problematic.

The possibility of such rewards or punishments points to another cardinal concern: the need to protect the privacy of research subjects. Those who participate should know whether anyone, including the researcher, knows who they are and whether their identity will be revealed to the research sponsors or in publications about the research. Here it is important to distinguish between anonymity and confidentiality. **Anonymity** means that the identity of subjects is not known. **Confidentiality** means that the researcher knows, or can figure out, who the participants are but will not reveal that information to anyone.

Research subjects selected through random-digit dialing or in mall-intercept studies may truly be anonymous, but in many cases the researcher does know, at least initially, the identity of participants and can match them with the information each one provides. When that is true, researchers ensure confidentiality by separating names and other identifying information from the data they collect from subjects. In quantitative research, findings are usually presented as aggregate data: "Seventy-five percent of the men agreed . . . ." In qualitative research, and occasionally in more quantitative studies, subjects are often identified in research reports only through general, nonspecific labels: "Three middle-aged men said . . . ."

Names or other identifying information should be used only if there is full, informed consent. On occasion, when pseudonyms are used, they should be chosen carefully both so that the identity of the subject cannot be inferred and so that someone uninvolved in the research cannot be mistaken for a participant. To lessen the danger of confusion, any use of pseudonyms should be stated clearly and prominently in all research reports.

Although anonymity is generally preferable, confidentiality is acceptable so long as subjects are not misled about the degree of privacy being offered. Deception is perhaps the most serious ethical problem, particularly when it involves misleading people about risks or benefits. However, sometimes withholding some information or engaging in some deception may be necessary in order to protect the integrity of the work. In such cases, ethical treatment requires telling people, up front, the general nature of the study and then **debriefing** them afterward so they understand what has occurred and can, if necessary, be given access to resources for coping with problems that may arise because of their participation in the study.

Working with Sponsors. Those who commission or fund research often have a vested interest in the study and/or the findings it produces. Therefore, some may put pressure on the researcher to do the research too quickly or to produce the desired results. They may also pressure the researcher to reveal the identity of those who participated or refused to participate or of those who are the source of the data that produced the "wrong" or otherwise undesirable results.

But designing a study to produce the "correct" results may, in fact, produce findings that are wrong. Acting on that wrong information can have consequences far more severe than learning the "bad news" could possibly have. Consider, for example, what might happen if a company that wanted "research" to confirm the wisdom of marketing a new product were to act on the basis of findings from a survey of company employees whose jobs may depend on going ahead with the project.

Researchers have their own vested interests. They may want or need the job. Quite naturally they want to please their employer or sponsor, and of course they too may hope for certain results. However, caving in to client demands or to their own desires is unethical. It is also short sighted. Any failure to follow general ethical guidelines in designing studies is likely to become known. Even if the results of the study do not lead to acting on the basis of incomplete or wrong information, the image and credibility of both the client and the researcher can suffer irreparable harm.

Analyzing Data. Even when a research project is properly designed, problems can occur. In some cases these problems are unavoidable; others can be avoided. Avoidable problems may stem from mistakes that accidentally happen during the course of an investigation. But they can also result from unethical behavior.

It should go without saying that researchers should not make up or falsify data, but that does happen. The risk of fabricated data is probably greatest when assistants are hired to do the work, especially if they are paid by the completed interview in survey research or paid for each document they examine for studies employing content analysis as the research method. Falsifying data can happen accidentally as a result of making mistakes in recording information, or it can be done purposely in an effort to get the desired results.

Principal investigators can catch most problems of made-up or falsified data by monitoring or spot checking their assistants' work. But principal investigators have also been caught engaging in both kinds of unethical behavior. Their lapses come to light most often when the findings look "too good to be true" or when other researchers discover they cannot replicate results.

Making up or otherwise falsifying data is quite serious because it undermines the value of research by striking at the very core of the scientific method. However, ethical problems that stem from inappropriate decisions about how to collect, manipulate, or analyze data and report findings are probably much more common. Examples of these problems include combining measures without checking the quality of their new composite measure, using the wrong statistic or failing to interpret it properly, and glossing over or selectively reporting findings.

Reporting Findings. Some clients and journals may actually encourage ethical lapses in handling and reporting data by accepting only studies that report statistically significant findings or ones where the data support the research hypotheses. In these situations the remedy is a firm reminder that importance and significance are not synonyms. Research that fails to confirm a hypothesis may provide more useful insight into a problem than work that gives the expected answer.

Whether problems with the handling and reporting of data arise as the result of outside pressures on the researcher or from a failure to understand how to use the chosen methods or report findings properly, the problems can be caught only if other researchers, uninvolved in a study, can evaluate the way the study was designed and conducted. Therefore, researchers must take care to provide a complete, accurate report of their methodology: what they were looking for, how they collected the data, and how they analyzed them. In doing so, the goal is not to teach anyone how to do the work. Rather, it is to provide enough information about each point in the study where another researcher might make a different decision so that others can follow the investigator's logic and reach an independent judgment about the overall quality of the work.

Quality Control

Even the best study is prone to error. But not all error is the same. Random error may have little effect on the findings, but systematic error can distort it in ways that may make "findings" meaningless.

Random error is sometimes called "noise." Like static or background sound that interferes with your ability to hear your favorite music on the radio, random error in research may also make it difficult to find answers. However, random error is no more likely to work in favor of or against any particular finding than static is likely to block out all the rock music but not the easy-listening tunes.

Random error is just something that "is" or something that happens over which the researcher may have little control. It can be the result of individual differences or the result of things that happen accidentally or unexpectedly during an investigation.

As a result of random error, results may vary. This can have repercussions in individual cases, but error in one direction is usually offset by error in the opposite direction. When results are aggregated, as they usually are in social science research, random errors tend to offset each other. Therefore, however "wrong" the results may be in some cases, they are unlikely to vary in systematic ways that will bias the findings.

As an example of random error at work, let's pretend you have been hired by a health-care facility to study the links between health and cognitive ability. As part of that investigation, you decide to look at the relationship between weight and reading. For convenience, you hire two research assistants to collect the information you need from students enrolled in a research methods class at a nearby college.

Being a careful researcher, you decide each student will be weighed six times using six different scales. You will then take the average weight as the "true" weight. To measure "reading ability," you will give two tests. Each test will have students read several short passages, but one will measure reading ability by asking students to answer a series of short-answer questions about the passages they have read. The other will require students to solve a problem and also write a short essay drawing inferences from the material they have read. Again, you will average each student's scores to produce scores for reading ability.

You, of course, find that students' weights differ greatly. One student may weigh only 98 pounds; another may weigh 350 pounds. You also find variability on the test scores. Both sets of results are "noisy"—there is wide variation that is the result of individual differences among the students in the class. That "noise" is not really error. Both students' weights and their reading abilities really do vary.

But you also find a different kind of noise. The six scales produced six different weights for each student; some of the students got very different scores on the two tests they took. Again, the results were noisy. And this noise is the result of error.

In the case of measuring weight, the scale that produced the heaviest reading did so for all students; the one that showed the lightest weight did so for everyone. If only one scale had been used, **systematic error** in determining students' weights would have occurred. That is, that scale would have biased the measurement by producing scores that were consistently too high or too low. Here, however, that would not matter much because the biasing would have been consistent across all students—all of them would have weighed "too much" or "too little."

In this study, the "wrong" weights would matter even less. Because six measurements were taken using six scales, some weighing "heavy" and others weighing "light," the error involved in determining each student's true weight is actually random error. The heavy readings are offset by the light readings so that the measurements are not biased in one direction or the other.

The test results, however, appear more troubling. Arranging the test scores in rank order shows that most students who did well on one test did well on the other; most who did poorly on one did poorly on the other. But some students did unexpectedly well or unexpectedly poorly on one or the other of the tests.

Some of that may be the result of random error—things over which the researcher has no control. One "poor" student might have guessed unusually well. A "good" student might have had a bad day. If the number of students who did unexpectedly well on one test closely matches the number who did unexpectedly poorly, random error is most likely at work. Unusually good or bad scores may make it difficult to assess the true knowledge of individual students, but for the class as a whole, the unexpectedly good or bad scores will offset each other.

But there may also be systematic error at work in our study. This kind of error may or may not stem from the researcher's desire to produce a certain kind of results, but it does tend to bias the findings in one direction or another.

Suppose, for example, two research assistants weighed the students. One weighed half the students early in the morning and also instructed those students to take off their shoes; the other weighed the remaining students after dinner and let them leave their shoes on. At the extremes, these differences wouldn't matter much. The 98-pound student would almost certainly weigh less than almost anyone else in the class; the 350-pound student would most likely still be the heaviest. But for students nearer the class average, a systematic error in weighing has been injected that could easily affect who is among the heavier students and who is among the lighter ones.

Systematic error could have an even greater effect on test scores. Perhaps the five heaviest students are all football players, and the tests were administered just an hour before they were scheduled to leave for the big game to decide the league championship. Almost certainly their minds would be elsewhere; their scores would be systematically lowered. Or suppose all of the passages on the essay/problem-solving test came from physics and engineering. Science and math students would almost certainly score higher on that test than would journalism students.

Or suppose that each of our two research assistants administered the test again to half the students. If each assistant responded to queries about the exam questions from a couple of light students and a couple of heavier ones, those attempts to help would inject some random error into the test scores. But if one assistant provided more or better instructions, scores would most likely be systematically better in that group than in the other group. In this case, if there were also weight differences between the two groups of students, the impact of systematic error would be compounded.

As the preceding examples illustrate, some random error is virtually inevitable; some systematic error may creep in. Some error comes from problems with the reliability of the research; some are validity problems.

Eliminating all error may be impossible, but quality control is possible. Although we have much more to say about reliability and validity in subsequent chapters, ensuring the quality of a research project begins with a concern for the reliability and validity of the study. In any given project, some trade-offs between reliability and validity may be necessary. But quality control requires striving for the best balance between them. Therefore, some understanding of reliability and validity is necessary in order to design the best possible study in any given situation.

Reliability

If a study is reliable, the measurements or observations used to collect data are stable, and the findings will be reproducible. Therefore, reliability depends partly on the quality of the measurements and/or observations and partly on creating and then following clear procedures throughout the study.

For measurements or observations to be reliable, they must be stable. That is, they must work the same way with all subjects in the study and also with other subjects in studies conducted at different times and in different locations. If the measures or observations are not stable, the findings will not be reproducible. However, even if the measurements or observations are reliable, reproducing the results may still be impossible.

Almost everyone would agree that weighing people should be a reliable way to determine weight. But that might not always be true. In earlier eras, butcher shops would advertise "Honest weight; no springs" because scales using springs are not reliable. Springs stretch so that, over time, scales with springs give progressively heavier readings for quantities of meat that are really the same weight.

In our study, each scale was a reliable instrument for determining weight. Although some produced heavier readings than others, they did so consistently. But our measurement of weight was not really reliable because there were no consistent rules for doing the weighing. Some people were weighed in the morning when most people usually weigh less; others were weighed later in the day when people tend to weigh more. Some were weighed with shoes on, others with shoes off. If the same procedures are not followed with all subjects in a study, then that study cannot be reliable.

To determine whether the tests used to measure reading ability are reliable, one would need to check them using procedures that are described in the chapter on measurement. However, even if the tests are a reliable instrument, their use in this study was not reliable because the assistants who administered them did not use the same procedures.

Even though a reliable measurement or observation technique should produce stable results and the findings from using them should be reproducible, it would be a mistake to believe that replicating a study will always produce exactly the same findings. Results are stable and replicable if using the same measures in the same way at the same time with the same subjects will always produce the same results. But that, of course, is an impossibility. One cannot weigh a person or give that person a test several times at the same time in order to determine whether each weighing or each administration of the test produces the same score.

If we conducted our study again next year with the same students, both weights and test scores would very likely differ. People gain or lose weight. After another year of schooling, their reading ability may improve, and so their test results might be higher. If we did the study at the same time but using a different group of students, their weights and test scores would probably also be different from those of students from the research methods class. In spite of differences in weight and scores, each of the studies might find the same relationship between weight and reading ability, but the studies might also show that the relationship between weight and reading ability differs among groups. If that were the case, we would have to reject or modify our original assumption about the relationship between weight and reading ability, but it would not mean our research was unreliable.

Reliability is related to accuracy and precision, but it is not the same thing as either one. A measurement or observation technique that produces accurate results will very likely be reliable, but a reliable technique may be inaccurate. The more precisely one attempts to measure something, the more reliability may suffer.

Note that in our study, each scale indicated a different weight for each student. None of the scales may have indicated the students' true weight. Although each scale may have been inaccurate, the results from using any scale or from using the average of all six readings produced a reliable measure of weight so long as the same procedures were used to weigh each student. The same is true of our measurement of reading ability. Neither test may have given a true indication of students' ability, but if administering the tests the same way to all subjects produces stable results, the tests are reliable.

Some of the scales may have weighed students to the nearest pound, and some to the nearest quarter pound. Those scales weighing to the nearest quarter pound would produce readings more precise than would the scales weighing only to the nearest pound. But each would be equally reliable, so long as the same rules for estimating amounts in between actual readings were used. However, on our tests of reading ability, an attempt to score the essays on a scale of 0 to 100 would appear to create a more precise answer than would scoring them simply as "excellent," "good," "average," "fair," or "poor." However, those doing the scoring could more likely agree on which papers are excellent and which ones are merely good than they could on whether a paper deserves a 93 or a 90. Therefore, using the 100-point scale would very likely be less reliable than using the simpler 5-point scale.

Reliability does not guarantee the accuracy or precision of measurements. Reliability simply means that any differences among subjects or from one replication to the next are not the result of problems with inconsistencies in the way measures work or in the way the study was conducted.

Validity

Although reliability is a necessary component of quality control, it is not enough. Quality control also requires a concern for the validity of the work. Reliability ensures that results will be reproducible. Validity offers assurance that the findings are real and not the result of some artifact in the way the study was conducted.

Internal validity addresses the question of whether the findings really answer a study's research question or test its hypotheses. **External validity** addresses the same concern more indirectly by asking whether the findings apply only to that study or whether they can be generalized—whether they will hold true—for other subjects, in other locations, and/or at other times.

Like reliability, validity is partly a matter of measurement/observation and partly a matter of how a study is conducted. One part of internal validity, sometimes called **measurement validity,** addresses the question of whether the individual measures used in the study are really measuring what they are supposed to be measuring. Another part looks to other factors that might affect the data. This part of internal validity is sometimes called **causal validity** because problems here can lead a researcher to conclude erroneously that scores on one measure are associated with or cause scores on some other measure when, instead, the findings stem at least partially from research procedures. Because this kind of internal validity problem is often idiosyncratic to a particular study, it affects the ability to generalize. Thus this kind of internal validity problem also raises problems with external validity.

The research teams of Thomas Cook and Donald Campbell (1979) and Donald Campbell and Julian Stanley (1963) have identified many factors that affect a study's validity. These factors are listed and explained in Box 2.1. Some of them can be controlled or eliminated by the researcher; some cannot be. More detailed information about checking the internal validity of measures can be found in the chapter on measurement. More about external validity is in the chapter on choosing subjects. The chap-

> **Box 2.1.** *Factors affecting validity*
>
> **Threats to Internal Validity**
>
> **History.** Events that occur during the study that may affect the results of the study. One example would be a public scandal involving a politician that makes news during the course of a study on attitudes toward government officials.
>
> **Maturation.** Changes in subjects' biological and/or psychological characteristics during the course of the study that may affect the results. Over the short term, subjects may grow tired, bored, or hungry. Over the long term subjects may actually grow older and wiser or their health and cognitive ability might decline.
>
> **Testing.** Effects occurring when the same or similar measures are used two or more times with subjects during the course of a study. A first test or pretest may sensitize subjects to the material and result in improved scores on a subsequent posttest. For example, just by virtue of having taken the SAT once, without any additional preparation, you are likely to improve your score on a retake.
>
> **Instrumentation.** Use of different measures and/or changes in the measuring "instrument" during the course of the study. Changes in measures can occur by using differently worded questions with different groups or in the pretest and in the posttest with the same group of subjects. Changes can also happen if equipment wears out or if a researcher does not use the same scoring or recording procedures consistently throughout a study.
>
> **Statistical Regression.** The tendency for individuals people who receive very high or very low scores on a test to score closer to, or "regress" toward, the mean on a subsequent test. That is, most people who score below average the first time they take a test will score somewhat higher if they take that test or a similar one again; those who scored above average will tend to score somewhat lower on retesting.
>
> **Selection.** Effects caused by the way subjects are chosen to participate in a study and/or the way they are assigned to treatment groups. If a researcher chooses too many or too few subjects with a particular characteristic, findings may be affected by the number of subjects in the study who possess certain characteristics. In any study of children, for example, scores on measures of cognitive or physical ability will almost always be lower for younger children than for older ones.
>
> **Mortality.** Effects attributable to subjects who drop out of a study before it is completed. This can happen through death, but more often people drop out because they move or because they decide they no longer want to participate in the study. As with selection, if too many people drop out or if those who drop out differ from those who remain in the study, mortality can bias the findings.
>
> **Threats to External Validity**
>
> **Selection.** Using subjects who are not representative of a larger population. Generalizing results to a larger population is impossible unless subjects are chosen using one of the probability sampling techniques described in Chapter 4.
>
> *(continued)*

Mortality. Loss of representativeness caused when subjects drop out of a study. Even when subjects are initially chosen using probability sampling techniques, the final sample may be unrepresentative of the population if too many subjects or too many subjects with a certain characteristic fail to complete a study.

Reactive Effects of Testing. Effects caused by sensitization through a pretest or simply from being a part of a study. For example, some people may score poorly because they suffer from test anxiety; others may score unusually well because they like being "on stage." Many people who know they are being observed or studied will change in ways that make them noncomparable to the population to which they are being compared. This is sometimes called the "guinea pig effect."

Reactive Effects of Research Arrangements. Any effects caused by the way research is conducted. People are very likely to behave differently in a laboratory setting than they would ordinarily behave. Subjects may, for example, more willingly comply with orders or try to adjust their behavior or their answers to questions in order to please the researcher. This is another kind of "guinea pig" effect. Outside the lab, people will respond to almost any change in routine in ways that make it easy to mistake that normal change for a real effect. This is the "Hawthorne effect," named for the location of an early study that illustrated the problem.

Multiple Treatment Interference. Using multiple manipulations or measures that interfere with or reinforce each other. Here, the assumption is that each one is independent of the other, but in reality they may cancel each other or produce an additive effect. This problem is most common in experiments, but it can occur with other methods.

Interaction Effects. Effects caused by a combination of selection and some other threat to internal validity such as history, testing, or instrumentation. Subjects selected from a town near a military base may be more sensitive to world news than those living in a rural area where fewer people may be directly affected by what is going on outside the United States; older adults may perform more poorly than college students on a pencil-and-paper test because they are less accustomed to being tested.

ters on individual research methods also discuss both internal and external validity. However, our study of the relationship between weight and reading ability nicely illustrates some common validity problems.

In that study, the researcher used scales to weigh students. Quite obviously the scales will measure weight. Therefore, the measurement has good internal validity. The researcher is measuring what she thinks she is measuring. Still, a problem may appear: Some students were weighed wearing shoes; others were not. Within each group, some undoubtedly wore heavier clothes than did others. Therefore, at least part of the measurement of students' weight is really a measurement of what they are wearing. As used in this study, the internal validity of the measurement may be good, but it could be improved.

But as weight is used in this study, a more fundamental problem with internal validity may occur than that caused by students' wearing different clothing. Recall that the main purpose of the study was to determine whether a relationship exists between a person's health and that person's cognitive skills. In this part of the study, weight seems to be a surrogate measure for "health." Although there almost certainly is some connection between weight and health, some people who are much heavier or much lighter than their "ideal weight" may be quite healthy. Others who are at or near their ideal weight could be quite ill. Used alone, weight as a measure of health is questionable. Internal validity would improve if additional measures such as blood pressure and percentage of body fat were added.

Slightly different internal validity problems crop up in the tests to determine reading ability. Even though having students read passages and then answer questions about those passages will measure reading ability, other factors will influence some portion of students' scores. For the football players, for example, things going on in their lives almost certainly depressed their scores. For a few other students, familiarity with the subject matter may have improved their scores. For others, part of the score on the essay/problem-solving test might come from writing ability rather than reading ability.

Even though a researcher has no control over what is happening or has happened in a subject's life, in this case internal validity might be better if the tests had been given on a different day. It could be improved by adding a greater variety of passages to the tests so that familiarity with one passage would very likely be offset by lack of familiarity with another and by developing a grading scheme that would exclude writing skills. But even then, problems might occur.

Again, recall that the purpose of the study called for assessing cognitive ability. Certainly reading ability is one kind of cognitive ability, but it is not the only kind. Therefore, internal validity could be further improved by including measures that tap other kinds of cognitive skills.

But even if all possible steps were taken to improve the internal validity of the measures, a problem with external validity would still be present. Because a subject's college major or events happening in a subject's life may influence test scores as much as or more than body weight does, the results from this small-scale study might not be typical for all students. They also might not apply to people who differ from the student subjects in lifestyle, age, socioeconomic status or other demographic characteristics.

Unless we can rule out other factors that may affect internal validity, external validity will be uncertain at best. Even if the measurement validity is good, we will not be able to generalize our results unless we can rule out the threats to internal validity that stem from research procedures. However, even with good controls for causal validity, findings may still be specific to the subjects who were studied unless they are truly representative of some larger group. This can occur only if subjects were selected using the probability sampling techniques described in Chapter 5.

Even if we took every step possible to improve the reliability and internal validity of our study of the relationship between health and cognitive ability, the study would still score low on external validity. We could not generalize the findings to other people who were not a part of the study. But that does not mean that the study would be worthless. Future replications using other groups as subjects can, over time, provide evidence as to whether the findings from small-scale studies such as this one exhibit any external validity.

Main Points

- Research methodology refers to the overall strategy for conducting research, whereas research methods are the individual techniques that researchers use in a given study.
- In research, a wide gap exists between the ideal and the possible because of practical limitations and ethical concerns. Moreover, the results of the best study may be less than perfect because some error will always be present. This error may be either random or systematic.
- Every study involves two types of cost: money and time. The types and extent of cost vary from one research design to another.
- Ethics involves decisions about what is right, moral, and just in a given situation. In order to protect people, researchers strive to minimize risks and maximize benefits.
- Most professional organizations have established codes of ethics that provide useful guidelines for making ethical decisions. In addition, research studies conducted by researchers who work for institutions that receive federal funds must receive consent from an institutional review board for any study involving human subjects.
- For the protection of human subjects, participation must be voluntary and based on informed consent. Informed consent refers to telling potential research participants about all aspects of the study before they agree to participate.
- In some cases subjects are guaranteed anonymity; in others, they are promised confidentiality.
- One of the most serious ethical problems is deception. If it is necessary to withhold information about the study in order to protect the integrity of a research project, subjects must be told the general nature of the study up front and debriefed afterward.
- When conducting research for a sponsor, many ethical dilemmas may arise. Researchers need to make sure that they follow ethical guidelines and do not capitulate to client demands or to their own desires.
- Researchers have an obligation to report their findings fully and honestly and to avoid any type of misconduct or fraud.
- All studies are prone to errors. Although random error is not likely to affect research findings, systematic error can seriously affect and distort findings.
- A study is reliable if the measurements or observations used to collect data are stable and the findings can be reproduced.
- Reliability depends partly on the quality of the measures/observations used in a study and partly on creating and following clear procedures throughout the study.
- A study is valid if the findings are "real" and not just an artifact of the way the study was conducted.
- The two main types of validity are internal validity and external validity.
- Internal validity refers to whether the findings really answer a study's question or test its hypotheses. One part of internal validity is measurement validity—do the measures used in the study really measure what they are supposed to be measuring? The other part of internal validity is often referred to as causal validity—can a researcher conclude that scores on one measure really influence scores on another measure?
- External validity refers to whether the findings from a particular study can be generalized to other subjects, to other locations, and/or to other times.

Terms to Know

research methodology
research method
research ethics
institutional review board
code of ethics
informed consent
voluntary participation
anonymity
confidentiality
deception
debriefing

random error
systematic error
reliability
validity
internal validity
external validity
measurement validity
causal validity
threats to internal
    validity

Questions for Discussion and Review

1. Interview several faculty members who conduct research to find out how they determine how long a study will take and how much it will cost.
2. Find out how your college estimates overhead for research projects. Also obtain copies of calls for research proposals from the federal government and from several nongovernment sources, and compare their provisions for providing funds for direct and indirect or overhead costs.
3. Arrange the ethical guidelines discussed in the chapter in terms of how important you believe them to be. Now arrange them according to how difficult you believe it would be for researchers to follow each ethical guideline. Which of the ethical guidelines are most likely to interfere with adhering to the scientific method? Explain your reasoning.
4. Use one of the search engines on the web to look up *professional code of ethics*. Examine the ethics codes for various professional organizations. In what ways are they similar? Different?
5. Obtain a copy of your university's Institutional Review Board (IRB) guidelines. How do they compare to the code of ethics in Appendix A and to other codes? In what ways do the guidelines help researchers meet the ethical concerns described in the text?
6. What is the difference between anonymity and confidentiality? All things being equal, which would you prefer to guarantee and why?
7. For each of the following research projects, identify the ethical flaw(s) and develop a strategy for eliminating or alleviating the problem:
   a. In order to investigate the quality of instruction at large universities, a researcher persuades instructors to provide him with essays students have written for assignments in beginning and advanced writing classes. The researcher evaluates the essays and then shares his evaluation of individual students' writing competence with the students' instructors.
   b. People who are called as part of a telephone survey are told their responses will be anonymous; however, callers are told to write the phone number on the cover page for the questionnaire. The researcher has a list of names and the corresponding phone numbers.
   c. As part of a study of media effects, a researcher recruits subjects for an experiment by telling them they will be asked to watch excerpts from televi-

sion news programs. Those who agree to participate are shown clips in which well-known television anchors and lesser-known ones report stories about well-known political leaders. Some of the stories cast the political leaders in a favorable light; others allege wrongdoing. All of the stories were made up by the researcher.
   d. As part of an investigation of communication networks in large organizations, a researcher rigs the computer system to intercept e-mail messages for content analysis.
8. Distinguish between reliability and accuracy, accuracy and precision, reliability and validity, internal validity and external validity.
9. Can a study or measure be reliable and not valid? Can a study or measure be valid and not reliable? Explain and justify your answers.
10. In what way(s), if any, do random error and systematic error affect the reliability and validity of a study?

Readings for Discussion

Asch, S. E. (1956). Studies of independence and conformity: A minority of one against a unanimous majority. *Psychological Monographs, 70*(9). This is a classic study that also raises serious questions about the guinea-pig effect.

Katzer, J., Cook, K. H. and Crouch, W. W. (1998). *Evaluating information: A guide for users of social science research*, 4th ed. Boston: McGraw-Hill, pp.73–107. The authors provide an easy-to-understand discussion of the two types of error, measurement, reliability, and validity as well as many thought-provoking questions for review and discussion.

Milgram, S. (1965). Some conditions of obedience and disobedience to authority. *Human Relations, 18*(1):57–76. In this classic the author thoughtfully analyzes the implications of researcher demand and the use of deception in his study of subjects' willingness to administer electric shock to others. See also Milgram, S. (1974). *Obedience to authority: An experimental view.*, New York: Harper and Row.

Pallone, N. J. and Hennessy, J. J. (1995). Deception, fraud, and fallible judgment. In N. J. Pallone and J. J. Hennessy, eds., *Fraud and fallible judgment: Varieties of deception in the social and behavioral sciences.* New Brunswick, NJ: Transaction Publishers, pp. 3-20. The authors describe problems that arise when researchers or their clients are also advocates for a particular outcome.

Singer, E. (1984). Public reactions to some ethical issues of social research. *Journal of Consumer Research, 11*(1):501-509. The author presents findings from telephone interviews with a national, random sample of 1016 subjects.

Smith, S. S. and Richardson, D. (1983). Amelioration of deception and harm in psychological research: The important role of debriefing. Journal of Personality and Social Psychology, *44*(5):1075–1082. The authors evaluate the ethical and practical implications of common practices.

References and Resources

Beauchamp, T. L., Faden, R. R., Wallace, R. J. Jr. and Walters, L., eds. (1982). *Ethical issues in social science research.* Baltimore: Johns Hopkins University Press.

This book contains useful chapters on privacy, the use (and misuse) of deception, and related issues in research on human subjects.

Campbell, D. T. and Stanley, J. C. (1963). *Experimental and quasi-experimental designs for research.* Chicago: Rand McNally College Publishing Company, pp. 5–6. These pages provide the authors' succinct identification and explanation of factors jeopardizing internal and external validity.

Cook, T. D. and Campbell, D. T. (1979). *Quasi-experimentation: Design & analysis issues for field settings.* Boston: Houghton Mifflin. Chapter 2 in this classic work provides a thorough discussion of the concepts of internal and external validity.

Greenberg, B. and Garramone, G. (1989). Ethical issues in mass communication research. In G. H. Stempel III and B. H. Westley, eds., *Research methods in mass communication*, 2nd ed. Englewood Cliffs, NJ: Prentice-Hall, pp. 262–289. The authors discuss the nature and origin of ethical dilemmas and strategies for resolving them.

Kimmel, A. J. (1988). *Ethics and values in applied social research.* Newbury Park, CA: Sage. The final chapter on the place and effect of personal values in decision making is particularly thought-provoking.

Mann, C. and Stewart, F. (2000). *Internet communication and qualitative research: A handbook for researching online.* Thousand Oaks, CA: Sage. This text examines ethical and practical considerations surrounding the conduct of computer-mediated qualitative research using materials collected from the Internet.

National Commission for the Protection of Human Subjects of Biomedical and Behavioral Research. (April 18, 1979). *The Belmont report: Ethical principles and guidelines for the protection of human subjects of research.* FR Doc. 79-12065. This is the basic government report leading to the adoption of standards for conducting research using human subjects.

Phillips, J. J. (2000). *The consultant's scorecard: Tracking results and bottom-line impact of consulting projects.* New York: McGraw-Hill. The author gives practical advice to both consultants and clients for getting the most from their relationship.

Rose, R. G. (1994). Psychological consultation to business. Odessa, FL: Psychological Assessment Resources. This book has good chapters on needs assessment, collecting the right information through research, and on presenting findings to clients.

Salmon, B. and Rosenblatt, N. (1995). *The complete book of consulting.* Ridgefield, CT: Round Lake Publishing. This is a step-by-step guide, complete with legal forms, for those who want to establish a consulting practice.

Wimmer, R. D. and Dominick, J. R. (2000). *Mass media research: An introduction,* 6th ed. Belmont, CA: Wadsworth, pp. 29, 30, 39, 186, 327, 329–330, 357–358, 360, 426. The authors present current information on the rates research firms charge for various kinds of research.

3

Planning Research Projects

GOOD RESEARCH REQUIRES carefully considering every phase of a project from initial conceptualization through the analyzing and reporting of findings.

Although it is convenient to think of the process as proceeding through a number of phases, it would be a mistake to conceive of that process as a linear one. More often it is an iterative one that can begin almost anywhere. Information and ideas uncovered in one phase have an impact on other phases. That impact may force a reconsideration of decisions about how to proceed that previously seemed firm.

Regardless of where the process begins or how it proceeds, the process will most likely progress from a preplanning phase to one involving the actual design for the research. The goal is to develop a research methodology that will achieve the best possible balance between reliability and validity while simultaneously taking into account both ethical concerns and budgetary and time constraints.

Preplanning

Research begins with an idea. But whether that idea is one that occurs to the researcher, is imposed by the researcher's job description, or is handed to the researcher by a client, the idea is often a very fuzzy one. That is particularly true in applied research because of the potential for miscommunication between the researcher and the research sponsor. Therefore, this phase requires open, candid discussion between the investigator and the sponsor in order to set clearly agreed upon parameters for the study and then decide how to proceed.

Setting Parameters

Much time, money, and effort can be wasted on research that ultimately is meaningless unless the researcher and the research sponsor come to an agreement in three areas. They must share the same understanding of the problem to be investigated, the purpose for the study, and the amount of time and money that is available for it.

Define the Problem. It seems obvious that designing a study that will produce useful findings requires establishing the problem to be investigated. Yet applied researchers

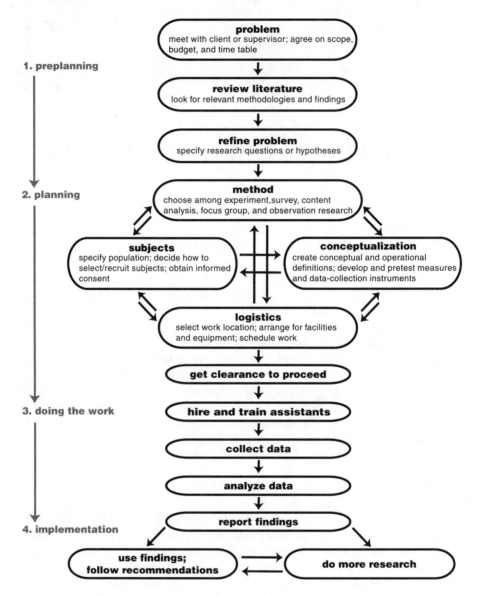

Figure 3.1. Planning and doing a research project

may initially find themselves in the position of having little guidance as to the nature of the problem they are asked to investigate.

A research firm may be hired to do a newspaper readership survey or to investigate complaints about the resolution of insurance claims. Staff researchers might be asked to evaluate the effectiveness of an advertising campaign; a nonprofit organization might call on its public relations practitioner for help in developing strategies to increase public support for the organization's goals. At first glance, the purpose of each of those projects seems clear. But in reality, each is subject to multiple interpretations.

As a practical matter, it makes a real difference whether the underlying problem that prompted the request for the readership survey is a perceived need to attract new subscribers or to satisfy current readers. If the latter, it also matters whether the newspaper wants to know how subscribers evaluate the paper or how they may react to proposed changes in content or delivery.

Similarly, it matters greatly which publics are of most concern to the nonprofit organization and whether the need is to find out what services will increase support or which issues most concern supporters and nonsupporters. In the case of the advertising researcher, it matters whether the client is more interested in whether the campaign initially increased awareness, knowledge, or purchase of the product or whether it is still having any or all of those effects. And in the case of the insurance complaints, it makes a difference whether the need is to know who complains, what they complain about, or whether there are differences in the complaints lodged against traditional insurance companies and against health maintenance organizations (HMOs).

In addition to raising questions about just what the client wants to know, the initial assignments also leave unanswered questions about the appropriate **level of analysis**. In the readership study, the client very likely needs an analysis at the individual level—readership by individual people.

The nonprofit organization might want to know about individuals, but data analysis comparing the opinions of several publics to each other and to organization priorities might also be appropriate. The study of the effect of the advertising campaign might also require analysis at the individual level, but it might be equally or more useful to conduct the analysis at the regional level—that is, design it so that it is possible to analyze data to show how effective the campaign is in different parts of the country.

For the insurance study, the client might also want to know what kinds of people most often complain or about regional differences in complaints. But the client might also be interested in an analysis at the level of the dyad—how often both doctors and their patients complain about the same claim. There might also be reason to determine whether differences occur between the kinds of complaints lodged against health maintenance organizations and traditional insurance companies, which would require data analysis at the organizational level.

Determine the Purpose. Some studies may be needed simply for surveillance purposes—to see what is going on. Others may help a company identify problems or opportunities or decide what they should do. In other cases, research may help determine how well goals are being achieved.

In the field of public relations, for example, practitioners frequently conduct **strategic research** to help them determine objectives and plan campaigns. They also do various kinds of **audits** to evaluate their communication and help clarify their organization's issue positions or determine the organization's image. They identify relevant groups, or publics, and investigate the opinions of those publics about the organization and their issue positions in comparison to those of the organization. **Situational analyses** help them monitor the environment, gather background information from the media and from their publics, and then use that information to decide what to do and how to do it in order to achieve organizational goals. They conduct **formative research** to help decide what to do and then to monitor, evaluate, and improve the campaign. At the conclusion of a project, they conduct **summative research** to learn whether campaign goals have been met.

Although those from other fields do not routinely use the same language to describe their work, they conduct or commission the same kinds of research. Technical writers

> **Box 3.1.** *Levels of analysis in research*
>
> The level (or unit) of analysis is the major entity of social life on which a research question focuses. The most common levels of analysis are the following:
>
> **Individual Level.** Focus is on the individual people such as adults, students, doctors, or, in some cases, individual social artifacts such as newspaper articles, funeral notices, or minutes of meetings. A study examining the relationship between watching television news and intention to vote is an example.
>
> **Interpersonal Level.** Focus is on the interactions between two or more people such as a doctor-patient dyad or a triad such as a love triangle. A study that examines the self-disclosure strategies of romantic couples is an example.
>
> **Group Level.** Focus is on social groups such as families, a club, or a task force. A study that explores changes in family composition and communication styles over time is an example.
>
> **Organizational Level.** Focus is on formal social organizations such as a corporation or church congregation. A study that examines the level of electronic monitoring of employees in Fortune 500 companies and in small businesses is an example.
>
> **Geographical Level.** Focus is on a specific geographical unit such as a town, census tract, state, or country. A study that examines regional differences in the popularity of country music is an example.

and educators conduct research to find out what kind of teaching techniques and training programs people want and need and **usability studies** to find out whether training programs, manuals, or other educational materials are effective. Publishers and broadcasters conduct or commission studies to find out what their audiences want, whether they are satisfied, and how they may react to change.

Advertising agencies and their clients use research to determine people's knowledge and opinions about companies and their products or services, to identify consumer wants and needs, and to monitor trends in consumer behavior through **tracking studies.** The results help them monitor, evaluate, and improve their product and their position relative to their competition.

Both the nature of the problem that triggers the need for research and the way the findings will be used help determine the kind of information that a researcher must collect. However, the question of why research is needed also has implications for the method of data analysis and the format for reporting the findings. Some studies may be exploratory or descriptive; others must provide explanations. Still others may need to be designed in order to show causality or to help predict the future.

Exploratory studies provide some initial information about a topic or problem. Their purpose is simply to show "what is out there." **Descriptive** ones tell "how things are." They give a snapshot of reality, whereas **explanatory** studies must answer questions about why things are the way they are or how things work. All three may be designed using the deductive or inductive model; data analysis may be qualitative, quantitative, or a combination of the two, depending on the preferences of the researcher and the client. If quantitative analysis is used, fairly simply univariate or bivariate statistics are often sufficient.

Box 3.2. *Basic categories of statistics*

In quantitative research, statistics are used to summarize large data sets. Statistics may be classified as univariate, bivariate, or multivariate according to the number of variables or measures used in calculating them and further categorized to indicate whether values cluster together or spread apart.

Statistics may also be categorized as descriptive or inferential according to how they are used. Because they summarize data on just one measure or variable, univariate statistics are usually considered descriptive. Bivariate or multivariate statistics may also be used descriptively to show patterns among characteristics or inferentially to make judgments about the strength and statistical significance of those patterns in a population.

I. Univariate Statistics. These univariate statistics summarize the distribution of scores on a single measure or variable. Most commonly they are used descriptively.
   A. Measures of Central Tendency. These statistics show how values on a single variable cluster. They include the **mean (average), median,** and **mode.**
   B. Measures of Dispersion. These show how individual scores on a single measure scatter across the set of all possible values. Some researchers place the **percentile** in this category, but the most widely recognized and used measures of dispersion are the **range** and **standard deviations. Variance** is used to calculate the standard deviation, but it is rarely reported separately.
II. Bivariate Statistics. These may be used descriptively to summarize data sets by showing patterns between two variables or measures or between two data sets on the same variable or measure or inferentially to test a hypothesis. When used with probability sampling, they may also be used inferentially to draw conclusions about whether the relationship is likely to hold true for an entire population.
   A. Measures of Association. These statistics show the relationship, or correlation, between two variables. The most common of these statistics are **chi square, Cramer's V, phi, Kendall's tau, Spearman's rho,** and **Pearson's r.**
   B. Tests of Difference. These statistics show whether there is any difference between two sets of scores. The most common ones include the **Z-Score, t-test,** and **analysis of variance (ANOVA).**
III. Multivariate Statistics. These make it possible to account for additional variables. Although they may be used descriptively, they are most often used inferentially to test a hypothesis or draw conclusions about a general population based on findings from a probability sample of subjects.
   A. Measures of Association. These show the relationship when three or more variables must be accounted for. Common ones include **partial correlation** and **regression. Factor analysis** may also be included here.
   B. Tests of Difference. These show whether there is any difference between sets of scores, while taking account of or holding constant scores on additional variables. The most common of these are elaborations of the basic ANOVA such as **analysis of covariance (ANCOVA), k-way or multiple analysis of variance (MANOVA),** and **multiple analysis of covariance (MANCOVA).**

However, if making **predictions** about the future—about "what will happen if"—or if showing a **cause-and-effect** relationship is the goal, the deductive model is most appropriate. Quantitative data analysis is virtually a requirement.

Causal studies also require tighter control over extraneous factors that could affect the outcome than is necessary for studies intended to serve other purposes.

For the exploratory, descriptive, and explanatory ones, it could well be enough to show that a particular relationship exists. For predictive ones, it could be enough to show that if the "cause" is present, one will very likely get the "effect." However, the cause-and-effect relationship might not be real. In many studies, there may be a **third variable problem** in that some other phenomenon is affecting or causing what at first seems to be a causal relationship.

In order to confirm a cause-and-effect relationship, a researcher must be able to show four things:

1. Temporal and/or geographic proximity exists between the presumed cause and the effect.
2. The cause always comes before and not after the presumed effect.
3. An empirical relationship exists between the cause and effect in which, for example, an increase in the cause consistently correlates with either an increase or a decrease in the effect.
4. Other possible causes for the presumed effect have been ruled out.

Identify Constraints. Designing a research project almost always requires making compromises between what ideally should be done and what can reasonably be done in any given case. Ethical considerations impose some limits on every project, but what is feasible more often depends on the available resources. Before the design process can proceed, it is crucial to determine how much money and time are available for the work. It is also important to identify, as soon as possible, what space, equipment, and support staff may be available.

Ideally, for example, our newspaper and public relations researchers might want to design studies that called for face-to-face interviews with 2,000 people, but that would be impossible on a budget of $200. At most several hundred telephone calls might be possible but only if the researchers could recruit callers who did not have to be paid and then only if the work did not have to be done quickly.

Similarly, our advertising researcher might want to conduct a tracking study that met the requirements for showing a cause-and-effect relationship, but she might have to choose a cheaper or quicker alternative because of time or budgetary constraints. The researcher investigating complaints about insurance claims might know that ideally he should collect information directly from HMOs, insurance companies, and their clients in every city and state as well as information from public records. However, even with a staff of willing workers and an unlimited budget, he would have to find a much quicker, if less comprehensive, way to collect data if he needed the findings for a congressional hearing next Friday.

As the examples suggest, some kinds of research are more expensive than others; some take longer. Some may require a particular kind of space or equipment that must either be available or attainable within budget limitations. Some projects can be accomplished by a single researcher, whereas others may require using available staff or hiring extra workers.

Reviewing the Research Literature

Once the basic parameters for the research project have been established, the pre-planning process becomes one of gathering information to help determine how to proceed. Even though this step takes time, omitting it can be costly.

Scientific research builds on the work of others. By reviewing the literature, you can find out what is known and unknown about the problem and its possible solutions. You may also be able to find out whether others have been faced with the same problem and, if so, how they have tried to solve it and how successful attempted solutions have been. If this information is extensive, a literature review may make further investigation unnecessary.

In most cases, however, you will find that others may have faced similar problems but not exactly the same one you are facing, or you may find they have investigated the same problem but in a setting different enough from your own that findings are not directly transferable. Occasionally you may even find that no one seems to have conducted research that addresses your concerns. In all of these cases, the value of the literature review is that it can provide methodological guidance.

Even previous work that seems only tangentially related to the problem at hand can provide ideas for choosing data-collection strategies and techniques. Because methods that have been used successfully in previous investigations are likely to work again, borrowing ideas from previous work saves the time that would otherwise have to be invested in developing original strategies and techniques. This kind of borrowing from previous research also increases the chances that the new study will be reliable and that the measures will possess acceptable internal validity. Finding a similar study and then replicating or partially replicating it also contributes to external validity by allowing findings to be placed in a broader context than would otherwise be possible.

Potentially useful information and ideas may be published almost anywhere. With the increasing availability of computerized access to the full text, abstracts, or indexes of many different kinds of publications, it is relatively easy and generally useful to begin by doing a broad computer-based search just to find out what may be available.

Depending on what that kind of search turns up, it may be necessary to go to the library to examine books or look through journal articles or compilations of abstracts in order to narrow the list of potentially relevant works to those that appear most useful. In other cases, it may be necessary to examine studies on other topics in order to find useful information or ideas that may lie buried within them.

For the literature review, it is usually most beneficial to concentrate on primary sources rather than on secondary ones. In **primary sources,** researchers report on their own work. In doing so, they usually provide enough information about their research methodology to make it possible for a reader to judge the quality and relevance of the work. In **secondary sources,** an author reports on someone else's work. Therefore, the information in these reports is akin to hearsay evidence. Secondary sources are more likely than primary ones to omit useful information or contain errors and ambiguities.

Although dissertations, theses, and some books qualify as primary sources, journal articles are more widely used by applied researchers. To find these articles, it is most efficient to first search the social science journals devoted to communication research. Some of the most generally available of these are listed in Box 3.3. If searching those journals turns up little relevant information, the search may be extended to journals from other fields or to the trade literature.

Box 3.3. *Communication research journals*

Advertising

Journal of Advertising
Journal of Advertising Research

Applied

Applied Communication Research
Journal of Applied Communication Research
Journal of Applied Communications

Education

Communication Education
Education Technology Research and Development
Journalism Educator
Marketing Educators' Journal

Journalism and Mass Communication

Communication Research
Howard Journal of Communication, The
Journal of Broadcasting and Electronic Media
Journal of Communication
Journal of Mass Media Ethics
Journal of Media Economics, The
Journal of Media Management
Journalism and Mass Communication Quarterly
Journalism Monographs
Mass Communication and Society
Newspaper Research Journal
Television and New Media

Marketing and Consumer Behavior

Journal of Consumer Research
Journal of Customer Service in Marketing
Journal of Marketing
Journal of Marketing Research

Public Relations

Public Relations Journal
Public Relations Quarterly
Public Relations Research Annual   (continued)

Speech Communication

Communication Quarterly
Communication Research Reports
Communication Studies
Human Communication Research
Human Relations
Quarterly Journal of Speech
Southern Communication Journal
Western Journal of Speech Communication

Specialized Topics

Behavior Research Methods, Instruments and Computers
Business Communication Quarterly
Communication and Cognition
Communication and the Law
Information Society, The
Journal of Health Communication
Journal of Media and Religion
Journal of Non-verbal Behavior
Political Communication
Public Understanding of Science
Technical Communication

Trade magazines, newsletters, and industry reports can be either primary or secondary sources depending on the article and the publication's policy. These sources should not be overlooked because they often provide very current findings from applied research projects and/or useful information about current problems and ways they are being addressed within an industry. However, even those articles published in the trade press that qualify as primary information sources rarely provide the kind of historical context or methodological detail found in books and journal articles. Because of the proprietary nature of the applied work reported in many of these sources, they may also provide only sketchy information about the findings. Moreover, the quality of the information available in the trade literature may be more suspect.

In contrast to articles published in the trade literature, articles published in research journals pass through a peer-review process before they are published. This process of submitting work to a panel of qualified researchers guards against seriously flawed research finding its way into the literature. However, it is always a good idea to perform an independent evaluation before relying too heavily on the work of others, regardless of where it may have been published. For information on evaluating research literature, see Appendix B.

Refining the Project

Although reviewing the literature can be a time-saver, the review can also turn up so many good ideas that researchers find themselves wanting to do everything. As a result, what may have started as a fairly simple project can easily mushroom into a research program that would take years to accomplish. Therefore, at this stage it is a good idea to revisit the purposes and goals identified earlier in order to differentiate between those things it is crucial to find out, those that would be useful, and others that would be merely interesting to know.

Time and the availability of resources will determine how much it is possible to do. However, creating a prioritized list of objectives and then submitting that list to the client or sponsor will help keep the project on track. If the client agrees that those priorities are appropriate, the researcher can concentrate on converting the purpose statement into a series of research questions and/or hypotheses.

Research questions are more common in applied research; hypotheses are more common in basic research than in applied studies. However, hypotheses are appropriate whenever the literature or personal experience provides clues as to the likely results of an investigation. Research questions are appropriate in cases where the investigator does not know what to expect.

Research questions are single sentences that specify exactly what the investigator hopes to learn. Research **hypotheses** are single sentences that tell specifically what the investigator expects to find. Although some research questions may ask about a linkage between or among phenomena of interest, hypotheses must link the terms to show the expected relationship between them.

As an example, our newspaper researcher might decide that it would be interesting to know how much time people spend with various kinds of content and which content types they think are most important, but the first priority is to find out who subscribes to the paper, who reads it, and how subscribers and nonsubscribers, readers, and nonreaders evaluate its quality.

If the client agrees, the researcher can then concentrate on turning those priorities into a series of research questions that will guide data collection and analysis. These research questions might be purely descriptive. That is, they might simply specify the kind of information that must be collected. As an example, one research question might be:

Q_1: What are the demographic characteristics of newspaper subscribers and readers?

However, research questions may also be relational. These questions ask whether two terms, each specifying a kind of phenomenon, are linked. An example of a relational research question would be:

Q_2: Is there a relationship between subscribing to the local paper and the way people rate it for quality?

or even more specifically:

Q_3: Do newspaper subscribers rate the quality of the newspaper higher than nonsubscribers rate it?

However, from reading the literature or through personal experience or simple logic, the researcher might be in a position to turn research question 3 into a **two-tailed,** or **nondirectional,** hypothesis:

H: There will be a significant difference in the way newspaper subscribers and nonsubscribers rate the quality of the newspaper.

With more information, he might create a **one-tailed,** or **unidirectional,** hypothesis:

H: Subscribers will rate the quality of the newspaper significantly higher than will nonsubscribers.

Research questions are answerable, but any answer will do. For the question linking newspaper subscribing and evaluation, both "yes" and "no" function satisfactorily. In contrast to the research questions, hypotheses are testable. Because they specify an expected linkage, they can be supported or proven false. Therefore, a null hypothesis will always be associated with every hypothesis. In contrast to the hypothesis that predicts a relationship, the **null hypothesis** predicts there will be no relationship. For both the one-tailed and two-tailed hypotheses predicting a relationship between subscribing and evaluations, the null hypothesis would be:

H_o: There will be no significant difference in the way subscribers and nonsubscribers rate the quality of the newspaper.

Finding support for the null hypothesis disproves the main hypothesis. But a one-tailed hypothesis may also be refuted by showing support for an **alternative hypothesis.** An alternative hypothesis calls attention to the possibility that the relationship might be just the opposite of what the researcher expects. For the one-tailed hypothesis about newspaper subscribing and newspaper evaluation, the alternative hypothesis would be:

H_a: Subscribers will rate the quality of the newspaper significantly lower than will nonsubscribers.

The exact wording of research questions and hypotheses will depend on the purpose of the study and on the state of knowledge about the problem that inspired the research. The important point is that relational research questions and research hypotheses should be worded so that the terms in them clearly indicate phenomena that can be measured or observed and that the specified relationship among those terms clearly points to temporal or causal order among them.

These terms become the variables used for data analysis. They may be classified as independent or dependent depending on their relationship to each other in the hypothesis or research question. The **dependent** variable is what the researcher is trying to explain. If a relationship exists between the independent and dependent variables, the observed or measured value of the dependent variable changes systematically with changes in the value of the **independent** variable.

In experimental research, the independent variable is the presumed cause; the dependent variable is the effect. In other kinds of research, the variables may sometimes be thought of as suggesting cause and effect. Just as often, the independent variable is the one whose value predicts the value of the dependent variable or that, in a temporal sense, precedes, or is antecedent to, the dependent variable. Therefore, the independent variable is sometimes called a **criterion** or an **antecedent** variable.

But sometimes the order among the variables is unclear. In the newspaper researcher's hypothesis, "subscribing" is the independent variable, and "quality" is the

> **Box 3.4.** *Common forms for hypotheses*
>
> Hypotheses (and relational research questions) may be worded differently depending on the research purpose and the state of knowledge about a research problem. The most common wordings take one of four specific forms:
>
> **If . . . then (Conditional) Statements.** These statements indicate that if one condition exists, another condition will also exist:
> H: If a person subscribes to a newspaper, that person will give it high marks for quality.
>
> **Continuous Statements.** These statements indicate that as one term increases (or decreases), the other term will systematically increase (or decrease):
> H: The longer people have subscribed to a newspaper, the higher they will rate it for quality.
> H: As people's income increases, their ratings of the quality of the newspaper will decrease.
>
> **Difference Statements.** These indicate that one phenomenon, or variable, differs in terms of the other phenomenon, or variable:
> H: There will be a significant difference in the way newspaper subscribers and nonsubscribers rate the quality of the newspaper.
> H: People with high levels of education are less likely than people with low levels of education to give the newspaper high marks for quality.
>
> **Mathematical Statements.** These state the hypothesis as an equation of the general form $Y = (f)X$, or "Y is a function of X":
> H: Evaluations of newspaper quality are a function of subscribing to the paper.

dependent variable. He hypothesized that whether or not a person subscribes precedes the evaluation. However, he could have hypothesized that evaluation precedes the decision to subscribe: People who give the paper high marks for quality will be more likely to subscribe than will those who judge its quality to be poor.

In multivariate research, developed by elaborating on a simple hypothesis, any variable that influences or comes before the variable identified as the independent one may also be referred to as an **antecedent** variable. Others that may influence the basic relationship between the independent and dependent variable are referred to as **intervening** variables if those variables are to be taken into account during data analysis. They may also be used as **control** variables in cases where the researcher wants to eliminate their influence.

To elaborate on the hypothesis linking subscribing and perceptions of newspaper quality, the researcher might add one or more additional variables to specify contingent conditions:

H_1: Among newspaper subscribers, older people will give the newspaper higher marks for quality than will younger people.

H_2: Among subscribers, long-time residents will give the newspaper higher marks for quality than will newcomers.

Deciding How to Proceed

The final, complete list of hypotheses and/or research questions serves as a guide for planning, collecting, and then analyzing data and reporting the findings. However, at this stage it may be necessary to reflect on that list and decide whether new research is really needed and, if so, whether this is a do-it-yourself project or one that requires help.

The newspaper researcher and the researcher assigned to find out about the effectiveness of a particular advertising campaign will almost certainly have to collect new data specific to their clients, but the public relations practitioner might find that there is enough published information on the salience various groups attach to important issues to make further investigation unnecessary. More likely, however, she might find that those published studies did not analyze the data or report findings in ways that fully satisfy her organization's needs. If that is the case, she might be able to accomplish her purpose by doing a secondary analysis using data sets that may be downloaded from the Internet or that may be available from other researchers.

A **secondary analysis** of pre-existing data sets should not be confused with relying on secondary sources. Although pre-existing data sets may not include all of the information you might want or the data may not have been collected exactly the way you would have collected it, you will have access to the original investigator's methodology and all of the raw data. This makes it possible for you to check the original investigator's work in a way that is impossible when relying on secondary sources. With access to the data set, you have complete control over how you analyze the data and how you interpret the findings.

Box 3.5. *Commonly used data sets available online*

General Social Survey: an annual personal interview survey of U.S. households conducted by the National Opinion Research Center (NORC). Since 1972 the survey has gathered data on demographics, attitudes, and behaviors (http://icpsr.umich.edu/GSS99).

Inter-University Consortium for Political and Social Research: provides access to survey data to students and researchers at universities belonging to the consortium. The data archive includes data from over 9,000 studies in more than 130 countries (http://www.icpsr.umich.edu).

National Election Studies: data on voting, public opinion, and political participation. (http://www.umich.edu:80/~nes/).

U.S. Bureau of the Census: demographic and other information gathered by the federal government (http://www.census.gov).

Other data sets can be found by searching the web using "social science data sets" as the key term. Also check out Data on the Net at http://odwin.ucsd.edu/idata/. This is an extensive, well-organized searchable and browsable database with links to websites that have numeric data sets ready to download and to websites of distributors and vendors of data that offer searchable data catalogs. The site is developed and maintained by the Social Science Data Collection, University of California, San Diego.

Even though the researcher investigating complaints about HMOs and insurance companies might not find pre-existing data sets that he could use to perform a secondary analysis, he might find much raw data available online in chatroom discussions or in documents available through government or industry websites. As with the case of the public relations researcher, finding pre-existing information or data sets simplifies the planning process by virtually eliminating many methodological options. But as with the newspaper and advertising researchers, it does not guarantee that the project can be carried out successfully without outside assistance.

At this stage both the newspaper researcher and the researcher investigating complaints might know how to proceed but realize that in-house resources are insufficient to allow them to collect and analyze the data in a timely manner without outside help. The public relations researcher might know how to access, download, and analyze individual data sets but needs help deciding whether a metanalysis would be appropriate and, if so, how to combine different data sets and analyze the data from that combined set. If a small advertising agency needed the data on campaign effectiveness to satisfy a client, the person charged with gathering that information might need help developing an appropriate methodology and then collecting and analyzing data.

In any of those situations, help can be obtained from research firms or independent consultants listed in city phone books under headings such as "research services." In many places, help may also be available from university research centers or from faculty in departments of journalism/mass communication, advertising, marketing, sociology, political science, or economics.

Outside experts may differ greatly as to services offered and areas of expertise. Some may specialize in one or two methodologies, whereas others have the capabilities to do almost any kind of research. Some may provide help with planning or data analysis; others will also do the work of data collection.

They may also charge very different rates or have very different capabilities as to the number of projects they can undertake at any given time and how fast they can provide the results to a client. In choosing the appropriate consultant or firm, cost is obviously a consideration, but so are expertise, quality, and speed of service. Therefore, careful checking is necessary in order to find the right help.

Planning

If, during the preplanning phase, it becomes obvious that you will need help with the actual research, you may want to contact several universities, research firms, or consultants and then hire outside help before planning the study. But with or without outside help, at this stage you should be ready to develop your research plan.

Developing the plan consists of three basic parts—choosing a method, developing a methodology appropriate to that method, and then working out the logistics for doing the work.

Although it is convenient to think first about choice of method before considering other facets of the planning process, developing the research plan may not proceed in that order. The basic problem may allow using any one of several methods or virtually dictate the choice of a particular method, but frequently methodological considerations—the availability of research subjects or the need for a particular kind of information—must be considered before choosing a method. In either case, the method or methodology may have to be rethought as budgetary or ethical considerations come to light. In devising the logistics, such as where or when the work will be done, it may become apparent that the initial plan is unworkable.

Choosing the Method

Researchers have their choice of five basic methods: experiments, surveys, content analysis, focus groups, and observation.

The model for scientific research is the **experiment.** In the classic experiment, subjects are divided into a test group that receives a stimulus or treatment (such as exposure to a particular message) and a control group that is not exposed to that stimulus. The ability to compare data collected before and after exposure to the stimulus and from those exposed and unexposed to it make this the method of choice whenever it is important to show cause-and-effect relationships.

Surveys are usually the best option for collecting large amounts of data from many people. In the true survey, a random sample of people, called respondents, are asked for their answers to questions asked in exactly the same way, in the same order, and with the same response options. This makes surveys the method of choice for determining the distribution of opinions and behaviors throughout a population.

A **content analysis** is a systematic investigation of messages in fixed format—a survey of documents, broadly defined. As such, it is appropriate only for studying what is in a document. It cannot be used alone to find out why that message is there or how people react to it.

Focus groups are small group discussions facilitated by a moderator trained to encourage participation and keep the conversation on track without leading it in a particular direction. Tapes or transcripts of the discussion become the raw data that are analyzed to produce findings. The focus group is a method of choice whenever there is a need to gather preliminary data, gain insight into people's thinking, or explore more fully the results from other kinds of research. Therefore, the focus group can be a useful component of evaluation research.

With **observational research,** the researcher gains access to a group or a particular location to find out firsthand how people behave in a natural setting. By watching and talking to people in their natural setting, this kind of research, when carefully done, can produce a better understanding of reality than any other method.

Detailed information on each of these methods and their most common variants can be found in Chapters 6 through 10.

Although each method has its own strengths and weaknesses, for some projects only one method may be a viable option. For others, several possibilities may exist. In those cases, there is rarely a right or wrong choice. More likely you will make your choice among options that are better or worse for a particular project.

The advertising researcher will very likely choose some variant of the experiment because that method is best for determining cause-and-effect relationships. But surveys and focus groups would both be options if the researcher later found it necessary to investigate why the advertising campaign was not as effective as it could have been or if the researcher wanted to gather ideas for a new advertising campaign.

The newspaper researcher and public relations practitioner will very likely choose the survey because that is the only method that will allow generalizing the findings beyond those people included in the study; however, if they were more interested in gathering ideas for how to increase subscriptions or improve support for organizational goals than in the generalizability of the findings, focus groups could be a good choice.

Given the constraint imposed by a tight deadline, content analysis may be the only option for the researcher investigating complaints about insurance, but with more time surveys and focus groups would become alternatives.

Although experiments and surveys are usually thought of as methods for quantitative research, under some circumstances both may lend themselves to qualitative analysis. Content analysis lends itself equally well to both quantitative and qualitative data analysis. However, quantitative analysis is inappropriate for data gathered by using focus groups or observation. With those methods, using statistics beyond an occasional percentage may introduce a false sense of precision or invite generalizing or making predictions beyond what the data allow.

Developing the Methodology

Once the basic method has been chosen, the methodology can be worked out. The methodology consists of the techniques or procedures for choosing subjects for the study and the ways in which the phenomena under investigation will be recognized, categorized, or measured. The goal should be to produce a written methodological statement that is clear and complete enough that another researcher could use it as a guide for replicating the study or, if need be, completing the work as the researcher intended to do it.

Choose the Subjects. In research, the subjects serve as the source of the information needed to test the hypotheses or answer the research questions. The key to choosing subjects wisely is to first determine who or what will be able to provide the most useful information at a reasonable cost in terms of time and money. In most cases, this means gathering information from a relevant group of people or, when content analysis is the method of choice, from relevant documents. But it may also require taking into account who will be willing or able to provide information or what documents are accessible to the researcher. Where many possibilities exist, you may need to impose geographic or temporal boundaries in order to limit the possible subjects to the most relevant ones.

For example, the newspaper researcher could reasonably assume that people living within the paper's circulation area would be appropriate subjects and that selecting subjects from that group would include both subscribers and nonsubscribers who can provide meaningful information.

The public relations researcher could also reasonably assume that people could provide meaningful information. But if the organization sponsoring the study were a regional or national one, the researcher would have to define the relevant publics more broadly than just those people living in the city where the organization has its headquarters. If one goal of the research were to compare supporters' and nonsupporters' interests to those of the organization, enough nonsupporters would most likely be available in the general population, but finding enough supporters might require choosing some subjects from the organization's membership or contribution lists.

The advertising researcher, however, could not readily assume that people would be able to provide meaningful, accurate information about purchasing habits over time. For such information, the researcher could try to get access to sales figures from stores selling the product, but some might not be willing to share those numbers. Therefore, the client company's own sales records would probably provide the best available information.

As the case of the advertising researcher illustrates, the source suggested by the level of analysis and the actual source of data may not be the same. That difference, as well as problems of access and availability, can also be seen in the study of complaints about HMOs and insurance companies.

For that project, the researcher wants to know about organizations, but information about them must come from people or from documents. Ideally the researcher might want to ask corporate officers about complaints they have received or examine internal documents from traditional insurance companies and from HMOs. But company officers might not cooperate; corporate records would very likely be unavailable to an outsider. If time were not a factor, the researcher could ask a large number of people about their experiences getting reimbursement for medical costs. But with a deadline looming, the subjects will necessarily be publicly available documents; however, because of rapid changes within the health-care industry, it might be appropriate to eliminate from the study any documents that are more than a few years old. With or without that limitation, information from them can be used for analysis at the level of the organization as well as to show similarities and differences in the complaints at the regional level or at the level of the individual complainant.

As the examples illustrate, the choice of appropriate subjects can be simple and obvious. But it can also be a complex, multistep process. Where many potential subjects are available for a study, you may need to limit their number. The advertising researcher would need to use all of the sales figures; the public relations researcher might be able to include all supporters but just some nonsupporters. The other researchers would also have to limit the number of subjects through sampling. Basic sampling techniques are described in Chapter 4. Additional method-specific information on choosing subjects can be found in Chapters 6 through 10.

Create Measures. Both the choice of method and the subjects available for a study have implications for the techniques that you can use to collect the information you need to answer the research questions or test the hypotheses. Creating these techniques requires defining the linked terms in the questions or hypotheses in ways that allow you to measure them. This process converts the terms into the variables used in data analysis.

The linked terms, as they are expressed in the questions or hypotheses, are concepts or constructs. A **concept** expresses an abstract idea formed by generalizing from particulars. A **construct** is a complex concept. It subsumes several concepts.

In our newspaper example, "subscribing" is a concept. It is unidimensional. "Quality" is a construct. It is multidimensional: People's perceptions of quality would most likely include judgments about the writing, reporting, photography, layout, accuracy, fairness, and perhaps even the delivery service.

Although constructs may be unique to a particular study and their meaning may not be readily apparent, the definition of a concept often seems obvious and universal. Nevertheless, defining them can be equally important. Creating clear definitions for concepts and constructs begins the process of linking the abstract world of theory to the real world where empirical research takes place. The definitions create the ground rules for measurement that, in turn, is the basis for data collection and analysis. More complete information about developing, testing, and using measures is given in Chapter 5.

Working Out the Logistics

Once the methodology is complete or nearly so, a few tasks remain before the actual research can begin. These include planning the work schedule and preparing and pretesting any materials needed for data collection.

The materials may include protocols to guide data collection by means of focus groups, observational research, and sometimes also for experiments; creation of test messages or other stimuli for manipulating the independent variable in experiments; questionnaires for use with surveys or to collect information in conjunction with experiments and focus groups; or coding sheets to use with a content analysis of documents or with transcripts from in-depth interviews, focus groups, or observations.

Preparing the needed materials may take anywhere from a few hours to months of hard work. But this is only a part of the job. The materials also must be pretested on a small group of subjects similar to those who will be used in the study in order to detect unclear questions, inadequate instructions, or other flaws that could seriously affect the reliability and validity of the study. In addition, it may be necessary to install and test the computer software that may be needed to facilitate data collection and analysis.

If you will use research assistants to collect data, someone will also need to train them in proper procedures. Otherwise differences among them will affect reliability and validity.

As work on this phase proceeds, decisions will have to be made about when to do the work. Many studies can be done at almost any time. For others, timing may be critical. The time by which data are needed by a client must be taken into account, but so must other factors. Seasonal differences in people's schedules may make certain kinds of subjects less available at one time of year than another or otherwise bias the results. You may have to time data collection to fit in between other projects to which in-house researchers or outside research firms are already committed or to coincide with the availability of secretarial support or of suitable facilities for conducting focus groups or laboratory experiments.

Doing the Work

All of the methodological and logistical details must be worked out in advance to keep problems that could affect the quality of the work to a minimum and to avoid unnecessary and potentially costly delays. With careful preplanning and planning, data collection should proceed fairly smoothly.

In collecting the data, the key to ensuring reliability is to make sure everyone connected with data collection adheres to the methodology developed in the planning stage. Having a written statement that everyone can consult as necessary helps guard against introducing artifacts into the data-collection and analysis process that could affect the reliability and validity of the work. Having such a statement to consult for details about how the work was done can also be useful in writing the final report or as a model to use in the planning process if the need should arise to replicate a study.

Although it is important to agree upon all procedures and to write them down in advance, things sometimes happen during data collection that make adjustments necessary. When that happens, changes should be introduced only as a last resort and then in a way that does the least possible damage to the overall reliability and validity of the study. Be certain to note these changes and the dates when they occurred in the written methodology so their effects can be detected in data analysis. Methods for detecting problems with subject selection and with measures can be found in Chapters 4 and 5.

If no changes in procedure were introduced during data collection or their effects were minimal, data analysis is primarily a matter of using the quantitative or qualitative procedures that are appropriate for the research method and that will provide suitable tests for the research hypotheses and/or answers to the research questions. Data-analysis procedures for use with each basic method can be found in Chapters 6 through 10.

Main Points

- Good research requires carefully thinking through every stage of a project from its initial conceptualization through the analyzing and reporting of findings.
- The research process is not always a linear one proceeding step by step through each of the phases of designing a project. Instead, ideas and information uncovered in one phase may affect other phases. This may force the researcher to reconsider decisions about how to proceed that may have seemed already set.
- The main goal of research design is to develop a methodology that will achieve the best possible balance between reliability and validity while taking into account ethical concerns and budgetary and time constraints.
- In the preplanning stage, a researcher sets the parameters for the study by clearly defining the problem to be investigated, determining the purpose of the study, and identifying constraints.
- When defining the problem, applied researchers need to determine exactly what the client wants to know and what the appropriate level of analysis is.
- Common goals or purposes of research include exploration, description, explanation, and prediction.
- In order to show a cause-and-effect relationship exists between two variables, a study must be able to show temporal and/or geographic proximity between the presumed cause and the effect, that the cause comes before the effect, that there is an empirical relationship between the two variables, and that other potential causes for the presumed effect have been ruled out.
- Once the parameters for a study have been set, the researcher should review the research literature. This may involve examining secondary sources as well as primary ones.
- The goal of the research has implications for how the researchers will analyze the collected data and how they will report the findings. The goal should be refined into a purpose statement that then can be converted into a series of research questions and/or hypotheses.
- Research questions are single sentences that specify exactly what the researcher wants to find out. Research hypotheses are single sentences that tell specifically what the researcher expects to find concerning the relationship between variables of interest.
- A researcher can develop two types of hypotheses: two-tailed, or nondirectional, hypotheses and one-tailed, or unidirectional, hypotheses.
- In a hypothesis, the dependent variable is the effect or other outcome a researcher is trying to explain. The independent variable predicts, precedes, or causes the change, outcome, or value on the dependent variable. If the independent variable

is viewed as preceding the dependent variable, it may be called a criterion or antecedent variable.
- If resources are unavailable to collect and analyze the data in a timely manner, a research firm or outside consultant may be hired to help plan and/or carry out the research.
- In the planning stage, the researcher develops the research plan, which involves choosing a method, developing a methodology appropriate to that method, and working out the logistics of doing the work.
- Sometimes the plan will be to use findings from previously published studies to answer the research questions or to use someone else's data to do a secondary analysis. But if the plan requires collecting new data, researchers have their choice of five basic methods, each with its own strengths and weaknesses. These methods are experiments, surveys, content analysis, focus groups, and observation.
- Once the basic research method has been chosen, the methodology needs to be worked out. The research methodology includes determining how to choose the subjects for the study and how to observe or measure the phenomena of interest.
- To measure the phenomena of interest in a study, it is necessary to define the terms in the questions or hypotheses. The terms as they are expressed in the questions or hypotheses are called concepts or constructs.
- Before data collection can begin, researchers must also work out the logistics for the study. This includes locating facilities, scheduling the work, and preparing and pretesting the materials that they will need for the actual data collection. It also may include training assistants on the proper procedures for collecting the data.
- In the final stage, the researchers do the work. That is, they collect the necessary data and analyze them using the appropriate quantitative or qualitative procedures.
- In collecting the data, the key to ensuring reliability is to make sure everyone involved in the data collection adheres to the methodology that was developed in the planning stage.

Terms to Know

level of analysis
strategic research
audit
formative research
summative research
tracking study
usability study
exploratory research
descriptive research
explanatory research
predictive research
cause and effect
third variable problem
research question
hypothesis

dependent variable
independent variable
criterion or antecedent
 variable
intervening variable
control variable
secondary analysis
research method
experiment
survey
content analysis
focus group
observation
research methodology
concept

one-tailed hypothesis
two-tailed hypothesis
alternative hypothesis

construct
null hypothesis

Questions for Discussion and Review

1. Explain the differences among descriptive, exploratory, explanatory, predictive, and causal research. When would each kind be most/least useful?
2. What are the four criteria necessary to show that a cause-and-effect relationship exists?
3. Look through a communication research journal until (see Box 3.3) you find examples of studies conducted at at least three different levels of analysis. For each article, identify the level of analysis. Provide evidence from the article to justify your conclusions.
4. Read the abstracts of each article in a recent issue of a communication research journal. On the basis of the abstract only, classify the research described in each article as primarily descriptive, exploratory, explanatory, predictive, or causal. Note any indication that the research focuses on other types of research questions. Explain why you classified each article as you did.
5. Explain why it is important to review the research literature before beginning any study.
6. Go to the library and search the literature on a communication topic of interest to you. What research methods have been used to study this topic?
7. What is the difference between a research question and a hypothesis?
8. Reexamine the articles you looked at for question 3. This time classify the articles according to whether the studies were designed to answer research questions or test hypotheses. In what kind of studies are the researchers most likely to set out to answer research questions, and when are they most likely to test hypotheses?
9. Convert each of the following research questions to a hypothesis.
    Q_1: How does the amount of use of the Internet for political information compare to use of newspapers for political information?
    Q_2: Is use of online media for political information related to age?
10. For an area of communication that interests you, create three research questions you could answer and three hypotheses you might test by conducting research.
11. For each of the hypotheses you created in question 9:
    a. Tell whether it is a one-tailed or two-tailed hypothesis.
    b. Tell whether the linked terms are concepts or constructs.
    c. Identify the independent and dependent variables.

Readings for Discussion

Datta, L. (2000). Seriously seeking fairness: Strategies for crafting non-partisan evaluations in a partisan world. *American Journal of Evaluation, 20*(2):251–263. The author presents strategies for planning and executing applied studies so that subsequent debates will focus on the findings instead of on the researcher's motives or the credibility of the work.

Davis, J. J. (1997). *Advertising research: Theory and practice.* Upper Saddle River, NJ: Prentice-Hall, pp. 18–32. In these pages the author discusses the planning process as it applies to advertising research.

Katzer, J., Cook, K. H. and Crouch, W. W. (1991). *Evaluating information: A guide for users of social science research,* 4th ed. Boston: McGraw-Hill. The chapters in parts 5 and 6 provide a thorough discussion of procedures for evaluating research as well as interesting end-of-chapter questions to guide the evaluation process.

Krough, K. S. and Lindsay, P. H. (1999). Including people with disabilities in research: Implications for the field of augmentative and alternative communication. *Augmentative and Alternative Communication,* 15(4):222–233. The authors present three scenarios to build a case for the theoretical and practical value of including people with disabilities among the subjects for applied research.

Setzer, R. A. (1996). *Mistakes that social scientists make: Error and redemption in the research process.* New York: St. Martin's Press. Chapters 2 and 3 address general problems in the planning process; Chapter 5 focuses on problems associated with relying on primary and secondary sources.

References and Resources

Becker, H. S. (1986). *Tricks of the trade: How to think about your research while you're doing it.* Chicago: University of Chicago Press. This is one of the standard works dedicated to research conceptualization and planning.

Blankenship, A. B., Breen, G. and Dutka, A. (1998). *State of the art marketing research.* Chicago: American Marketing Association. This book, written for managers and marketing executives who direct or work with researchers, explains methodological and technological developments that have revolutionized marketing research as it takes the reader step by step through the research planning and development process.

Booth, W., Colomb, G. G. and Williams, J. M. (1996). *The craft of research.* Chicago: University of Chicago Press. This is another classic work on the research process.

Bracken, J. K. and Sterling, C. H., eds. (1995). *Telecommunications research resources.* Mahwah, NJ: Lawrence Erlbaum Associates. This is an annotated, topical, indexed guide to the most important sources for research on electronic media.

Creswell, J. W. (1994). *Research design: Qualitative and quantitative approaches.* Thousand Oaks, CA: Sage. This textbook provides instructions for conducting the two basic kinds of research and guidance for blending or choosing between them.

Fink, Arlene. (1998). *Conducting research literature reviews: From paper to internet.* Thousand Oaks, CA: Sage. This book provides guidance for locating, interpreting, and analyzing relevant studies from many different published and unpublished sources.

Hedrick, T. E., Bickman, L. and Rog, D. J. (1995). *Applied research design: A practical guide.* In addition to covering the basics, this book has information on testing for feasibility and on trade-offs among precision, reliability, and validity.

Rubin, R. B., Rubin, A. M. and Piele, L. J. (1996). *Communication research: Strategies and sources,* 4th ed. Belmont, CA: Wadsworth. This is a standard introduction to locating information sources in the field of mass communication.

Sterling, C. H., Bracken, J. K. and Hill, S. M., eds. (1997). *Mass communications research sources*. Mahwah, NJ: Lawrence Erlbaum Associates. This is an annotated, topical indexed guide to 1,400 important print and Internet sources of research and other information about print and electronic media history, economics, content, audience research, policy, and regulation in the United States.

4

Choosing Subjects

ONE OF YOUR first tasks as a researcher will be deciding who or what holds the information you need to test your hypotheses or answer the research questions. Those people, documents, or other entities become the subjects for the study.

The first step in choosing subjects is to define an appropriate population. The **population** is the set of all **elements**—people or things—from or about which information is needed. Although the terms can be used interchangeably without confusion, some researchers prefer to define a population as the set of all people from or about whom data are needed and use **universe** to describe a set of relevant documents, organizations, or other elements.

Only rarely, however, will all possible elements be defined as the relevant population. In most cases it is both necessary and appropriate to define the **target population** much more narrowly so that its exact boundaries more directly address the purpose for a given study.

Conceivably a researcher investigating declining audience shares for network television might define the population as all people living within the United States, but most likely that population would be too broad: It would include infants and toddlers as well as many others on the margins of producers' and advertisers' concerns. Therefore, for one purpose, a researcher might define the target population as "all adults ages 18 to 49" and for another as "all households with at least one child under 18."

Where the relevant population is relatively small, it may be possible to collect data from all members of the population. If data come from all units within the population, the study is based on a **census**—data collected from everyone or everything within a target population. However, where the population is large, as it would be in most studies of a media audience, collecting information from all individuals or households would be too time-consuming and costly to be feasible. In such cases it will be necessary to choose just some people or households for inclusion.

For including just some subjects, researchers have two choices. They may decide to use probability sampling or nonprobability sampling.

With either method, the first step is to specify the **sampling frame**. A sampling frame is an actual list identifying all of the individuals, households, documents, or other elements specified by the definition of the target population. Where no list is available, the sampling frame indicates a procedure, such as random-digit dialing, for locating relevant elements.

For **probability sampling,** a subset of a population is chosen from the sampling frame in accordance with scientific, statistical procedures so that the sample will be representative of the target population as a whole. In **nonprobability sampling** representativeness is not ensured because the subset or sample is chosen for other reasons and according to other criteria.

The choice of probability or nonprobability sampling depends on several factors. These include the amount of time and available resources for the project and its purpose.

Probability sampling is often more time-consuming and expensive than nonprobability sampling. However, it is the most appropriate choice whenever the purpose of the research is to draw inferences about the population as a whole. For that to be possible, the sample must be relatively large, but it can be only a small proportion of the entire population. Moreover, it must be possible to convince those chosen for the sample to participate in the research project. Therefore, probability sampling is often used for survey research and for content analyses.

In identifying a sampling frame to use with probability sampling, the goal is to find and then use one that neither **overregisters** nor **underregisters** elements in the target population. A sampling frame that includes all or most of the elements in the target population but also includes elements that are not part of that frame suffers from overregistration. One that misses many relevant units suffers from underregistration.

Overregistration or underregistration of the sampling frame is not a major concern in nonprobability sampling. Samples may be quite small. Because representativeness cannot be guaranteed, nonprobability sampling is most appropriately used in research projects where drawing relatively firm conclusions about the population is not the goal. But it can also be used in cases where time and money may be too limited to allow using a large random sample or in situations where it may be difficult, if not impossible, to choose a large sample of subjects and get them to participate in the study. Therefore, it is often the method of choice for laboratory experiments, focus groups, and for observational research.

Probability Sampling

Probability sampling is the process of selecting individuals, households, documents, or other elements from a target population in such a way that each element in the population has an equal and known chance of becoming a part of the sample. That means the sample must be chosen randomly (i.e., according to the laws of chance). The researcher should not know a priori which elements will be selected. Neither the researcher nor the subjects should be able to influence in any way the decision about who or what gets included in the sample.

Where those conditions are met, the sample should match the population on all relevant characteristics within certain limits that can be calculated. The number or proportion of those characteristics within the population or universe is called a **parameter.** The number or proportion within the sample is referred to as a **statistic.** If, for example, 48 percent of the people in a population of 10,000 people are men, the 48 percent is a population parameter. In a probability sample drawn from that population, the percentage of men would be a statistic. That statistic might or might not exactly match the percentage of men in the population; however, we can infer the percentage of men in the population with a certain precision and degree of confidence.

Sampling Theory

The ability to infer, or state a knowledge claim, about a population based on results from a sample rests on the laws of **probability**. Before we begin our research, we cannot know the parameters for all of the relevant characteristics of our population. Nevertheless, the **central limit theorem** tells us that the mean values for repeated samples drawn from a population will cluster around the true mean value for any given characteristic. That is, *for any characteristic we wish to measure, the mean values found from repeated sampling will tend to converge, or cluster, around the actual mean value for the target population.* Plotting these sample averages or means will produce a **normal** or **bell-shaped curve** around the population parameter.

You can easily test this basic theorem by first imagining a universe of all possible coin tosses. From personal experience and simple logic, you know that half the time the coin will come up heads; half the time it will be tails. But personal experience also tells you that you can't really count on every other toss producing a heads. Sometimes you will get several tails before you get another heads.

With that background, test these assumptions from probability theory by graphing the results of 50 samples, each consisting of 10 coin tosses.

Perhaps in the first sample of 10, you got three heads and seven tails. Another time your sample might have turned up only one heads or only one tails, but gradually you saw that you got three or seven heads more often than one, two, nine, or ten; four or six heads more often than fewer or more heads and five heads even more often. Even repeatedly drawing very small samples from a population of infinite size supports the theory. The plot of the results began to look pretty much like a bell.

From your basic statistics class, you recall that when the resulting plot is a normal or bell-shaped curve, 95 percent of all samples will fall within a specified range—one standard deviation—of the true population mean. Using that knowledge, you can calculate the **sampling error.**

For a bell-shaped curve, the actual **dispersion** in the estimate of a population mean is a function of sample size. Therefore, the **standard deviation** is based primarily on the

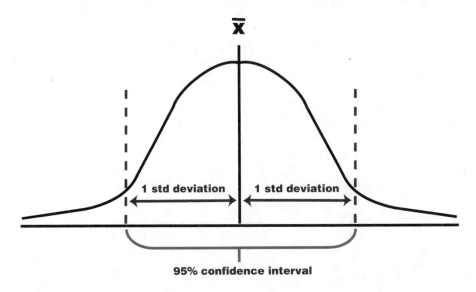

Figure 4.1. Sampling distribution and 95% confidence interval

number of units in the sample. Where we are dealing with percentages, as is most often the case in communication research, the formula for the standard deviation is as follows:

$$\sigma = \sqrt{\frac{p(100-p)}{n}}$$

In our experiment, p is the percentage of heads, expressed as a decimal, for any one set of coin tosses, and n is the number of tosses in a sample. Therefore, using the results from our first set of coin tosses, $p = .3$ and $100 - p = .7$; $n = 10$.

With one more piece of information—the **confidence interval**—we can state the likely error, or range of deviation from the true population parameter. The most common confidence interval is the 95 percent level. Its value in the equation is 1.96. Therefore,

$$\text{Sampling Error} = 1.96\sqrt{\frac{p(100-p)}{n}}$$

Plugging in numbers from our first set of 10 coin tosses, we get:

$$\text{Sampling Error} = 1.96\sqrt{\frac{(.3 \times .7)}{10}}$$

or

$$\text{Sampling Error} = 1.96(.145) = .284$$

Expressed as a percentage, that's a sampling error of 28.4 percent.

Box 4.1. *Calculating sampling error and confidence interval for mean scores*

Most often sampling error is calculated using percentages. But if you are working with averages, you may need to know the possible range within a population for the mean score you calculated for subjects in your sample. To do that, you must use this formula:

$$\text{Sampling Error} = (\text{desired confidence level}) \frac{SD}{\sqrt{n}}$$

where,

$$\text{SD (standard deviation)} = \sqrt{\frac{\Sigma(x-\bar{x})^2}{n}}$$

Confidence Interval = (mean score in sample ± sample error)

Example: mean = 25; SD = 6.9; n = 100; confidence level = 1.96
Sampling Error = 1.96(6.9 / 10) = 1.35
Confidence Interval = 25 ± 1.35 or (23.65 – 26.35)

Box 4.2. *Sampling errors based on sample size for the 95 percent confidence Interval*

Sample Size	50/50 split	40/60 split	30/70 split	80/20 split	90/10 split
25	20.0	19.6	18.3	16.0	12.0
50	14.2	13.9	13.0	11.4	8.5
75	11.5	11.3	10.5	9.2	6.9
100	10.0	9.8	9.2	8.0	6.0
150	8.2	8.0	7.5	6.6	4.9
200	7.1	7.0	6.5	5.7	4.3
250	6.3	6.2	5.8	5.0	3.8
300	5.8	5.7	5.3	4.6	3.5
400	5.0	4.9	4.6	4.0	3.0
500	4.5	4.4	4.1	3.6	2.7
600	4.1	4.0	3.8	3.3	2.5
800	3.5	3.4	3.2	2.8	2.1
1,000	3.2	3.1	2.9	2.6	1.9
2,000	2.2	2.2	2.0	1.8	1.3
5,000	1.4	1.4	1.3	1.1	.85

How to read this table: For a characteristic that we expect to be about evenly split in the population (e.g., 50/50 split), a sample of 100 would yield a sampling error of no more than ± 10%, at the 95% confidence level. A "50/50" split means the population is relatively varied. A "30/70" split means it is less varied; most people have a certain characteristic; a few do not. Unless we know the split ahead of time, it is best to be conservative and use a 50/50 split.

Sampling error does not mean that something is necessarily wrong with our sample. Sampling error tells us how closely a sample statistic may match the corresponding population parameter and how confident we can be about that finding. Calculating and reporting sampling error simply acknowledges the reality that a sample may not perfectly reflect the target population.

In the case of the sample of coin tosses, our confidence interval tells us that in 95 out of 100 samples, the true number of heads in the population will be within 28.4 percent of the number of heads we found in our sample of 10 coin tosses. That is, the true percentage of heads in a population of an infinite number of coin tosses might be as low as 1.6 percent (30 percent – 28.4 percent); it could be as high as 58.4 percent (30 percent + 28.4 percent).

If five heads had turned up in the first sample, the sampling error would be even larger: .309 or 30.9 percent. In fact, the case of a 50-50 split produces the greatest uncertainty. Therefore, reports of sampling error are usually based on that worst-case scenario.

With a sample of 10, it wouldn't really matter whether we found one, three, or five heads. The sampling error would be so large that we really could not draw any firm inferences about the true number of heads in our population of all possible coin tosses. But note: If our sample were larger, our sampling error would be lower.

If we assume half heads and half tails, the sampling error for a sample of 100 coin tosses would be only .098 or 9.8 percent; for a sample of 1,000 tosses, it would be .031 or 3.1 percent.

Sampling error depends on the size of the probability sample. It does not depend on the size of the population. But here, there are caveats.

A sample may not be a true probability one if the sample is so large relative to the population that the last units chosen have a much greater chance of being included in the sample than did the first ones chosen. Most statisticians do not believe this poses a problem unless a sample is more than 20 percent of the population. But if the sample approaches or exceeds 20 percent of the population, it is usually better to use a census rather than to sample. If a census is not feasible, then one should make generalizations with extreme caution because the sample may really be a nonprobability one.

Moreover, regardless of how the sample was chosen or of its size, the sample finally available may not be a true probability one if too few subjects, and particularly too few with certain characteristics, are available or agree to participate in the study. As we shall see in Chapter 7, low response rates due to difficulties contacting people in a sample or to their refusal to answer questions is an increasingly severe problem in survey research. A high nonresponse rate will decrease external validity.

Public Opinion Quarterly is a good source for current research findings on problems related to sampling and on sampling techniques and their application.

Types of Probability Sampling

Probability sampling allows a researcher to use powerful statistical tests to analyze data collected from a relatively small number of subjects and then generalize the findings to the target population from which the sample was drawn.

For selecting a probability sample, the four basic techniques are simple random sampling, systematic random sampling, stratified random sampling, and cluster sampling. Each technique approaches the need to ensure representativeness in different ways, making it better suited for some purposes than for others. Because each of these techniques will work with all of the basic research methods, the information about them presented in this section is generally applicable. Method-specific adaptations will be described in subsequent chapters.

Simple Random Sampling. This is the basic, and perhaps ideal, technique because it truly ensures that each unit in a sampling frame has an equal and known chance of being included in the sample. Today it is commonly used in public opinion polls, marketing, and audience research.

For these studies the target population is usually all households or all individuals in the nation or in a particular geographic region. The most commonly used sampling frame is the set of all telephone numbers, and the technique for sampling is random-digit dialing.

With this technique, phone banks used to contact subjects are programmed to dial numbers with specified area codes or prefixes. The area codes and prefixes can be selected to produce a national sample or to include only phone numbers from a specific geographic region. Because numbers with the same prefix are usually assigned during a particular time period and to people living in a particular part of a city or county, prefixes can also enhance the probability of finding households with certain demographic or lifestyle characteristics.

As a sampling method, random-digit dialing has many advantages. With the appropriate equipment in place, drawing the sample is quicker and easier than with most other methods. Moreover, the researcher does not have to know ahead of time how many people are likely to agree to participate in the study and adjust sample size accordingly. The system simply continues to select phone numbers until the specified sample size is achieved.

For many purposes, phone numbers are the best available sampling frame. There is, for example, no master list of all businesses, households, or adults living in the United States, but most businesses and households do have telephones. Therefore, phone numbers are the only possibility for studies where all businesses, households, or individuals are the target population. However, even in cases where that is true, problems with overregistration and underregistration inevitably arise, which may be more or less severe depending on the purpose of the study. In contrast to sampling directly from phone books, random-digit dialing includes unlisted numbers but at the expense of also including many nonworking numbers and ones that may be irrelevant for a particular study—for example, businesses when the interest is in households.

Whether the phone number is used as a way to sample households or individuals, the method will almost certainly overregister middle- and upper-socioeconomic status households and underregister ones of lower socioeconomic status. Although almost everyone has access to a phone, middle- and upper-class families are more likely to have multiple phone lines; less affluent households either may not have a phone or may share one. For that reason and because their households tend to be somewhat larger, lower-socioeconomic individuals as well as young adults sharing an apartment or living in group settings such as university dormitories may be underrepresented relative to older, more affluent individuals.

Because of the problems with overregistration and underregistration associated with random-digit dialing, researchers work from specialized lists whenever possible. In addition to the enrollment, employee, customer, subscriber, and contributor lists maintained by schools, businesses, nonprofit organizations, and government agencies, many of the larger research organizations create, maintain, and then make available to their clients specialized lists reflecting various interests, lifestyles, and psychographic and demographic profiles.

Many of these lists are computerized using software that has a random-number feature. Where this is the case, the computer can quickly select a random sample of any size from the entire list or from some specified portion of it: a random sample of all men or of all patrons who contributed more than $500 in the last year, for example.

Technologically assisted sampling is the norm today, but using either of two methods for manually drawing a simple random sample will work so long as there is a master list that identifies all elements in a target population. The first is the classic "pull a name out of a hat" method; the second makes use of a table of random numbers.

With the "pull a name out of a hat" method, all elements in the sample frame are separated from each other so that information about just one element—a name and phone number, for example—is on its own separate slip of paper. The slips are then placed in a large container and thoroughly mixed. The number of slips required for the sample are then drawn individually from the container.

> **Box 4.3.** *Excerpt from a random-number table*
>
> | | | | | | |
> |---|---|---|---|---|---|
> | 10097 | 32533 | 54876 | 39292 | 74945 | 74339 |
> | 08422 | 02051 | 68607 | 47048 | 86799 | 50500 |
> | 60970 | 14905 | 76520 | 80959 | 91655 | 70593 |
> | 82406 | 99019 | 12807 | 66065 | 31060 | 85269 |
> | 63573 | 98520 | 33606 | 27659 | 76833 | 29170 |
> | 09732 | 88579 | 25642 | 88435 | 73988 | 02529 |
> | 16631 | 35006 | 96733 | 38935 | 31624 | 78919 |
> | 20206 | 64202 | 76320 | 19474 | 69298 | 54224 |
> | 35552 | 75366 | 20801 | 44437 | 19746 | 59846 |
> | 92325 | 87801 | 82732 | 35083 | 35970 | 36207 |
> | 34095 | 20969 | 90700 | 99505 | 58629 | 16379 |
> | 40719 | 55157 | 64957 | 35749 | 58104 | 73211 |
> | 42791 | 87338 | 20468 | 18062 | 03547 | 20468 |
> | 21438 | 13092 | 71060 | 22132 | 45799 | 52390 |
> | 69074 | 94138 | 87637 | 91976 | 04401 | 10518 |
> | 21615 | 01848 | 76938 | 09994 | 51748 | 19255 |

The random-number technique is a bit more complex. To use it:

1. Number all elements on the master list.
2. Decide how many elements you want in the sample.
3. Decide which digits in each entry in the random-number chart you will use if the number in your sample is smaller than the five digits used as entries—for example, if you want a sample of 500, you might use the first three digits, the middle three, or the last three. It doesn't matter so long as you always use the same three.
4. Randomly pick as a starting point one of the entries in a table of random numbers. That number identifies the element on your sampling frame that will become the first one in your sample.
5. Select additional units for your sample from numbers you encounter in the random-number chart as you proceed through it systematically. You may go down columns, across rows, or diagonally. It's your choice so long as you use the same method consistently.

Systematic Random Sampling. Although the manual methods for selecting a random sample work well, both take a lot of time and involve a lot of work. Therefore, if the list used as the sampling frame is long and neither random-digit dialing nor computer-generated sampling is an option, most researchers use systematic random sampling.

To construct a systematic random sample:

1. Determine the number of units in the list that is used as the sampling frame.
2. Determine the appropriate sample size.
3. Divide the number of units in the list by the desired sample size to determine the **skip interval.** This skip interval is sometimes also called a **sampling interval.** This interval is usually indicated as k.

4. Identify a starting point by randomly selecting a number between 1 and *k* by using a random-number chart, pulling a number out of a hat, or asking someone unconnected with the research to name a number between 1 and *k*.
5. Draw the sample by selecting the unit identified as the starting point and then every *k*th unit after that. To speed up the process of selecting the units, you can use a ruler to measure the distance between the starting point and *k*, and use that distance as the skip interval.

As an example of how systematic sampling works, imagine you need a sample of 400 members of your organization. The membership list contains 10,000 names. Dividing 10,000 by 400 produces a skip interval of 25. Using a random-number chart, you select 17 as the starting point. The 17th person on the list is the first member included in your sample; counting down the list, you also include members 34, 51, 68. . . . But all that counting could become tiresome, so you take out your ruler and measure the distance between the first name on your list and name number 17, between name number 17 and number 34, and between 34 and 51, and take the average of those distances—let's say that's $3^{1}/_{4}$-inches. From then on you simply measure, putting into your sample one name at every $3^{1}/_{4}$-inch interval.

Estimating the number of units is more difficult, but the same procedure works equally well with sampling from a telephone book. Using it avoids the problem of including businesses when you want households because you can sample only from the white pages. If some businesses are listed in the white pages, they can be excluded from the sample while still maintaining its essential randomness. To do this simply follow pre-established procedures such as alternately selecting the first eligible number directly above and below the *k*th number on the list. You can use a similar procedure to avoid households when you want businesses or to exclude phone numbers whose prefixes indicate they lie outside a target area.

If it is important to have a sample that includes unlisted numbers, you can arbitrarily add a number, usually 1, to the last digit of each phone number so that phone number 111-1111 becomes 111-1112 and 111-2222 becomes 111-2223. Although this "add 1" technique lets you recover some unlisted numbers, it also creates some problems. Some of the adjusted numbers will not reflect the target population; others will not be in service. Therefore, you will have to adjust the sample size upward to ensure getting the desired number of cases.

With systematic random sampling, all units in the population do not have an equal and known chance of becoming part of the sample once the skip interval and starting point have been selected. However, initially all units did have that equal and known chance of being included. Therefore, using this procedure creates a sample with all the attributes of the simple random sample, so long as there is no biasing order to the units on the list. How well a systematic random sample represents the target population is more a function of the quality of the list than it is of the sampling procedure.

But be aware that, even with a good list, this method may not work very well if names are entered on that list in some order (such as chronologically—perhaps according to the time people joined an organization) that may lead to people with some characteristics having a greater chance of being in the sample than is true for people with other characteristics. Systematic random sampling from a telephone book works because there is no reason to presume that people or businesses whose name begins with "A" differ in any fundamental way from those whose names begin with any other letter.

Stratified Random Sampling. For simple random sampling and systematic random sampling, the sampling frame may be a single list or it may be found in several lists that can be combined. In either case, both methods work equally well when the purpose of the research is to study an entire population. But it may not work as well if the population is heterogeneous on certain characteristics important for a particular investigation. This is particularly true if a study requires examining each of those characteristics separately and especially if one of those characteristics occurs much less frequently than others.

To understand the potential problem, imagine that in our test of sampling theory we had been selecting jelly beans from a giant candy jar instead of sampling coin tosses from an infinite universe of all possible coin tosses. Whereas there were only two coin characteristics of interest—heads and tails—our jelly beans might come in eight different flavors.

If the same number of jelly beans of each flavor were present in the universe of all jelly beans, a perfect sample of 400 jelly beans would have 50 of each flavor. But samples are rarely perfect, so we very likely would have somewhat more than 50 jelly beans of some flavors and fewer than 50 of others. That's fine for studying the population of all jelly beans, but the relatively small number of jelly beans of each flavor could cause data-analysis problems.

Those problems will be even more severe if the jelly beans of each flavor are not equally represented in the target population—if, for example, 50 percent of the jelly beans are red and 20 percent are green, but there are only 4 or 5 percent of each of the other six colors. In such a case, you can easily imagine drawing a sample that would have only a few orange jelly beans or maybe even none at all. That might not be much of a problem if your only interest were in the population as a whole. But it would be a problem if you also needed to compare the different flavors of jelly beans to each other or if you were really most interested in learning about those orange jelly beans.

Even though worrying about studying jelly beans may be a bit fanciful, the basic problem is a common one. A community may have many more Democrats than Republicans; colleges may have more freshmen than seniors and more undergraduates than graduate students; companies typically receive more letters of complaint than letters praising their products or services; and, of course, nonprofits usually receive more small contributions than large gifts from their patrons.

In any of those cases and many similar ones, the way to avoid the problems created by too few members of some group of interest is to use stratified random sampling. In this procedure, you use either simple random sampling or systematic random sampling to select samples of equal size from separate lists, each representing a stratum or characteristic of interest.

Using stratified random sampling makes it possible to draw meaningful inferences about each stratum. It also facilitates comparing one stratum to another. But because members of one stratum may be overrepresented or underrepresented relative to their presence in the sample as a whole, inferences based on the entire, stratified sample may be misleading.

If, for example, we used data from a sample of 200 likely voters, stratified so that there were 100 Republicans and 100 Democrats in the sample, and then compared the two groups, we would most likely find that Republicans are much more likely than Democrats to favor the Republican candidate for governor. From this we might reasonably conclude that Republicans and Democrats have very different candidate preferences. But if we had used that procedure in a state where Republicans outnumbered

Democrats by a margin of two to one, it would be unreasonable to conclude that the election might go either way simply because equal numbers of respondents said they support and oppose the Republican candidate.

We could have avoided the problem of drawing potentially misleading conclusions about the population as a whole by selecting twice as many names for the Republican stratum as for the Democratic one. But that kind of **disproportionate likelihood sampling** might have required us to use a larger sample than we could afford, so we could also handle the problem through sample weighting.

Sample weighting is a statistical procedure that adjusts the distribution of characteristics of subjects in a sample to more nearly reflect their distribution in the target population. To use sample weighting with our stratified sample of 100 Republicans and 100 Democrats, we would divide the percentage of Republicans in our sample (50 percent) by their percentage in the state as a whole (66.6 percent), do the same for the Democrats (50 divided by 33.3), and enter the results (.75 for Republicans, 1.5 for Democrats) into the weighting option in whatever computer program we are using to analyze our data. This procedure should produce results more nearly reflecting the true distribution of characteristics within the population than would be true if we had used the unweighted sample.

Nonprobability Sampling

Although drawing inferences about or generalizing findings to a target population is possible only when subjects are chosen using one of the probability sampling techniques, nonprobability samples play an important role in applied communication research.

They are commonly used in formative research to gather ideas for planning public relations or advertising campaigns or as part of evaluation research to see how well program goals are being met. They are also used in business and by media researchers for surveillance purposes whenever they are considering making a change or want some idea of how relevant publics are receiving their practices. Journalists, too, use them to select settings or choose sources for their stories.

Types of Nonprobability Sampling

Each nonprobability sampling technique has its own strengths and weaknesses. But one or more of them is an appropriate method for choosing subjects for laboratory experiments, focus groups, and observational research. In these kinds of research, working with large enough random samples is usually not feasible. In any case, generalizability is rarely the goal.

Quota Sampling. This method for choosing subjects most closely resembles probability sampling. In fact, pollsters striving for generalizable findings routinely used this method before the laws of probability and their application to sampling were well known.

To use this method, researchers first identify some obvious demographic characteristics within a target population that will very likely translate into differences in knowledge, opinions, or behaviors and then seek out subjects with those characteristics in rough proportion to their presence in the target population.

If gender were the only characteristic of interest, a researcher would have little trouble finding approximately 50 men and 50 women who would be willing to watch and react to a pilot for a new situation comedy. But if the researcher also wanted to take race into account, finding men and women from the major racial groups in proportion to their numbers in the population would be more difficult. If she also wanted to test the pilot with various age groups and with different education and income levels and from various religious backgrounds, determining the proper quota for each combination of characteristics and then finding subjects with that combination might become almost impossible.

Even if the researcher found the appropriate quota of subjects by gender, race, age, education, income, and religion, she would have no way of calculating the sampling error and then generalizing the subjects' opinions about the television program to the population as a whole.

Finding the appropriate quota of subjects with identifiable characteristics may ensure a more representative sample than other nonprobability sampling techniques might produce, but it cannot ensure that the sample is truly representative on unobservable characteristics. Therefore, quota sampling is most useful for those situations where there is a need to increase the likelihood of finding diverse viewpoints but no real need for true generalizability.

Purposive Sampling. Purposive sampling is much like quota sampling in that the researcher has some special reason for choosing subjects. However, that purpose is rarely to mimic characteristics present in a target population. Instead, the goal is to choose subjects who can be expected to provide useful information.

A public relations researcher might choose to gather information about issue priorities from the organization's top financial backers to use for program planning purposes; educational software developers may try to determine how well suited a new educational program is for use in different settings by testing it in just a few inner-city and suburban schools. And, of course, journalists use purposive sampling when they seek out the governor, the chair of the House and Senate finance committees, and a couple of business and labor leaders to comment on a new tax proposal.

Because the samples are often small and can be assembled quickly and easily, the method is well suited for use with focus groups and in-depth interviews designed to monitor the environment, gather information for use in early stages of campaign development, or as a reality check on a company's plans or on recommendations and conclusions suggested by findings from full-scale research projects. Purposive sampling is also very useful for selecting sites for observational research and in-tact groups such as schools or classrooms for use with experimental research.

Snowball or Network Sampling. Purposive sampling is most useful when the researcher knows or can find out and then locate those subjects who are most relevant. Snowball sampling accomplishes the same purpose in situations where the relevant subjects are unknown.

In this form of sampling, the investigator need identify only one subject and then use that subject to help find others who might be of interest. In this way, the sample size "snowballs" until those who have been drawn into it begin to point to others who already are part of the sample or are repeating information provided by others already included in the network.

This is the method commonly used by anthropologists and ethnographers when they find an informant who serves as their guide and entry into the culture they are studying. But the method works equally well in communication research when there is a need to map a community, track down opinion leaders, or locate power brokers. Thus, advertising and marketing directors use it to find the teenagers whose tastes influence others to buy certain brands of clothing. Issue-oriented organizations use it to find out where power really lies within a legislature or who legislators listen to; and businesses may use it to find out how information circulates through a company.

Although the method is most often used in conjunction with survey research or as part of an observational study, the method can also be used with content analysis. By finding one relevant document such as a book, journal article, or report, the investigator can use clues within that document such as footnotes or passing references to people, places, or other documents to track down more information.

Convenience Sampling. Unlike quota sampling and purposive sampling where an investigator seeks out subjects for a reason that is closely related to the purpose of the research, with convenience sampling the only goal is to find a number of research subjects quickly and easily. Instead of seeking out people with certain characteristics that reflect population demographics or who have relevant information to offer, a researcher using convenience sampling takes information from whoever is available.

The most obvious use of this method is the "person-on-the-street" interview long favored by journalists as a way of finding sources to comment on some topic. Although this kind of **straw poll** is popular both with those who are quoted and with a media audience that enjoys learning about other people, the shortcoming is that, no matter how many people are interviewed or quoted, their views cannot be generalized to the entire population. They are more often a reflection of where and when the interviewing was done than they are of widespread sentiments.

As a gauge of popular opinion, findings based on convenience samples are quite often misleading. However, convenience samples can be used appropriately. One such use is the **mall intercept.** In this method, researchers interview every k^{th} person entering a store or shopping mall. Of course, the people selected this way are not a random sample of all consumers. Moreover, those who shop on Tuesday morning or Thursday evening may be very different from each other and also from those shopping Saturday afternoon or Sunday at noon. But by sampling at various times, the procedure can provide very useful information about consumer-related matters such as people's shopping behaviors, product preferences, or opinions about the quality of goods and services at that particular site.

The mall intercept is also useful for finding people who can be directed to a mall auditorium where they can watch and respond to proposed advertising campaigns or test products. Used this way, the mall intercept becomes a convenient way to find subjects who are immediately available for focus groups and laboratory experiments. It is very much like the long-standing practice of using members of service organizations or students enrolled in "the large introductory class at Enormous State University" as subjects for experimental research. As with those experiments, the results cannot be generalized, but the sample is appropriate for determining possible effects of a particular stimulus.

Self-selection Sampling. In the convenience sample, the researcher finds subjects; in a self-selecting sample, the subjects choose themselves. Therefore, findings based on this kind of sampling can never be generalized to a target population. Indeed, as a gauge of public opinion, the findings may be even more misleading than those based on straw polls.

That is often the case with **telephone call-in polls.** In these polls, readers, listeners, customers, or constituents are invited to call a toll-free number to register their views on the issue of the day. The results are then tabulated and publicized, often in ways that suggest they are a measure of public opinion. However, as is the case with unsolicited phone calls or mail, those with strong feelings are most likely to register their views. Of those, people with complaints or who strongly oppose some position are more likely to select themselves into the sample than are those who are quite satisfied or who support a particular measure.

Although call-in polls can easily be used in misleading and unethical ways, calling for volunteers is a widely used accepted practice for locating people to participate in focus groups or experiments. Offering people a chance to select themselves into a sample by means of a questionnaire printed in a newspaper or magazine can be a good way to find out whether there is interest in a particular topic or whether there will be strong opposition to proposed changes or courses of action. It is also one way to establish rapport with a public and create good will.

Multistage Sampling

The sampling techniques discussed in the previous sections are usually sufficient for locating subjects when lists are available or there is some other way to gain direct access to an appropriate target population. But when the units on the best available list do not match the level of analysis or the units from which information must be collected, accessing the target sample may require using multistage sampling.

Superficially multistage sampling resembles stratified random sampling. However, in stratified sampling, the lists used for sampling from each stratum are for different subsets of the same target population and are capable of directly reaching members of that population. With multistage sampling, repeated samples are drawn from a series of sampling frames that do not reflect the same target population. Instead, each stage more narrowly defines that population until it can finally be accessed in the last phase of sampling.

Although multistage sampling is often treated as a probability sampling technique, some phases may use nonprobability sampling techniques. Therefore, whether findings based on subjects selected through multistage sampling can be generalized to a target population and, if so, to what population, depends on when in the process probability sampling is used.

If probability sampling is used for all stages, the results are, of course, readily generalizable. However, if nonprobability sampling is used in initial stages and random sampling techniques are used in later phases, findings can be generalized, but only to those populations chosen through the use of probability techniques. If probability sampling is used in the early stages and nonprobability sampling in later ones, the results will not be generalizable.

To understand why multistage sampling might be necessary, consider the selection of subjects from the white pages of a phone book or via random-digit dialing. Those phone numbers are, of course, assigned to households, each of which might have multiple residents. As a way of minimizing sample biases related to gender and age differences in phone-answering behavior and also in willingness to participate in research

To select households:

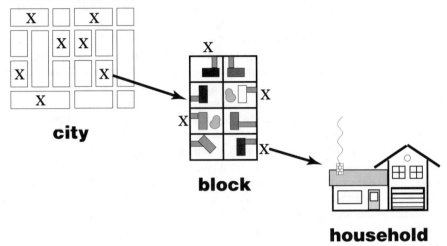

To select documents:

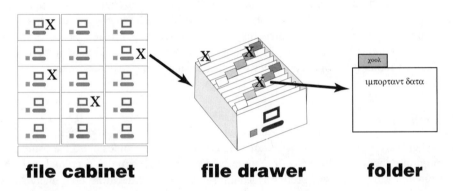

Figure 4.2. Multistage probability sampling

projects, researchers often use one or more **screens** to sample a second time within multiple-resident households.

Common screens include questions to identify and select into the sample the person who last celebrated a birthday or to choose alternately from men and women or from among the oldest male, youngest male, oldest female, and youngest female. Although using one of these screening techniques may require making additional phone calls if the appropriate person is not immediately available, this second stage of sampling usually produces a sample that is more truly representative of the target population than could be obtained without the use of a screen.

Using screening questions works well whenever choosing only one person or other unit from each of the elements identified in the last available sampling frame is appropriate. But sometimes it may be desirable to use **multistage cluster sampling** to choose more than one subject from each unit in the last available sampling frame. Depending on the purpose of the research, these clusters of subjects may be a census, a probability sample, or a nonprobability one.

For studies of news coverage, researchers typically use multistage cluster sampling. They also use it to find and select members of populations for which there are no master lists or who might be difficult to locate through all-purpose lists such as phone numbers.

As an example, an organization might choose to examine trends in news coverage of environmental issues. For that study, a researcher might choose to study coverage by elite media—network television newscasts, the major news magazines, and newspapers such as the *New York Times* and *Washington Post*—during the last 10 years. With the media and a time frame selected, the researcher might then sample weeks or days within each of those 10 years before finally choosing a set of stories for analysis.

To learn even more about the climate of opinion on environmental issues, the organization might seek information about issue salience among environmental activists. But, of course, there is no master list so it would first be necessary to identify relevant organizations such as the Sierra Club and Audubon Society and obtain lists of their chapters. With the lists in hand, the researcher could then use either probability or non-probability techniques to sample chapters and finally members of those chapters.

To find people who are not members of environmental organizations but who might support their causes, the researcher might first turn to commercial firms specializing in profiling zip codes by the demographic and psychographic characteristics of people living in each zip code. From one of those lists, the researcher could choose appropriate zip codes, then turn to maps to select blocks within each zip code, and then select respondents by using a **reverse directory.** Reverse directories provide the same information as a phone book, but the list is organized alphabetically and then numerically by address rather than alphabetically by a business name or personal surname.

Sample Size

Whenever collecting data from an entire population is impossible, choosing an appropriate technique for selecting some subjects for inclusion is a crucial step in designing a study that will produce meaningful results. But deciding how many subjects to include is equally important. Although time and budgetary constraints may sometimes make it necessary to work with samples that are smaller than the ideal, the goal should be to get a sample that is big enough for the purpose of a particular study but not so big as to add unnecessarily to the time and cost for the study.

Decisions about sample size depend primarily on the:

- homogeneity or heterogeneity of the population
- expected variability on key measures
- the way data will be analyzed and used

Although techniques for choosing an appropriate sample size differ somewhat depending on whether the sample will be a probability or nonprobability one, population size is not a factor unless the sample size approaches 20 percent of the population. If that happens, sampling may not be appropriate.

Probability Samples

For probability samples, the sampling error is the key. But considered in isolation, it will not provide enough information to make a wise choice.

Because sampling error decreases with sample size, one might assume that, with sample size, "the bigger, the better." But note that as sample size increases up to about 600 subjects, sampling error decreases rather rapidly. After that, sampling error decreases much more slowly. Also recall that sampling errors are normally calculated assuming the worst-case scenario—a 50-50 split in responses on a key measure. Therefore, for some projects 3,000 subjects might be overkill; for others, 3,000 subjects might not be enough.

For studies where external validity is important, samples under 100 are almost always too small. However, samples in the 200 to 600 range are usually quite adequate if:

- subjects are fairly homogeneous on key independent variables
- you can expect separation greater than that 50-50 split on key dependent variables
- all of your data analysis will be based on data obtained from the entire sample

The more of those conditions that apply, the smaller a sample may be; the fewer that apply, the bigger it will have to be.

To illustrate the problem, imagine a pollster draws a random sample of 600 people, stratified so there are 100 from each of six congressional districts in the state. Those 600 respondents may be enough to determine with some degree of confidence who is leading in a statewide election campaign for governor, U.S. Senator, or president. But in hotly contested statewide races, the final margin of victory is likely to be so small that it would fall within the margin of error even if the pollster had chosen to use a sample of 3000.

Similarly, 100 respondents from each district is enough to say something meaningful about differences in candidate preference among the districts. However, the sampling error associated with a sample of 100 people from a district is so large that determining who might be leading in a race for U.S. Representative becomes impossible except in those districts that are so overwhelmingly Republican or Democratic that one party's candidate almost always wins with 60 percent or more of the popular vote.

But even in those districts, the adequacy of a sample of 100 will depend on choice of statistics for data analysis. Although we have more to say about particular statistics in subsequent chapters, some statistics such as analysis of variance are designed to detect differences even when sample size is quite small. Measures of association such as chi-square and Pearson's r require much larger samples, especially if a particular value for a variable of interest is rather rare or if there are many possible values on those variables—just a few orange jelly beans but eight possible flavors of jelly beans as opposed to just heads and tails for coin tosses.

As with the orange jelly beans in our earlier example, both the sample of 100 within a congressional district and the statewide sample of 600 would very likely be inadequate for finding members of the Temperance Party. It might also be inadequate for examining religious differences in political opinion and voting behavior because within any community there typically is more religious diversity than political diversity.

By increasing the size of a simple random sample, you would very likely enhance your chances of getting some Temperance Party members and at least a few members of many different religions into your sample. But even a sample of several thousand probably would not include enough Temperance Party members to include them as a distinct group in your data analysis or enough members of a number of distinctly different religious traditions to allow you to make comparisons across religious groups.

Because there are lists that identify voters by their party affiliation, you could select a random sample stratified by political party. But because there are no master lists identifying people by their religious affiliation, you most likely would have to use multistage cluster sampling in order to select enough representatives from each of the religions of interest to allow you to make those kinds of comparisons.

To achieve the same precision, however, a multistage cluster sample would have to be larger than would be necessary for a stratified random sample. The reason for this is related to the concept of heterogeneity. Stratified random sampling assumes, and is appropriate when, there is greater variability (i.e., heterogeneity) on the dependent variable between strata than there is within a stratum. In multistage cluster sampling, as might be used to first choose religions and then choose people, there will probably be variability both within and between the clusters. But if heterogeneity on the dependent variable between clusters is very large compared to the heterogeneity within the cluster, sampling error can be considerable, and it may very well increase with each subsequent sampling stage.

Nonprobability Samples

Because there is no way to calculate the sampling error for a nonprobability sample, time and cost factor more heavily into decisions about how many subjects should be included in a nonprobability sample than they do for probability ones.

In making decisions about sample size, the type of sampling and the way you will analyze and use data should also come into play. But because generalizability is not the goal for investigations based on nonprobability samples, the number of subjects selected is typically quite small.

With the true self-selecting sample such as a telephone call-in poll, sample size is purely a matter of how long the lines are open, how many people choose to call in during that period, and then how many responses one might be able to record or analyze. Purposive samples may also be self-limiting. How many subjects depends on the researcher's judgment about how many have relevant information and how many the researcher can include given time and budgetary constraints.

For convenience samples used in connection with laboratory experiments, two intact groups such as classrooms or organization chapters or as few as 20 volunteers could be enough for a preliminary study or for a replication of earlier work. However, additional groups from more different settings would be necessary for a fuller investigation. Many more than 20 volunteers would be required for investigations where risks associated with the possibility of getting a few idiosyncratic subjects into a study or missing subjects with important characteristics are high, as they might be in pretesting an expensive campaign or determining the likely effects of warning labels.

The same is true of recruiting subjects for focus groups. Two groups of six to twelve people each would be enough to gather some useful information, but adding more groups recruited from different locations would provide at least some protection against missing important information or putting too much weight on what might be idiosyncratic responses.

For both experiments and focus groups, quota sampling is one way to introduce diversity while simultaneously minimizing the risk of recruiting atypical subjects. However, it is important to recall that choosing even large numbers of subjects in this way does not guarantee that the sample will be representative of the population as a whole.

Where convenience sampling, quota sampling, or snowball sampling are used in connection with projects where representativeness is appropriate, the only safeguard is to follow the practice of ethnographers: keep selecting subjects until new additions to the sample are no longer providing information that differs from that gathered from other subjects previously selected into the sample.

Validating a Sample

No sample is likely to be perfect. With a probability sample, there will always be sampling error that is usually calculated using a 95-percent confidence interval. That means that even with perfect sampling technique, there are 5 chances out of 100 that the sample may be a bad one—the subjects selected will not be representative of the population from which they were drawn. In addition, an accidental bias may have crept into the sampling technique. A high nonresponse rate can also cause problems because subjects from that sample who actually were available for a research project often are not like the ones who are unavailable. Therefore it is always a good idea to check the sample for obvious problems. For random samples of individuals intended to represent the population as a whole, the best check is to compare the demographic characteristics of subjects included in the sample to the characteristics for the population. If the sample represents the population, the figures for such characteristics as gender, race, age, and income should match, within the limits of sampling error, the figures from the U.S. census or a similar data set.

A similar check may be performed on samples representing other populations so long as information about the population is available or can be obtained for one or more measures for which information about the sample was collected. Many organizations, for example, will be able to provide data on the gender, age, education level, or race of employees or members that can be used to validate a sample.

If the figures from a sample for any of those characteristics do not match, even after the census data have been adjusted to account for differences in the time periods when the two sets of information were collected, the sample and sampling procedures should be examined in an effort to locate the problem.

When there is no obvious reason for the problem but the discrepancy between sample data and census data is large, the first step to solving the problem is to draw a supplementary sample and test it against both the census data and the data from the original sample. If data from the two samples match within the limits of sampling error, the problem most likely lies with the census data. However, if the new data match the census but not the original sample, very likely that first sample just happened to be a bad one. If that is the case and the discrepancy is large, the only real solution may be to draw a new sample and begin again. But in most cases, that should not be necessary.

Most problems of this kind, as well as ones where the source of the bias cannot be identified, can be handled by using the sample weighting techniques discussed earlier in this section. Weighting is frequently necessary to make a sample more nearly reflect the strength of racial, ethnic, or religious minorities within a community. Sample weighting can also be used in cases where there are some subjects with a particular characteristic in a simple or systematic random sample but not quite enough to draw meaningful conclusions about that subgroup.

Where the source of sample bias can be identified—perhaps, for example, too few zip codes in middle-income neighborhoods got into the initial stage of multistage cluster sampling—the problem may also be handled by drawing a supplementary sample and adding those subjects to the initial sample. Alternatively, the researcher could randomly select clusters with certain characteristics from the original sample to create a new, smaller stratified random sample.

The same procedure can be used to check quota, convenience, and self-selecting samples. But where representativeness on a few variables is a strong indicator that a probability sample is truly representative of the population from which it was drawn, there is no similar guarantee that a match on a few obvious characteristics between members of a nonprobability sample and the general population also means the sample will match the population on other characteristics.

Purposive and snowball samples are usually validated by having an outside expert—someone who is familiar with the type of research but unconnected to the project—check the sampling procedure and the sample itself for reasonableness.

Main Points

- A population consists of all possible elements of whatever a researcher is interested in studying.
- A census is based on collecting data from all elements in a population, whereas a sample is based on collecting data from a subset of elements in a population.
- The first step in sampling is to specify the sampling frame. A sampling frame is a list that identifies all of the elements specified by the definition of the target population. In cases where a list is not available, the sampling frame indicates the procedure that the researcher will use to locate the relevant elements.
- A probability sample is chosen from the sampling frame using scientific, statistical procedures. Therefore, it will be representative of the target population as a whole.
- Simple random, systematic random, stratified, and cluster samples are all types of probability samples.
- Nonprobability samples do not assure each population element has a known chance of being included in the sample. Therefore, they lack representativeness.
- Quota, purposive, snowball, convenience, and self-selected samples are all types of nonprobability samples.
- When the units on a list do not match the level of analysis or the units from which information must be collected, multistage sampling may be necessary. In multistage sampling, repeated samples are drawn from a series of sampling frames.
- Whether the findings from a multistage sample can be generalized to a target population and, if so, to what population depends on whether and when probability sampling is used in the process.
- Deciding which type of sample to draw depends on time, resources, and the purpose of the research.
- Data based on a properly drawn probability sample are generalizable to the sampled population within the limits of sampling error. Sampling error is the difference between sample statistics and population parameters. The size of the error is based largely on the size of the sample.
- Using the central limits theorem and confidence intervals makes it possible to tell how close a sample statistic may be to the population parameter and how confident we can be about our findings.

- The decision about the sample size needed for a particular study depends primarily on the homogeneity of the population of interest, the expected variability on key measures, and the way data will be used.
- The most common way to validate a sample or to check it for biases is to compare the demographic characteristics of the subjects included in the sample to known characteristics in the population.

Terms to Know

population
element
universe
target population
census
sampling frame
probability sampling
nonprobability sampling
overregistration
underregistration
parameter
statistic
central limit theorem
normal curve
sampling error
confidence interval

simple random sampling
systematic random sampling
sampling interval
stratified random sampling
sample weighting
quota sampling
purposive sampling
snowball sampling
convenience sampling
straw poll
mall intercept
self-selection sampling
multistage sampling
multistage cluster sampling
reverse directory
screen

Questions for Discussion and Review

1. What is the difference between a census and a sample? When would it be most appropriate to use a census? A sample?
2. Distinguish between a probability sample and a nonprobability sample, and explain the advantages and disadvantages of each.
3. When is stratified random sampling more efficient than simple random sampling? Under what conditions would it be advantageous or even necessary to use disproportionate stratified sampling? To use cluster sampling?
4. Would using a reverse telephone directory as a sampling frame be likely to overregister or underregister elements from a population? Explain your answer.
5. Find an example of a probability sample described in a journal article. What was the population (or universe) of interest? What was the sampling frame? What type of probability sampling was used? Note any strong or weak points in how the sample was actually drawn. Considering the population of interest and the sample drawn, how confident are you in the ability of the researcher to generalize the findings to the population based on the sample?
6. When might it be more appropriate to use quota sampling rather than purposive sampling? Under what conditions would it be advantageous or even necessary to use snowball sampling? Convenience sampling? A self-selecting sample?

7. Find an example of the use of nonprobability sampling in a newspaper story or magazine article. What type of sampling was used? How confident are you in the conclusions drawn based on the sample? Would it have been possible and/or advantageous to use some other form of sampling?
8. Evaluate the following statement: "The bigger the sample size, the better." Take into account the factors to be considered when determining the sample size for a study and how each factor affects decisions about sample size.
9. Search the web for a *sample size calculator*. Plug various population sizes and confidence levels into the appropriate places. How does the size of the population affect the needed sample size? How does the degree of confidence desired affect the sample size?
10. For each of the following, indicate the type of sample you think is the most appropriate, explain what you would use as a sampling frame, tell how many subjects you would want in your sample, and justify your choices:
    a. A study to determine who buys DVD players
    b. A study to determine the demographic makeup of people who listen to country-western radio stations
    c. A study to compare the number of hours worked per week by minority and majority men and women at your college
    d. A national survey of Democrats' attitudes toward potential candidates for the next presidential election
    e. An in-depth study of computer hackers
    f. A study examining the use of the Internet by college students in the United States
    g. A study to determine newspaper readership of people who visit museums
11. You are doing an internship at a local radio station. As part of your internship, the station manager tells you to interview every senior citizen in the city of 25,000 to find out which of the station's shows draws the largest number of listeners over the age of 65. But having taken a research methods course, you suggest to the station manager that selecting a probability sample and conducting careful interviews with those people selected will yield better information with less interviewing time. The manager then tells you to develop a plan to convince the station owners how you might do this.
    a. What is the population for this project? Try to define the population as specifically as possible.
    b. Could you use random-digit dialing to sample this population? What would be the advantages and disadvantages of using this technique?
    c. Three sampling frames that you might use are (1) the city directory, (2) the local telephone book, or (3) the Social Security Administrations list of those in the county who receive benefits. What might be the advantages and disadvantages of each? Can you think of any other sampling frames that might be useful?
12. Draw a snowball sample of people who watch pro-wrestling on TV. Ask friends and relatives to locate a first contact, and then call or visit this person and ask for names of others. Stop when you have identified a sample of 10. What problems, if any, did you encounter? How would you proceed if you had to draw a much larger sample? How representative would your sample be of all those who watch pro-wrestling?
13. Use the website for the General Social Survey, http://www.icpsr.umich.edu/gss, to locate the 1990 GSS sample. Then answer the following questions about the

1990 GSS sample. Hint: Use the search engine to find information on the sampling design for the 1990 survey.
   a. What is the population of interest?
   b. What type of sampling procedure was used?
   c. What was the sample size?
   d. What was the response rate?
   e. What is the sampling error for the sample used in this survey?
   f. What would the sampling error be if you wanted to analyze data from only the men in the sample?
14. Using census data (http://www.census.gov), validate the GSS sample by comparing the 1990 GSS to the 1990 U.S. Census on several different demographic characteristics (e.g., age, sex, race, or education).

Readings for Discussion

Atkin, D. J. and LaRose, R. (1994). Profiling call-in poll users, *Journal of Broadcasting and Electronic Media, 38*(2):217–227, Spring 1994. This study examines potential demographic and lifestyle biases in call-in polls as well as the potential for rigging results.

Baxter, L. A. and Bullis, C. (1986). Turning points in developing romantic relations. *Human Communication Research, 12*:469–494. The article illustrates the use of snowball sampling.

Poindexter, P. M. and Lasorsa, D. L. (1999). Generation X: Is its meaning understood? *Newspaper Research Journal, 20*(4):28–36. This article illustrates use of plus-one telephone sampling and methods for validating a sample.

Richmond, V. P. and Croskey, J. C. (2000). The impact of supervisor and subordinate immediacy on relational and organizational outcomes. *Communication Monographs, 67*(1):85–95. The authors combine samples to examine dyadic relations.

Segal, M. N. and Hekmat, F. (1985). Random digit dialing: A comparison of methods. *Journal of Advertising, 14*(4):36–43. Research findings have implications for choosing among methods.

Stamm, K., Johnson, M. and Martin, B. (1997). Differences among newspapers, television, and radio in their contribution to knowledge of the contract with America. *Journalism and Mass Communication Quarterly, 74*(4):687–702. The authors make creative use of multistage sampling to find radio listeners.

Streib, G. D. and Poister, T. H. (1999). Assessing the validity, legitimacy and functionality of performance measurement systems in municipal government. *American Review of Public Administration, 29*(1):107–123. The authors pay special attention to validating their sample.

Wearden, S., Fidler, R., Schierhorn, A. B. and Schierhorn, C. (1999). Portrait vs. landscape: Potential users' preferences for screen orientation. *Newspaper Research Journal, 20*(4):50–61. The study makes practical use of mall-intercept sampling.

References and Resources

Barnett, V. (1991). *Sample survey principles and methods.* New York: Oxford University Press. This is a thorough, sophisticated work.

Berry, S. H. (1985). Field sampling in data collection for evaluation research. In L. Burstein, H. E. Freeman and P. Rossi, eds., *Collecting evaluation data: Problems and solutions.* Beverly Hills, CA: Sage, pp. 107–120. This is a good source of information on choosing subjects for assessments.

Kalton, G. (1983). *Introduction to survey sampling.* Beverly Hills, CA: Sage. The chapter on stratification in this thorough, highly technical work is especially useful.

Kerlinger, F. N. (1986). *Foundations of behavioral research,* 3rd. ed. New York: Holt, Rinehart, and Winston. This classic work contains thorough discussions of probability theory, sampling, and the concept of randomness.

Walker, J. T. (1985). *Using statistics for psychological research.* New York: Holt, Rinehart, and Winston, pp. 169–172. These pages give a good overview of the central limit theorem.

Wilburn, A. J. (1984). *Practical statistical sampling for auditors.* New York: Marcel Dekker. This book, written for auditors and accountants, is a good source of information on sampling work and work products and sampling to detect rare events.

5

Creating Measures

MEASUREMENT IS SIMPLY the process of assigning numbers, amounts, or other labels to things we observe or experience. Each of us does it every day. When, for example, we say we jogged two miles before breakfast before driving to work in our Chevrolet or tell someone a class is hard or a TV show is awful, we are using measures to record, analyze, and make sense of our world and then share our reality with others.

But some of the measures, such as distance and make of car used in the preceding example, would have meanings that would be the same for everyone. Even people who would normally use kilometers would understand and accept distance measured in miles. They would also understand what we mean when we call our car a Chevrolet, and so they would be able to recognize and compare it to other cars.

But some of the other measures are much more ambiguous. What does it really mean to say a class is hard or a TV show is awful? The difference between the way we create and use measures like those in our daily lives and the way researchers would do it is in the care that goes into the process.

Conceptualization

In research, measurement begins with the creation of definitions for the terms—the concepts or constructs—linked together to form the research questions or hypotheses that guide the research process. These conceptual definitions must fully describe what the researcher means by each term. To be useful, these definitions should:

- denote the essential qualities or meanings of the concepts or constructs
- exclude dimensions of the concepts or constructs that are irrelevant or nonessential for the purposes of the study
- be clear, precise, and complete

As an example, let's consider a research question related to that "awful TV program." A client might want to know whether advertising on great programs is more effective than advertising on awful ones—about the relationship between program quality and advertising effectiveness.

To help the client find the answer, we would have to agree on conceptual definitions for "quality" and for "effectiveness." Quality might be judged on production values or on the acting and writing. But it could also be judged on the basis of subject, educational content, portrayal of prosocial values, relative absence of sex and violence, audience reaction, or number of awards received. By "effectiveness" we might mean improved corporate image, increased sales, or just the number of people the advertisement will reach.

Our conceptual definitions of quality and effectiveness might be somewhat arbitrary. They could involve any or all of those possibilities, but they should be comprehensive. That is, they should include all of the attributes that help define the terms in ways that are useful for a particular purpose and that will distinguish the terms from other similar ones.

The key to creating the conceptual definition is to ask and then answer a series of questions:

- How have others defined this term?
- What else could it mean?
- What do I mean by this term?
- Why do I mean that?

Let us assume that the researcher and client agree that the conceptual definition of effectiveness is increased sales of the client's products. They might also agree that a quality program is one that portrays prosocial values and is relatively uncontaminated by scenes of sex and violence. However, to define quality that way, they would also have to further specify what they mean by prosocial values and what counts as a scene of sex or violence. But with that done, they could go on to the next phase in the measurement process.

Operationalization

The conceptual definition points the way to the second step in the measurement process: creation of the operational definition. Where the conceptual definition specifies the meaning of a term, operationalization specifies the procedures or rules for describing its observable characteristics. The operational definition should retain as much of the conceptual definition as possible. That is, there should be a good **conceptual fit** between the specified meaning of a term and what will serve as an indicator of that term in the real world.

For measurement purposes, researchers have three basic operational procedures available to them. They may measure by manipulation, through observation, or by using self-reports.

Manipulation

This procedure is used most often in experimental research and then only for the independent variable. In its most common form in communication research, an investigator artificially creates one or more versions of a message to serve as an indicator of the concept of interest.

As an example, our researcher might create two television programs: one conforming to the agreed-upon definition of a quality program and one violating that definition by containing much sex and violence and very little prosocial content. For data-analysis purposes, the high-quality version might be coded as "2" and the other program as "1" to indicate "low" or "no" quality. If our researcher had created several programs instead of just two, again absence or very low quality could be coded as "1," and other versions could be given higher numbers to reflect the relative amount of quality in each one. If versions of a manipulation are simply different (e.g., a violence version and a sex version), each version would be assigned its own number, but the numbers would simply be numerals. Numerals are useful for data analysis, but they have no real meaning.

As an operationalization procedure, manipulation has the advantage in that it gives the researcher complete control over the independent variable. Because the researcher creates the indicator, it should reflect the underlying concept exactly. The downside, however, is that creating a manipulation that is a pure, complete indicator of that concept and of nothing else can be a very time-consuming and difficult task. Moreover, the specially created stimulus may be somewhat artificial. And, as the name for this operationalization procedure implies, it is always manipulative. In its common form, manipulation requires exposing subjects to stimuli that they otherwise would not receive. Therefore, the risk of researcher-induced unintended consequences always exists.

Observation

For both qualitative and quantitative research, operationalization through observation consists of creating rules for watching and then noting behaviors or physiological responses that signal categories or varying amounts of an underlying concept. Most often the researcher is the measuring instrument, looking for and then taking note of the cues that become the measures, but machines may sometimes be used as the watching and recording instrument.

This procedure is the primary technique for measurement in studies using observation as the research method, but it can also be used with other methods. It is appropriate for both the independent and dependent variable as well as for any antecedent, intervening, or confounding variables that may be of interest.

With the researcher as the instrument, the observer could immediately categorize or rank cues based on what can be seen or heard. For example, our researcher might observe and record which products shoppers at several stores select during the week before and after the ads run in programs of varying quality.

To study employee morale, a trained observer might score supervisors and their employees according to how often, during one working day, each approaches the other to initiate a conversation and then how comfortable, assertive, or fluent each appears to be during each encounter. However, the observable behaviors of supervisors and employees might also be recorded as **field notes**—written or taped narrative descriptions of a research setting and what transpires there. These notes would then become the raw stuff for subsequent measurement and analysis using the techniques described in the qualitative data-analysis sections of Chapters 7 through 10.

Human observers can detect and record behaviors, but they may also use instruments as the observer. A common example would be the use of computer software to record how many people visit a website or how many pages of the site they open. Similarly, humans may note and record indicators of message effects such as facial expression or body language, but they may also use instruments to detect and keep track of physiological cues such as heart rate, perspiration, or muscle contractions.

Operationalizing a concept through observational techniques is the best way to find out what people really do, but the procedure is ill-suited to learning why they do it. A trained observer may be able to detect cues that indicate a person's motives, opinions, or intentions, but there is always the danger the observer will misperceive or misinterpret what transpires. Instrumental techniques such as functional magnetic resonance imaging can be used to detect which parts of the brain are most active in response to certain kinds of stimuli, but even they cannot tell what a person is actually thinking. Moreover, both human observers and machines as observers are obtrusive; their mere presence may affect what is observed. Observation also raises ethical questions about privacy and subject autonomy.

Self-reports

This is the basic method for survey research, but it is also used to collect data from experiments and focus groups and in observational studies. Content analysis is one variant of it. As an operational procedure, self-reports are appropriate for all variables.

Most commonly, self-reports take the form of answers to specific questions. The questions may call for subjects to select from among a fixed list of response options that serve as indicators for each concept, or they may allow subjects to answer in their own words. Answers in subjects' own words may also be found in textual material such as diaries, letters, media accounts, or transcripts of interviews that a researcher can examine for indicators of various concepts. Indicators from textual material can be converted to measures according to the principles of content analysis described in Chapter 8. As with open-ended responses to specific questions, this variant requires the researcher both to ask the question that serves as an indicator and to answer it on behalf of the narrative.

The question-answer format is the staple for news reporting and public opinion polling. But applied researchers also use it to find out what people like or dislike about a product, why they do what they do, or what they want and need. For those latter applications, researchers may also examine self-reports in the form of complaint letters or turn to internal memos, reports, or minutes of meetings. For evaluations of software, manuals, or other documentations, they may have subjects record their thoughts in diaries as they try to use the material.

In our study of the relationship between program quality and advertising effectiveness, the investigator might select programs for the manipulation by having experts rate available programs for quality by using the agreed-upon conceptual definition as a guide. Alternatively, subjects might be shown programs, each with the same commercial embedded within it, and then asked to rate the programs for amount of prosocial content, sex, and violence. For advertising effectiveness, the researcher could either ask people who saw different programs how likely they are to buy the advertised products or analyze sales figures to find out how many products were sold before and after advertisements ran on programs of varying quality.

Self-reports are a good way to measure knowledge; they are also useful for learning about motives, attitudes and opinions, past behaviors, and future intentions. The strength of the self-report is that it gathers information directly from subjects who presumably are the experts on what they know, think, feel, experience, and do. The weakness of relying on questions and answers is that the technique assumes subjects are knowledgeable and that they can and will answer questions clearly, accurately, and

> **Box 5.1.** *Example of operational definitions for a concept*
>
> **Concept:** Television Viewing
>
> **Operational Definition 1: Self-report.** Television viewing is defined as the answer to the question: "Approximately how much time did you spend watching television yesterday?" Record answer in minutes/hours.
>
> **Operational Definition 2: Observation.** Television viewing is defined as being present in a room where a television set is turned on and someone is paying at least minimal attention to it. On their home visits, social workers will record whether a television set is turned on and, if so, who is in the room with the television. Those present will be considered "viewers." Each viewer will be coded in one of three viewing categories: (1) inattentive viewer (not paying attention to the television), (2) casual viewer (paying some attention; also doing other things), or (3) attentive viewer (paying full attention to the television).
>
> **Operational Definition 3: Manipulation.** Television viewing is defined as receiving information via a television. Subjects in Condition A (viewing) will be shown three stories presented as one segment of a local television news cast. Subjects in Condition B (nonviewing) will be exposed to the same information but as newspaper stories appearing together on the front page of a daily newspaper.

honestly. But that may not be the case if questions reflect on a person's competence, are highly personal, ask about socially undesirable attitudes or behaviors, or if the inquiry raises concerns about how the information might be used.

For other kinds of self-reports, similar problems may arise. But using open-ended narratives also has all of the problems associated with observational procedures in which the researcher/observer serves as both the measuring and the recording instrument. Moreover, using some kinds of narrative self-reports raises additional ethical concerns about privacy and autonomy of research subjects, especially in those cases where a report such as a diary was not produced specifically for a research project.

Levels of Measurement

Operationalization specifies the procedures for recognizing indicators for each concept of interest. It also specifies the rules for **measurement.** Measurement is the process of determining the existence, characteristics, size, or amount of some indicator for each research subject. Because each subject may fit into a different category or exhibit a different amount of an indicator, measurement turns concepts into the **variables** used in data analysis by assigning subjects to levels or categories for each indicator.

The way numbers are assigned can produce four different levels of measurement: nominal, ordinal, interval, and ratio. These levels are hierarchical. That is, when the levels are arranged in ascending order from nominal to ordinal to interval to ratio, each higher level has all the characteristics of the lower one(s).

> **Box 5.2.** *The four levels of measurement*
>
> | Level | Exhaustive, Exclusive Categories | Ranked Categories | Fixed Categories Distance Between | True Zero |
> |---|---|---|---|---|
> | Nominal | Yes | | | |
> | Ordinal | Yes | Yes | | |
> | Interval | Yes | Yes | Yes | |
> | Ratio | Yes | Yes | Yes | Yes |

Nominal Level Measurement

Nominal measures assign subjects to one of two or more categories. These categories do not represent amounts. Therefore, the numbers assigned to them are **numerals**—arbitrarily assigned tracking devices devoid of any mathematical meaning.

A typical example of nominal measurement would be categorizing subjects as male or female. For data-analysis purposes, all men might be coded as "1" and all women as "2," but the numbers could not be interpreted to mean that a woman is twice as much of a person as a man. Even when we use statistics with nominal level measures, the data are essentially qualitative. We could, of course, use the numerals to calculate an average score for gender. Finding that score to be 1.43 would tell us that slightly more than half of our subjects are men, but the average could not be interpreted to mean that the typical subject in the study is part female and part male.

To measure a variable at the nominal level, four conditions must be met:

- There must be at least two categories into which subjects may be coded.
- Categories must be equivalent. They must be subsets of the same thing.
- Categories must be exhaustive. There must be a place suitable for coding each subject; however, one category may be labeled "other" in order to provide a place for rare instances.
- Categories must be mutually exclusive. Just as there must be a place for every subject, there must be only one place where the subject fits.

To illustrate these points, let's pretend that, as part of our study of program quality and advertising effectiveness, we want to find out about people's favorite television programs. Categorizing programs as "comedy," "drama," "news," "prime-time," and "local" would fulfill the first requirement, but those categories would violate the requirement for equivalence: The first three refer to content, the fourth to time of day, and the last might refer to content or to point of origin. A subject who prefers local news could be placed in either or both of two categories—"local" and "news"; there would be no category suitable for the person who prefers talk shows. The categories are neither exhaustive nor mutually exclusive.

Ordinal Level Measurement

At this level of measurement, the categories represent amounts—more or less of some indicator. However, the amounts are imprecise. They simply rank or order subjects along some dimension, but the distance between any two adjacent categories is unspecified and most likely is not the same. There is no way to tell from the categories themselves how far apart any two categories are. Nor can we tell how much difference there may be among subjects who place themselves in or are placed in the same category.

Ordinal measurement is the most commonly used level in communication research. Typical examples would be measuring television use by asking subjects to tell whether they watch "a lot," "some," "a little," or "never" or measuring communication competence by giving each of a number of speakers a letter grade or by assigning each of them a number that would indicate their rank from most competent to least competent.

Using those examples, one person who says he watches television "a lot" may view 8 hours a day, every day; another who says she watches "a lot" may watch only a couple of hours on most days. Yet both will be coded the same and treated as equivalent for data-analysis purposes. Similarly, the judges ranking speakers for their competence might have given three subjects the same grade and ranked them 1, 2, and 3 even though they believed the subject ranked as 1 was much better than subjects 2 and 3, who were essentially of equal competence.

Interval Level Measurement

Interval level measurement has all of the features of nominal and ordinal measurement. With interval measurement, there is no true zero, but differences between the various amounts are fixed and equal. The numbers represent a standard measurement unit, called a **metric.** Because they have real meaning, it makes sense to add and subtract them.

Unfortunately, there are few true examples of interval level measurement in communication research or, for that matter, in any of the social sciences. But an everyday example of an interval measure is temperature. If we use the Fahrenheit scale, we know that there is the same difference in amount of warmth between 2° and 3° as there is between 28° and 29°; 3° is 1° warmer than 2°, and 29° is 1° warmer than 28°. But 0° Fahrenheit does not mean the absence of all warmth. The zero is not a true zero, so we cannot infer that 28°F is 14 times warmer than 2°F.

Ratio Level Measurement

Ratio level measurement adds the true zero to the characteristics possessed by interval measurement. With a true zero, it becomes possible to multiply and divide the numbers to produce results that have real meaning. On the Kelvin scale for measuring temperature, we can say with confidence that 28°K is 14 times as warm as 2°K because 0° K does represent an absence of all heat.

For a measure to be a ratio, the possibility of a true zero must exist, but it is not necessary to place any subjects in that category. Researchers are measuring age at the ratio level when they ask people how old they are and allow the subjects to respond with any age rather than requiring them to choose from among several age categories.

Asking subjects to tell how many hours and minutes they spend each day watching television or how many times each day they talk to clients on the telephone are examples of measurement at the ratio level. So are scoring a test to indicate the number of right or wrong answers and counting the number of words in a document or the number of newspaper articles on a certain topic.

Implications

Note that levels of measurement are not the same as the levels of analysis discussed in Chapter 3. Levels of analysis describe the subjects of interest; levels of measurement describe the scheme for recording presence/absence, amount, direction, or other characteristics for each subject on each indicator of a concept. Levels of analysis are important for deciding who to study and what comparisons to make in the data-analysis phase. Levels of measurement are important for deciding how to make those comparisons.

Different levels of analysis require using different statistics. **Nonparametric statistics** are designed for use with nominal and ordinal measures. **Parametric** statistics are intended for use with interval or ratio measures.

Because of the hierarchical arrangement of the levels of measurement, any statistic that can be used with lower levels can also be used with higher levels. But purists will argue that statistics designed for higher levels should not be used with lower levels of measurement. However, as a practical matter, distinctions between ordinal and interval or ratio measures may not be all that important. Researchers often use Pearson's r with ordinal-level data; they treat measures that technically are ordinal as if they are interval or ratio. Sometimes they even use nominal variables in regression analysis by creating a series of **dummy variables.** Directions for doing this are in Chapter 7.

Box 5.3. *Level of analysis and choice of statistics*

Level	Appropriate Statistics
For Univariate Analysis	
Nominal	Frequencies, mode
Ordinal	Frequencies, mode, median, range
Interval or Ratio	Frequencies, mode, median, mean, range, variance, standard deviation
For Bivariate Analysis	
Nominal Independent Variable and Nominal or Ordinal Dependent Variable	Cross-tabulation, chi-square, phi, Cramer's V
Nominal Independent Variable and Ordinal or Ratio Dependent Variable	t-test
Ordinal Independent Variable and Ordinal Dependent Variable	Cross-tabulation, Kendall's tau, gamma, Spearman's rho
Interval or Ratio Independent and Dependent Variables	Pearson's correlation coefficient (r)
For Multivariate Analysis	
Nominal	Cross-tabulation
Ordinal	Cross-tabulation
Interval or Ratio (Ordinal, with caution; Nominal with dummy variable analysis—see Chapter 7)	Multiple Regression, Analysis of Variance (ANOVA), factor analysis

Nevertheless, it is a good idea to employ the highest possible level of measurement in order to get the full benefit of statistical analysis. Parametric statistics are more powerful than nonparametric ones. They can more readily detect whether scores on one variable are systematically related to scores on another than can nonparametric statistics.

For each level of analysis, it is also a good idea to use more rather than fewer categories or levels for each individual measure. It is always possible to combine categories for data-analysis purposes, but once subjects have been grouped into a category during data collection, it is impossible to separate them into two or more categories.

In general, using more categories provides additional information without adding to the cost of data collection. Consider a typical dichotomous yes/no measure. Simply asking whether someone reads an organization's monthly magazine will separate subjects into two categories: readers and nonreaders. In data analysis, researchers sometimes treat this kind of question as a ratio measure because "no" does represent zero.

But statistics designed for use at the ratio level assume many responses are possible and that the categories for those responses will be **continuous.** Because there are only two discrete categories, a yes/no question will, at best, be a very weak ratio measure. Moreover, the "yes" category requires us to treat as equivalent people who eagerly read every issue and those who glance at the newsletter only occasionally. The yes/no dichotomous question tells us nothing about how much subjects may read. To get that kind of information, we could just as easily have created an ordinal measure by offering three options: "always," "sometimes," and "never."

To make even finer distinctions among readers we could add more categories. Instead of offering discrete categories from which subjects could choose, we could even create a **continuous** ratio measure by asking how many issues subjects read during the last year.

If our only purpose in asking that yes/no question about readership were to filter out the readers so that we could study nonreaders, just two response options would be fine; however, two would be less satisfactory if we really wanted to study the readers. With just "yes" and "no" as response options, either we would miss important information about readers or else we would have to make further inquiries in order to get the information we could have gotten from a single question.

At the same time, asking people how many issues they read might offer too many options. With 13 possibilities (12 issues, plus 0 for those who never read), the responses could be very inaccurate. Subjects might simply be guessing. In making those guesses, they would also be more likely to favor some possibilities and avoid others than they would be if there were fewer possible response options.

The tendency to favor some options and avoid others might not matter much if our sample were very large. But with smaller samples, we might well have very few subjects for some readership levels. Therefore, in order to use some of the statistics we might want to use for our data analysis, we would end up combining categories. Choosing the right number of categories depends partly on the purpose of the research and partly on the nature of the measure. However, sample size may also be a factor.

Measurement Scales

Some concepts have attributes or features that can be measured only at the nominal level. But to measure people's behaviors, opinions, or knowledge, researchers must use scales that do more than classify by type. As used here, a **scale** is any measure that taps strength, amount, direction, salience, importance, or some combination of these.

Technically, most scaling techniques produce ordinal level measures, but the measures can usually be treated as interval or ratio ones without any problem. Most of the scales and scaling techniques that are used in communication research were initially designed as fixed-response questions for use in conjunction with operationalization through self-reports; however, most can readily be adapted for use as observational measures.

Likert Scales

This scaling technique, developed by the psychologist Rensis Likert (1932), measures strength and direction of people's attitudes or opinions toward a person, event, issue, or other phenomenon by having subjects use a five-point scale to indicate how strongly they agree or disagree with a series of statements. A typical Likert measure might ask:

Do you strongly agree, agree, neither agree nor disagree, disagree, or strongly disagree that the company newsletter is a good source of information?

_____ strongly agree
_____ agree
_____ neither agree nor disagree
_____ disagree
_____ strongly disagree

The decision as to whether "strongly agree" or "strongly disagree" should be coded as 1 would be arbitrary. But once the researcher made that decision, other options would be given progressively higher numbers so that the final numbering would reflect both direction and intensity.

Because subjects who have never considered a topic or do not have strong feelings about it tend to give positive answers, researchers often work around this **positive-response bias** by wording some questions so that "strongly disagree" becomes the favorable answer:

Do you strongly agree, agree, neither agree nor disagree, disagree, or strongly disagree that the company newsletter is boring?

_____ strongly agree
_____ agree
_____ neither agree nor disagree
_____ disagree
_____ strongly disagree

As a variant of the classic Likert scale, researchers often substitute response options such as "support/oppose," "pleased/displeased," or "likely/unlikely" for the traditional "agree/disagree." They may also use only a three-point scale or expand the scale to include seven or even nine points.

Decisions about the number of response options depend on the number of subjects, how fine a distinction they may be able to make, and how many distinctions among subjects the project demands. Although it isn't easy to label each point on an expanded scale, using extra points is appropriate in cases where a relatively high proportion of all subjects might reasonably be expected to choose one option over all the others. If that option is likely to be the midpoint or if it is important to distinguish between possible supporters and possible nonsupporters, it could even be appropriate to eliminate the midpoint or neutral category by offering an even number of response options.

Semantic Differential Scales

Like the Likert scale, the semantic differential scale provides information on direction and intensity. But the semantic differential does so by having subjects rate the specified phenomenon on a number of seven-point scales bounded by words that are polar opposites or antonyms.

To create this measuring technique, Osgood, Suci, and Tannenbaum (1957) selected large numbers of word pairs and then had their subjects use those word pairs to evaluate people, occupations, organizations, and objects. In this way they were able to locate those pairs that tap three different dimensions of meaning: activity, evaluation/judgment, and potency.

Rather than using large numbers of word pairs to locate those dimensions, today applied researchers typically select just a few items from the original list. Using this method, a researcher might explore how employees feel about the company newsletter by presenting them with a small set of word pairs and asking them to circle the appropriate number:

good	1 2 3 4 5 6 7	bad
boring	1 2 3 4 5 6 7	interesting
useful	1 2 3 4 5 6 7	useless
necessary	1 2 3 4 5 6 7	unnecessary
unattractive	1 2 3 4 5 6 7	attractive

In this example some favorable items were placed on the left, next to the number 1, and some on the right, next to the number 7. This technique helps reduce response bias; it also makes it possible for the researcher to identify and remove from the study those subjects who appear to have chosen a number and then circled it for all items in the set instead of actually using each set of word pairs to evaluate the referent.

Although seven-point scales are typical, researchers may choose to use an even number of points to eliminate the midpoint or neutral position, or they may add or subtract points while preserving a midpoint. However, they rarely use fewer than five points or more than ten. One exception is the **Stapel Scale,** which uses just a single word along with five positive and five negative numbers plus a midpoint indicated as zero:

Think about the company newsletter. Then read each word shown below. Then circle a number to indicate how well that word describes the newsletter. The better the word describes the newsletter, the larger the positive number you should circle; the less the word describes the newsletter, the larger the negative number you should circle.

	+ 5		+ 5
	+ 4		+ 4
	+ 3		+ 3
	+ 2		+ 2
	+ 1		+ 1
boring	0	useful	0
	− 1		− 1
	− 2		− 2
	− 3		− 3
	− 4		− 4
	− 5		− 5

With that exception, however, scales with more than ten points require subjects to make finer distinctions than may be possible. When confronted with too many categories, many subjects will simply ignore the extremes, in the process turning what started out as an expanded scale into one that effectively has fewer points. At the same time, scales with fewer than five points often fail to provide enough information because they allow for so little variance in response.

Thurstone Scales

Although Likert and semantic differential scales may appear to be interval levels of measurement, the distance between response categories is not really fixed. To overcome this shortcoming, L. L. Thurstone (1929) developed the technique bearing his name to measure direction and intensity or salience.

To create a Thurstone scale, a researcher would generate several hundred statements relating to the referent being investigated. Fifty to 100 people, reasonably knowledgeable about the subject but unconnected to the research, would then be asked to categorize the statements independent of each other into eleven categories ranging from extremely favorable to extremely negative or perhaps extremely important to extremely unimportant. The researcher would then select those statements—usually at least twenty but certainly at least one from each category—that were coded most consistently by the judges. To calculate the mean score on each item, the researcher would then add together the values given to each item by each judge and divide by the number of judges. These statements would then be ranked according to those average scores, worded to allow for agreement or disagreement, and then submitted to subjects. Those subjects' scores on this Thurstone, or equal-appearing interval scale, would be determined by giving each positive response the value it received in the rating by judges before adding together the values for all Thurstone items.

Although Thurstone scaling finds frequent use in psychology and education research, it is less often used in other fields. Because of the amount of work involved, applied communication researchers may occasionally use a pre-existing Thurstone scale, but they almost never create their own.

Guttman Scales

A Guttman scale measures intensity or commitment to an opinion, cause, person, organization, or course of action by presenting subjects with a series of statements on a topic, which can be arranged in a hierarchical fashion. This technique, developed by Louis Guttman (1950), is also referred to as **scalogram analysis.**

With Guttman scaling, an organization could locate its most committed supporters by asking subjects to indicate whether they:

    _____ serve on a committee or hold an office in the organization
    _____ attend meetings or other programs sponsored by the organization
    _____ contribute money to the organization
    _____ read the organization's newsletter

Presumably, a person who holds office or serves on a committee would also check all of the other options; those who attend meetings or programs would probably also give money and read the newsletter, but many who would read the newsletter or give money might not give of their time. Therefore, the volunteers would be scored "4," attendees would get a "3," contributors a "2," and readers a "1" on a Guttman scale.

Common Guttman scales such as those measuring political activism, social distance, and prejudice can be found by consulting books devoted to measurement scales. However, researchers often create items specific to the purpose of their own investigation and then test them to see whether they create a Guttman scale by using the scalogram feature in computer programs designed for quantitative data analysis.

Feeling Thermometers

The University of Michigan Survey Research Center developed this technique primarily for use in political polling. Subjects are shown a picture of a thermometer with temperatures from zero to 100° or asked to imagine one. Typically they are then told that if they feel neutral toward the referent person, organization, position, or other object, they should select 50°; they should pick a higher number to indicate how much more "warmly" disposed they are toward the referent or a lower temperature for "cooler," less favorable feelings.

With such a large number of "degrees" from which to choose, data generated in this way lend themselves to analysis at the ratio level. However, measurement is really ordinal. Because the numbers on the thermometer have no defined meanings, it is impossible to tell whether one subject's selection of 53° is really different from another subject's selection of 65° or how much more favorably disposed subjects are to a referent receiving an average of 75° than to one receiving an average of 65°. Moreover, there may not be a true zero on this scale. Although subjects usually find it easy to relate to the thermometer and pick a number, they tend to avoid extremely high and low ones.

Ranking Scales

In contrast to the previously described measures that call for subjects to provide information on amount, intensity, importance, or some similar evaluative quality for a single referent, ranking scales provide a means for comparing two or more referents on one or more of those qualities.

In the most common version, subjects are presented with a list and told to prioritize items on it:

Please rank the following kinds of television entertainment programs from 1 to 5 to indicate how well you like each kind. Use 1 to indicate the kind of program you like best. Use 5 to indicate the kind you like least. Use every number between 1 and 5. Use each number only once.

_____ situation comedies
_____ action/adventure
_____ drama
_____ variety shows
_____ game shows

Even with clear instructions, some subjects will use the wrong number—5 instead of 1 in the example—to indicate their favorite. Some will not provide a number for each item on the list; others may insist on giving two or more items the same number or use fractions such as 2.5 to indicate ties. These problems can be minimized if the researcher records responses.

If subjects can see the list of items and have a bit of time to think about it, they can usually rank ten items without much difficulty. But if the researcher must read the list, most subjects will have difficulty coping with even five or six items. Therefore, researchers may use a forced choice variant. To use this technique with the list of five kinds of television shows, subjects would first be asked to name their favorite and least favorite kind. After using the appropriate numbers to record those responses, the researcher would then read the remaining three kinds of programs before asking which of those the subject likes best and least.

Another variant of the basic ranking technique is the **paired choice:**

Which kind of television program do you like better?

| _____ situation comedy | or | _____ variety show |
| _____ variety show | or | _____ drama |

If all possible pairs are presented, the researcher can use the set of comparisons to rank individual items.

Composite Measures

Concepts are usually **unidimensional.** However, there may be several acceptable ways to measure even those concepts that consist of just one basic aspect or component. We could, for example, measure "readership" by asking people whether they "always," "sometimes," or "never" read a single paper or each of several papers, how many days they typically read them, how much time they usually spend reading them, or whether they read a particular paper or each of those papers "yesterday."

Because each question provides slightly different information even though all are measuring essentially the same thing, using several measures is usually preferable. These related measures may be combined to create a single **composite measure.**

Using composite measures simplifies data analysis by reducing the number of variables the researcher must consider. Although composite measures will bury or obscure the information each individual measure provides, they are more likely than a single measure to capture fully the true meaning of an underlying concept.

Therefore having multiple measures for more complex concepts and constructs is even more important than it is for simpler concepts, but identifying individual items that may appropriately be combined can be a difficult task. Recall that, for internal validity, measures should be measuring what they are supposed to measure—nothing more, nothing less. Therefore, each measure should tap just one concept or construct and, for most purposes, just one dimension of it. But in contrast to a concept, constructs such as "credibility," "quality," and "effectiveness" are **multidimensional.** They consist of several aspects or components. Therefore, locating individual items and then defending their combination to create a new variable requires more than the simple logic we used to argue for combining different measures of readership into a new composite measure.

Measurement techniques such as the Likert, semantic differential, and Guttman scaling techniques described in the previous section were developed specifically to incorporate multiple measures into data collection and then locate items that can be combined to create higher-level composite measures.

Although "scale" often refers both to individual ordinal- or higher-level measures and to a composite measure created by combining them, both the composite measures and the set of items that they combine may also be called an **index**. In this context, an **index** or **scale** is any composite measure composed of a supportable combination of individual measures. Many examples of this kind of scale or index can be found in the research literature.

If there are no pre-existing scales or indexes suitable for use in a study or if the preexisting ones must be shortened or otherwise modified, support for combining items or making modifications may be obtained by first identifying some possibilities and then checking the **correlation** between each pair of items in that set of possibilities. Items that logically appear to tap the same dimension of a concept can probably be combined if there is also a positive and statistically significant correlation between them.

A positive correlation indicates that both measures are working the same way—subjects that rate high on one measure also rate high on the other; subjects that rate low on one measure rate low on the other. Researchers usually accept a significance level, indicated by "p," of .05, which means that there are only 5 chances in 100 that the relationship is not real—that the relationship may have occurred by chance alone. Occasionally, however, they may set the significance level higher or lower depending on sample size and on the consequences of being wrong about whether the relationship is real.

If a study includes many similar items, each addressing the same topic, you may use **factor analysis** to confirm that items identified from theory or by checking correlations really do tap the same dimension of a concept. Factor analysis may also be used in an

---

Box 5.4. *Sources for existing measures*

Bruner, G. C. and Hensel, P. J. (1996). *Marketing scales handbook.* Chicago: American Marketing Association.

Miller, D. C. (1991). *Handbook of research design and social measurement,* 5th ed. Newbury Park, CA: Sage.

Robinson, J. and Shaver, P. (1973). *Measures of social psychological attitudes,* 2nd ed. Ann Arbor: Institute for Social Research.

Robinson, J. P. (1998). *Measures of political attitudes.* San Diego: Academic Press.

Rubin, R. B., Palmgreen, P. and Sypher, H. E. (1994). *Communication research measures: A sourcebook.* New York: Guilford Publications.

Schuessler, K. (1982). *Measuring social life and feelings.* San Francisco: Jossey-Bass.

Shaw, M. E. and Wright, J. M. (1967). *Scales for the measurement of attitudes.* New York: McGraw-Hill.

> **Box 5.5.** *Creating a composite measure or index*
>
> **Steps**
>
> 1. Select potential items for use in the composite measure.
>   - Make this initial selection on the basis of theory, an exploratory factor analysis (see Box 5.6), or logic.
>
>   CAUTION: All potential items for inclusion in a composite measure must be related to the underlying concept. Every item included in the index should have face validity. Those without face validity should be excluded.
>   - Select only items that provide a high degree of variance. Use frequencies as your guide. For example, if the frequencies indicate that everyone (or almost everyone) gave the same answer to a particular question, there would be little or no variance on this item. Including it would contribute little to the new index.
> 2. Examine the bivariate relationships between the items.
>   - A composite measure should be unidimensional. All the items should be measuring the same concept. If the items reflect aspects or degrees of the same concept, we would expect the responses to the items to be correlated with each other.
>
>   CAUTION: Examine all the bivariate relationships. Be sure to use the appropriate statistic. Use the information in Box 5.3. as a guide.
>   - If the computer program you are using will calculate Cronbach's alpha, you can use the Cronbach's alpha procedure as an alternative to examining bivariate relationships. Using this strategy also eliminates the need for step 5: The value of Cronbach's alpha tells you whether combining the items will produce a valid index.
>
>   Note: You may eliminate this step and steps 3 and 5 if logic alone serves as a sufficient justification for combining items. For example, these steps would not be needed to create an index of "total newspaper use" by adding together scores on items measuring readership of three different newspapers or by multiplying the number of days a person reads the paper during a typical week by the amount of time the person spends reading the paper on a typical day.
> 3. Eliminate inappropriate items.
>   - Use the bivariate correlations or Cronbach's alpha as your guide. If an item is not correlated with other items, this is evidence that it is not measuring the same concept as the other items; it is not a valid item and should be excluded from the index. If the value for Cronbach's alpha is lower with an item included than it is if the item is eliminated, that item should be eliminated.
> 4. Combine appropriate items into an index.
>   - Use the "compute" function in a computer program such as SPSS or SAS to create the new index. Summing scores—adding together the values for each subject on each item that the bivariate analysis or Cronbach's alpha indicates may belong in this new measure—is most common; however, it may sometimes be appropriate to use other mathematical techniques such as multiplying the value on one variable by the value on another.
>
>   CAUTION: Before combining items into an index, make sure all of them are worded in the same direction. If, for example, some are worded so that a high value indicates the presence of the concept and others are worded so that a
>
> *(continued)*

low value indicates its presence, you will need to recode some items so that either high or low (but not both) indicates its presence.

Note: You may also have to decide what to do about any missing data. If there are relatively few cases with missing data, you might decide to exclude them from the index and from data analysis. On the other hand, if there are a lot of cases with missing data, you may want to use another strategy. Some of the available strategies include mean substitution, assigning the middle value to cases with missing data, and assigning values at random.

5. Validate the index.
   - The most common methods for doing this are to calculate Cronbach's alpha or to determine whether scores on the index you have created will successfully predict scores on other measures of the same concept.

Example: Construction of the index "political diligence"

1. Two items from a survey questionnaire appear to be related to the underlying concept of political diligence.
   - Media attention (mediattn) = Please tell me how much attention you paid to stories about the recent presidential election that you happened to come across in the news media. Did you pay a lot of attention, some attention, a little attention, or didn't you pay any attention to stories you happened to come across in the news media?
   - Media effort (mediaeff) = How much effort would you say you generally put into getting information from the news media about the presidential election? Would you say you put in a lot of effort, some effort, a little effort, or didn't you put any effort into finding information about the presidential election?

   Combining these items produces an index with face validity.

2. Examine the correlation matrix for mediattn and mediaeff:

	mediattn	mediaeff
mediattn	1.000	.477
		($p = .01$)
mediaeff	.477	1.000
	($p = .01$)	

   (Note: $n = 395$)

   The matrix shows how strongly each item is correlated with each other item. This matrix indicates that the correlation between the two items is sufficiently strong (.477) and statistically significant at the .01 level (two-tailed). Therefore, we are justified in concluding that mediattn and mediaeff are measuring the same underlying concept.

3. By using the reliability routine in our data-analysis program, we find that Cronbach's alpha for this index is .65. Combining mediattn and mediaeff produces a valid index.

4. Because there are only two items in our index, we cannot improve validity by eliminating any item. For our index, we must combine mediattn and mediaeff.

5. To create the index of political diligence, use the compute function to sum respondents' scores on each item. The scores on mediattn and mediaeff ranged from 1 (none) to 4 (a lot). Therefore, the range for our new variable, the index of political diligence, is 2 to 8.

Box 5.6. *Doing a factor analysis*

Overview

Mathematically, factor analysis is very complex. But conceptually it is rather simple. In effect, the procedure sorts variables by placing those that cluster with each other but not with other variables into a number of unidimensional sets or "factors." In factor analysis, an **eigenvalue** greater than 1 indicates a true factor. **Factor loadings** indicate which items from the set of all items are part of each factor; items that are part of a factor or dimension will typically have a value greater than .5 on one factor and a value much lower on other factors.

This ability to find meaningful clusters of items that **load** on the same factor makes it possible to do an exploratory factor analysis as a first step in locating items that may be combined to form a scale or index that is unidimensional (all items measure a single dimension of a concept). The procedure can also be used to confirm whether a scale or index is unidimensional or whether it is really multidimensional (items tap more than one dimension of the same concept). If a scale or index is multidimensional, the results from a factor analysis will help sort the items into meaningful clusters or factors.

Example

As part of a radio listening survey, respondents were asked to rate 18 different types of music on a scale of 1 to 5 where 1 meant that the respondent did not like that kind of music at all and 5 meant that it is one of the respondent's favorite types of music.

To help the station managers make programming decisions, we can do an exploratory factor analysis to see which types of music tend to cluster together. For this factor analysis, we will use Principle Component Analysis as the extraction method and Varimax as the rotation method. These extraction and rotation methods are commonly used; however, other methods are possible. Using different extraction and rotation methods can produce different results.

Using Principle Component Analysis with Varimax Rotation, the factor analysis produces six factors, indicated by an Eigenvalue of at least 1.0:

(continued)

exploratory sense to locate items that cluster together without first examining their correlations.

Quality Control

Using good measures is a fundamental requirement for producing research findings that will be useful and withstand public scrutiny. One of the quickest and easiest ways to find good measures is to search the literature for individual measures or indexes that have been thoroughly checked and have withstood the test of time. However, even widely accepted and used measures need to be checked to ensure that they will work as intended. Elapsed time since a measure was developed or differences in research purpose, population, or conditions under which data must be collected can render even excellent, widely used scales suspect in some situations.

Type of Music	Factor 1	2	3	4	5	6
Alternative	–7.17E–02	**.823**	–5.34E–02	.181	.156	–2.77E–02
Christian Contemporary	4.35E–03	5.64E–02	.256	–7.29E–02	**.860**	5.08E–02
Classical	**.691**	–7.79E–02	–4.08E–02	.202	.166	.169
Classic Rock	.348	–.255	.249	**.528**	9.06E–02	.333
Country	–.291	–.165	**.514**	–5.85E–02	.236	8.40E–02
Easy Listening	–6.57E–02	–.328	**.603**	6.66E–03	.218	7.16E–02
Gospel	.265	–6.37E–02	3.07E–02	–.139	**.849**	3.56E–02
Hard Rock	–6.70E–02	.278	–5.85E–02	**.849**	–8.18E–02	4.28E–02
Heavy Metal	–4.65E–02	.190	–.110	**.792**	–.178	.149
Jazz	**.748**	–.136	–.215	2.52E–02	.208	1.78E–02
Oldies	.213	–1.21E–02	**.757**	–8.98E–02	–9.60E–02	7.79E–02
Pop	–2.43E–02	.227	**.652**	–2.99E–02	.142	.219
Progressive/College/New Age	.152	**.798**	–2.32E–02	.172	–.125	3.35E–02
Rap	–6.77E–02	2.54E–02	–4.52E–03	.120	–.124	**.844**
Reggae	**.537**	.440	–2.71E–02	2.51E–02	–.194	.239
Soul	**.644**	.257	4.27E–02	–.160	0.02E–02	.293
Spanish	**.565**	.112	.290	–.178	–1.82E–02	3.27E–02
Urban/R&B/Dance	.151	–1.22E–02	.160	–1.82E–02	.122	**.848**

The factor loadings in the cells of this table indicate the extent to which a particular type of music correlates with a factor. Notice that each item has a loading on each of the factors, but each is the highest on one factor. Typically a loading of .5 or higher is considered to be important if it is not exceeded by the same items loading on another factor. These highest loadings on each factor are in boldface type.

Notice that the types of music that loaded heavily on factor 1 are classical, jazz, reggae, soul, and Spanish music. Typically all items that load on a factor have something in common. In this case we might suspect that these types of music may in some sense be considered "educational or highbrow." Typically we would expect that we could create a unidimensional scale by combining all items that load on the same factor into a single, composite measure. But in this case, we would definitely want to check further. To see whether the items that load together on Factor 1 really "belong" together, we would check inter-item correlations and also calculate Cronbach's alpha (see Box 5.7).

Because most measures are in the form of questions to be answered by research subjects or variants of them for use by observers, in most cases the checking involves examining both questions and response options to make sure they are unambiguous and unbiased. At this stage it is also important to examine the instruments that you will use to collect and record information to make sure they are easy to use and that the arrangement of items will not bias the results.

The researcher responsible for a project can do much of this checking, but some problems such as missing measures or language that may be unsuitable for certain subjects can be hard to detect. Therefore, it is usually a good idea to have someone who is familiar with research procedures but who is unconnected to the project examine the measures and data-collection instruments. It is also a good idea to pretest them with a small sample of subjects and/or observers similar to those who will participate in the actual research.

In subsequent chapters we have more to say about question wording and developing instruments for data collection. But once this preliminary checking is complete, it is time for more formal checks to assess the reliability and validity of the measures.

Reliability

At the most fundamental level, a measure that is reliable is stable and dependable. Using it over and over should produce essentially the same results. But measures may still be reliable—dependable—even if the results obtained by using the measure again and again differ so long as those differences can be predicted on the basis of changes over time in subjects or their situations.

A test of writing ability might be reliable even if you scored much higher the second time you took it than you did on the first try, so long as most people taking the test also improved over time, and the amount of improvement could be predicted on the basis of other factors such as the kind of cognitive development that is normal as young people mature or, perhaps, whether subjects were enrolled in a writing class during the interval between test administrations. But to be a truly reliable test, it should also provide an accurate assessment of writing ability.

Accuracy is easier to achieve with measures of knowledge or skill than with other kinds of assessments for which no real benchmark may exist. But even tests of knowledge or skill may not provide an accurate assessment. They may overestimate or underestimate ability. Even so, the test may be reliable if the results are stable and dependable—if they consistently overestimate or underestimate by the same amount so that the measure will relate to other measures in predictable ways.

It is important to note that accuracy and precision, though related, are not the same thing. Indeed, in the quest for reliability, a trade-off between accuracy and precision often occurs. Grading that test of writing ability on a scale of 0 to 100 would be more precise but probably no more accurate than simply rating essays as A, B, C, D, or F. Indeed, trained coders could much more consistently place essays into one of five categories than they could choose from among the 101 categories to attach an appropriate number to individual essays. Therefore, a more precise measure might actually be less reliable than a less precise one.

Because checking the reliability of measures can be a rather complicated matter, researchers have developed a number of standard procedures.

Concurrent Reliability. In practice, the reliability of a measure is most often assessed by checking the strength and statistical significance of its correlation with some other measure. If Pearson's r or another appropriate statistic indicates that scores on the two measures vary together as expected, the measure in question is presumed to be reliable because results from using it are dependable and predictable.

For the measure to be dependable, the correlation may be positive or negative. That is, high scores on one variable may go with high scores on the other, low scores with low scores, or high scores on one variable may consistently be associated—correlated with—low scores on the other. Here, the important point is not the direction of the correlation, but that there is a correlation and that it is what one would expect based on theory or findings from previous research.

Alternate Forms Reliability. This is a variant of the basic correlation technique. But because the goal here is to see whether the two measures or tests are equivalent—the results obtained with one measure or one test produce results that are similar to and consistent with those of another version—the correlations must be both statistically significant and positive. Subjects scoring high on one measure or test should score high on the other; those who score low on one should score low on the other.

Split-half Reliability. The consistency or stability of a measure is most commonly assessed by dividing subjects into two groups and then comparing the scores on a measure for one group to those for the other. If the measure behaves consistently, the scores for the two groups should be the same, within the limits of sampling error. This sampling error will be higher for these subsets than for the entire group because each subset is smaller than the group from which it was drawn.

If the data-analysis program you are using has a routine for random sampling, the simplest way to generate two groups is to let the computer create two random samples. However, you can also do it manually by numbering the subjects consecutively and then putting even-numbered subjects in one half and odd-numbered ones in the other.

In cases where it is also important to check for stability over time, subjects may be divided into two or more groups according to when data from them were collected. This technique is appropriate whenever data collection occurs over a period of days, weeks, or months. It is also useful for determining whether some extraneous factor that occurred during data collection had any effect on subjects' responses.

Test-retest Reliability. To use this technique, you administer the same test to the same subjects at two different times. In addition to providing a test for stability and consistency, this procedure can also help detect the effect of some other factor. Therefore, it is the basis for experimental research.

In a slightly different version, the test-retest procedures can be used to pretest items before they are included in formal data collection by administering them to several groups that are similar to those who will be the subjects for the research. If all groups score similarly on the measures, the measure would appear to be consistent and therefore reliable. If the scores among groups are different but those differences vary systematically with other measures, such as time of day or the demographic characteristics of those taking the test, then the measure is dependable.

Cronbach's Alpha (Coefficient Alpha). The basic correlation technique and the split-half and test-retest can all help check the reliability of individual items or composite measures. Like them, Cronbach's alpha, also referred to as coefficient alpha, is based on correlations. However, it is used exclusively to check the reliability of a composite measure. It is an inter-item correlation summary statistic that estimates the total variance of the set of scales or individual measures and their individual contribution toward the new composite measure.

Calculating Cronbach's alpha can be a laborious task, especially with large samples or for composite measures that consist of many individual items. However, most computer programs have a feature that will calculate it for you.

> **Box 5.7.** *Formulas for statistics for assessing the reliability of measures*
>
> **Cronbach's alpha**
>
> $$\text{Cronbach's alpha} = \frac{k}{k-1} \times \left(1 - \frac{\text{sum of } s^2 \text{ of each test item}}{s^2 \text{ of the total test score}}\right)$$
>
> k = the number of items in the index
> s^2 = the variance
>
> **Scott's pi**
>
> $$\text{Scott's pi} = 1 - \frac{p_o - p_e}{p_e}$$
>
> p_o = percentage of observed agreement (expressed as a decimal) of two judges coding the same item independently
> p_e = percentage of agreement expected by chance = sum of squared percentages (expressed as a decimal) in each coding category
>
> **K-R 20**
>
> $$\text{K-R 20} = \frac{k}{k-1} \times \left(1 - \frac{\text{sum of } P \times Q \text{ for each test item}}{s^2 \text{ of the total test score}}\right)$$
>
> k = number of items on the test
> P = proportion of subjects (expressed as a decimal) answering the item correctly
> Q = proportion of subjects (expressed as a decimal) answering the item incorrectly
> s^2 = variance in test scores

Intra- and Intercoder Reliability. As measurement instruments, humans are prone to making errors. Therefore, it is especially important to check for reliability whenever measurement is done through observation.

The easiest way to determine whether a single observer is reliable is to perform an intracoder reliability check. To do this, the observer would redo a portion of the work and then calculate the percentage of all items coded or rated the same in the two trials. For two or more coders, intercoder reliability can be determined in much the same way: simply calculate the percentage of agreement between pairs of observers.

However, simple percentage of agreement tends to overestimate reliability. Just by chance alone a judge will sometimes choose the same category on two trials, or two judges may agree with each other even if they are not paying attention or following directions. Therefore, researchers generally use **Scott's pi** because it factors out chance agreement.

Because most data-analysis programs do not have a feature for calculating this kind of reliability, researchers usually calculate it for only the most problematic measures by recoding or re-rating a random sample of 10 percent of the items that were previously coded. Typically, reliability is assumed if Scott's pi is at least .7, but if an item is particularly hard to judge, a reliability coefficient of .6 may be acceptable.

Test-item Analysis. Checking items that may be scored using dichotomous categories such as right or wrong, good or bad, or competent or incompetent may be done by calculating the correlation of a subject's score on individual items with the score the subject received for the total set of all test questions. To do this, the most common technique is the **Kuder-Richardson formula 20,** more commonly referred to as **K-R 20.**

K-R 20 works on the logic that if a subject gets the first item on a test, or for that matter any item on a test, wrong, that subject is likely to have a lower score on the entire test than a subject who got the question right. Therefore, if calculating K-R 20 indicates that is not the case, the item is deemed unreliable because it doesn't dependably discriminate between those who, for example, are more knowledgeable or competent and those who are less knowledgeable or competent.

As with Scott's pi, K-R 20 is usually not available as an option in data-analysis programs.

Validity

Validity is related to reliability, but they are not the same thing. A measure can be reliable but not valid. It cannot be valid if it is unreliable. Validity presumes reliability; external validity—the ability to generalize—presumes internal validity. Reliability guarantees that a measure is consistent, stable, and dependable; internal validity establishes

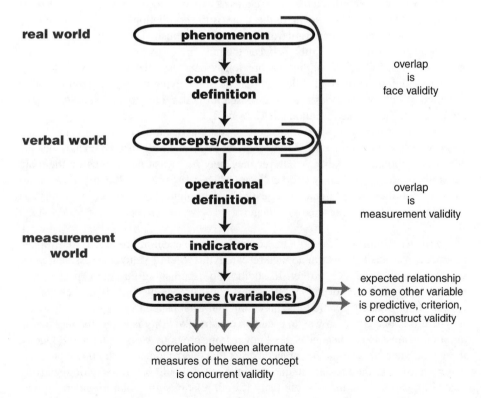

Figure 5.1. The relationship between measurement and validity

whether measures are actually measuring what they are intended to measure. To make a case for the internal validity of their measures, researchers use one or more assessment techniques.

Content Validity. Assessments of content validity rest on logic alone to determine whether a measure reflects the attributes or content of a particular concept as it is usually understood or may have been defined for a particular study. In its simplest form, a researcher claims **face validity.** That is, "on its face" the measure is a reasonable one. It appears to be measuring what it is intended to measure.

Because that kind of claim is so easy to make, face validity is the most common assessment technique. But because there is no external evidence to support the claim and there are no criteria for determining whether a measure is doing what it is supposed to do, it is also the weakest test for validity.

To buttress the claim for content validity, researchers may use **jury validation.** They call in a panel of experts or a group of subjects similar to those who will participate in the actual study and have them decide whether the measures are reasonable. This kind of **audit procedure** is commonly used with inductive, qualitative research. It is also used to help with development of measures for more quantitative, deductive research.

In comparison to the claim of face validity, jury validation provides more evidence for the quality of a measure. However, in essence it is simply face validity involving more judges.

Criterion Validity. This technique brings external evidence to bear on claims for validity by establishing that a measure relates as expected to some outcome, called the **criterion.** Therefore, it is essentially the same as the correlational tests for reliability.

Using this method, **concurrent validity** can be established by finding a statistically significant, positive correlation between the measure or test in question and another measure or test for which validity has previously been established. For example, concurrent validity of a proposed college-entrance test involving working in teams to build specified objects out of blocks can be assessed for both reliability and validity by testing it against the more conventional SAT or ACT.

Predictive validity exists if a measure or test can accurately forecast some future criterion. Although the face validity of both the ACT and SAT may be open to debate, both have predictive validity. Whatever they may be measuring, scores on them are generally predictive of ability to do college-level work. Similarly, the predictive validity of the building-block test can be assessed by determining whether scores on that test are positively related to success in college.

Construct Validity. Used more commonly in academic research to evaluate abstract constructs than it is in applied research, this kind of validity assessment takes criterion validity one step further. In addition to establishing that the expected relationship exists between or among measures, construct validity insists that claims for concurrent or predictive validity be based on well-established theory.

Therefore, one common way to establish construct validity is to test the measure in question against two other measures known from previous work to be valid. One of those other measures would be for a construct that theory indicates should be positively correlated with the measure in question; the other should be inversely or negatively related to the measure in question. Only if both relationships hold up under rigorous testing should the measure in question be accepted as valid.

Main Points

- Measurement refers to the process of assigning numbers, amounts, or other labels to things we observe or experience.
- The first step in measurement is to create a conceptual definition for the concepts in the research question and/or hypothesis that are guiding the research process.
- A conceptual definition should denote the essential qualities or meanings of the concept, exclude dimensions of the concept that are irrelevant or nonessential for the purposes of the study, and be clear, concise, and complete.
- Once a concept is clearly defined, the second step in the measurement process is to create an operational definition.
- Operational definitions specify the procedures or rules for recognizing and recording observable characteristics of a concept or construct.
- For measurement purposes, researchers have three different operationalization procedures available to them: manipulation, observation, and self-reports.
- There are four different levels of measurement: nominal, ordinal, interval, and ratio. These levels are hierarchical with each level possessing all the characteristics of those below it.
- Different levels of measurement require using different statistics. Nonparametric statistics are designed for use with nominal and ordinal measures. Parametric statistics are intended for use with interval or ratio measures.
- Because parametric statistics are more powerful than nonparametric statistics, it is usually a good idea to measure a variable at the highest level of measurement possible. It is also a good idea to use more rather than fewer categories or levels for each individual measure.
- A scale is a measurement instrument that taps strength, amount, direction, salience, importance, or some combination of these. Common scales in the social sciences include Likert scales, semantic differential scales, Stapel scales, Thurstone scales, Guttman scales, feeling thermometers, and ranking scales.
- Composite measures are created by combining several measures. They are useful because they can more fully capture the meaning of complex concepts and constructs than can any single indicator. They can also be used as a data-reduction technique.
- Factor analysis is a technique used to determine whether items in an index really tap the same dimension of a concept. It is also used for data reduction.
- A reliable measure is stable and dependable; a valid measure is one that actually measures what it is intended to measure.
- Concurrent reliability, alternate forms reliability, split-half reliability, and test-retest reliability are all procedures used to test the reliability of a single item or a composite measure.
- Cronbach's alpha is used exclusively to check the reliability of a composite measure.
- Scott's pi is used to assess intracoder and intercoder reliability for observational measures.
- K-R 20 is used to check the reliability of test questions for which there are right and wrong answers.
- Content validity, face validity, jury validation, criterion validity, concurrent validity, predictive validity, and construct validity are techniques used to assess the validity of measures.

Terms to Know

concept
conceptual definition
operationalization
conceptual fit
manipulation
observation
self-report
open-ended question
fixed question
reliability
variable
levels of measurement
nominal level of measurement
ordinal level of measurement
interval level of measurement
ratio level of measurement
exhaustive
mutually exclusive
nonparametric statistics
parametric statistics
dummy variable
continuous measure
scale
Likert scale
positive response bias
semantic differential scale
Stapel scale
Thurston scale

Guttman scale
feeling thermometer
ranking scale
paired-choice ranking scale
unidimensional
composite measure
multidimensional
index
correlation analysis
factor analysis
validity
concurrent reliability
alternate forms reliability
split-half reliability
test-retest reliability
intracoder reliability
intercoder reliability
Scott's pi
test-item analysis
K-R 20
content validity
face validity
jury validity
audit procedure
criterion validity
concurrent validity
predictive validity
construct validity

Questions for Discussion and Review

1. Provide a conceptual and operational definition for each of the following concepts: campus involvement, political participation, magazine readership, and media bias.
2. Search the literature for four research articles that focus on "television violence," "advertising clutter," or some other concept your instructor suggests. Is the concept defined clearly in each article? How similar are the definitions?
3. Use the Web to find information on "brand loyalty." Prepare a report in which you compare and contrast the way brand loyalty is conceptualized and measured in the different sources you find. Be sure to list your sources and the websites where you found them.
4. You have been hired to conduct research on negative advertising by candidates for political office. Create a conceptual definition for "negative advertising" and operational procedures using (a) manipulation, (b) self-reports, and (c) observation to measure negative advertising. Describe the advantages and disadvantages of each of these measurement approaches; tell when each would be appropriate and inappropriate.

5. Why is manipulation inappropriate as a measurement technique for dependent variables?
6. Why do researchers often use multiple indicators to measure a concept? Why do they create composite measures? What are the advantages and disadvantages of using single questions rather than multidimensional index scores from composite measures?
7. Why is it important for a researcher to consider the level of measurement of the variables in a study when planning a research project? When analyzing data?
8. For each question on the survey provided in Appendix C, identify the level of measurement. For each question, tell whether it would have been both possible and advantageous to substitute a question measuring the same concept but at a different level (such as ratio instead of ordinal).
9. You are interested in finding out how students at your school feel about a topic that interests you or about one your instructor suggests. Show how you might measure student opinion on this topic by using a (a) Likert scale, (b) semantic differential scale, (c) Thurstone scale, (d) Stapel scale, (e) feeling thermometer, and (f) ranking scale. What are the advantages and disadvantages of each of these ways of measuring opinion on the topic you chose?
10. Go to the website for the General Social Survey (http:www.icpsr.umich.edu/gss98) and explore the GSS code book. Write down the variable names, variable descriptions, and the response items available to the respondent for variables measured by using the Likert scale, semantic differential scale, and the feeling thermometer.
11. Compare and contrast face validity, concurrent validity, criterion-related validity, and construct validity. When is jury validation and/or an audit procedure appropriate for assessing validity?
12. Quite often a college student's grade point average (GPA) is taken as an indicator of intelligence or future success. Give some reasons why GPA may not be (a) reliable and (b) valid as a measure of each of these concepts: intelligence and future success. Suggest some alternative measures that may be more reliable and/or more valid indicators of these concepts. Explain how you could assess the criterion-related and construct validity of a GPA and one of the alternative measures you created for intelligence.
13. Examine the questions measuring attitudes toward abortion (abdefect, abnomon, abhlth, abpoor, abrape, absingle, abany) on the 1999 General Social Survey. Identify one other variable that you think should be correlated with the measures of abortion attitude. Extract these eight variables using the extraction program provided and analyze them using a data-analysis program (e.g., SPSS, SAS) to examine the (a) frequency distributions on the abortion items, (b) bivariate correlations between each pair of abortion items, and (c) bivariate correlations between each abortion item and the other variable you selected. Is the assumption that the seven abortion variables all reflect variations of the same concept supported by the frequencies and bivariate correlations? Would an index created by summing the values on these seven items be a reliable measure of attitudes toward abortion? (Hint: Calculate Cronbach's alpha.)
14. Use the information you developed in question 13 as the basis for combining individual questions to create a reliable composite measure of attitudes toward abortion. Examine the validity of this new index by checking the relationship between the index and some other variable(s) from the data set that you think should be correlated with abortion attitudes.

15. Look through recent issues of a communication journal, and find an article reporting on a quantitative, explanatory study. For that article, (a) state the research questions or hypotheses, (b) tell how the author(s) measured each concept or variable in the question or hypothesis, (c) identify the level of measurement for each concept or variable, (d) tell whether the author provides evidence for reliability and validity, and (e) give your own assessment of the reliability and validity of the work. In your report, be sure to include a full citation for the study you examined.

Readings for Discussion

Abernethy, A. A. (1991). Differences between advertising and program exposure for car radio listening. *Journal of Advertising,* pp. 33–40, April/May. This study makes use of several kinds of self-report measures.

Campbell, D. T. (1988). Definitional versus multiple operationalism. In D. T. Campbell, ed., *Methodology and epistemology for social science: Selected papers.* Chicago: University of Chicago Press, pp. 31–36. In this paper, a master of research design discusses the implications of methods for operationalizing concepts.

deBell, C. S., Montgomery, M. J., McCarthy, P. R. and Lanthier, R. P. (1998). The critical contact: A study of recruiter verbal behavior during campus interviews. *The Journal of Business Communication, 35*(2):202–203. The authors compare observational ratings of interviewer-interviewee interactions to ratings by the interviewees.

Fico, F. and Soffin, S. (1994). Covering local conflict: Fairness in reporting a public policy issue. *Newspaper Research Journal, 15*(4):64–74. The authors create a technique to measure balance in individual newspaper stories.

Frary, R. B. and Tideman, T. N. (1997). Comparison of two indices of answer copying and development of a spliced index. *Educational and Psychological Measurement, 57*(1):20–32. This study makes sophisticated use of concurrent validity as a basis for creating a new index.

Gaziano, C. and McGrath, K. (1986). Measuring the concept of credibility. *Journalism Quarterly, 63*(3):451–462. This study illustrates the use of factor analysis to isolate dimensions of a complex construct.

Johnson, R. L., McDaniel II, F. and Willeke, M. J. (2000). Using portfolios in program evaluation: An investigation of inter-rater reliability. *American Journal of Evaluation, 21*(1): 65–80. This study uses the basic intercoder reliability technique to assess the use of portfolio reviews as part of the evaluation of the Even Start family literacy program.

Rimmer, T. and Weaver, D. H. (1987). Different questions, different answers? Media use and media credibility. *Journalism Quarterly, 64*(1):28–36. This study examines the effect of operationalization decisions on research findings.

References and Resources

Blalock, H. M. Jr. (1982). *Conceptualization and measurement in the social sciences.* Beverly Hills, CA: Sage. In this work, the noted scholar addresses the need for generalizability, simplicity, and precision and the implications of trade-offs among them in social science research.

Chakrapani, C. (1998). *How to measure service quality and customer satisfaction.* Chicago: American Marketing Association. This handy source of measurement techniques also provides information on which techniques work best for various purposes.

Emmert, P. and Barker, L. L. (1989). *Measurement of communication behavior.* White Plains, NY: Longman. In addition to covering research basics, this book has excellent chapters on more specialized topics such as nonverbal communication measurement, interaction analysis, and physiological and unobtrusive measures.

Guttman, L. (1950). The basis for scalogram analysis. In S. A. Stouffer, L. Guttman, E. A. Suchman, P. F. Lazarsfeld, S. A Star, and J. A. Clausen, eds., *Measurement and prediction.* New York: Academic Press. This chapter is the developer's explanation of the theory and practice of Guttman scaling.

Likert, R. (1932). A technique for the measurement of attitudes. *Archives of Psychology, 140*:1–55. This is the original work explaining Likert scales.

Myers, J. H. (1999). *Measuring customer satisfaction: Hot buttons and other measurement issues.* Chicago: American Marketing Association. A top marketing researcher and an academic, Myers provides detailed guidance to researchers for large and small businesses and not-for-profit organizations for selecting, creating, and using measures and for analyzing data from self-reports.

Osgood, C., Suci, C. and Tannenbaum, P. (1957). *The measurement of meaning.* Urbana: University of Illinois Press. This is the classic work explaining the theory and creation of semantic differential scales.

Thurstone, L. L. (1929). Theory of attitude measurement. *Psychological Bulletin, 36*:222–241. This is the foundational work on Thurstone scaling.

Webb, E. J, Campbell, D. T., Schwartz, R. D. and Sechrest, L. (1999). *Unobtrusive measures.* Thousand Oaks, CA: Sage. This is an updated and expanded version of the classic work arguing for use of multiple and nonreactive measures.

Part III

Basic Research Methods

6

Experiments

AN EXPERIMENT INVOLVES exposing subjects to a stimulus under controlled conditions in order to determine the effect of that stimulus.

In an experiment, the stimulus, which serves as the independent variable, is manipulated. That is, the researcher creates one or more versions of a stimulus such as a public service announcement or other message and uses them as measures for the independent variable. The dependent variable is the effect. It can be measured through observation or self-reports.

To conduct the basic experiment, the researcher collects data on the dependent variable, then exposes one group of subjects to the stimulus, and compares their postexposure scores on the dependent variable to their pre-exposure scores and also to the scores from another group, the **control group,** which did not receive the stimulus.

Experiments can be relatively cheap, or they may be extremely expensive depending on the design of the study, the length of time over which data collection must take place, the number of researchers involved, sample size and the relative ease of finding subjects, whether those subjects must be paid in order to gain their cooperation, and the cost for creating the manipulation and for acquiring space suitable for doing the work.

However, the chief advantage of the experiment is that, of all the basic research methods, an experiment provides the strongest evidence for cause-and-effect relationships. Rival explanations for the effect can be eliminated or ruled out.

But in exerting the kind of control necessary to increase reliability and internal validity, external validity usually suffers. For the typical laboratory experiment, it is almost always impossible to convince a large random sample of subjects to participate in the study. Validity problems can also exist because the manipulation and the setting for the experiment may be artificial. Experiments also can present more ethical challenges than other kinds of research.

Communication Applications

For all their problems, experiments are a staple for some kinds of applied communication research.

- In advertising and public relations, the experiment is the method of choice for determining whether a campaign is likely to be or has been effective.

120   Basic Research Methods

- Public relations practitioners and issue advocacy groups use experiments to document both beneficial and harmful effects of media coverage.
- Reporters sometimes conduct field experiments to find out whether businesses and other organizations are complying with the law or to determine whether their practices are discriminatory.
- Businesses conduct experiments to find out whether programs they have implemented are working the way they were intended to work.
- Technical communicators conduct useability tests.

Useability tests, which investigate whether people can understand messages or follow directions to perform some task, are one kind of evaluation research. But broadly understood, evaluation research includes any type of study, qualitative or quantitative, that is designed to assess the impact of policies, programs, or messages. As the examples suggest, experiments are particularly useful for **evaluation research** because of their ability to detect true cause-and-effect relationships.

Although experiments are used less often in communication research than they are in other fields such as psychology, they are the basis for all scientific research. The logic underlying the experiment informs the design for surveys, content analyses, focus groups, and observation research.

The Logic of Experiments

By systematically manipulating the independent variable and eliminating or accounting for all extraneous factors or **confounds** that might also affect scores on the dependent variable, researchers make it possible for an experiment to show causation.

Causation

As we noted in Chapter 3, being able to show a causal relation requires four things, each of which is possible with an experiment.

1. There must be some temporal or geographic proximity between the presumed cause and the effect. This requirement is met by showing that subjects who are exposed to the stimulus exhibit the effect, but those who are unexposed to the stimulus do not exhibit the effect.
2. The cause must precede the effect. With an experiment, the researcher controls the time sequence. By taking measurements on the dependent variable before and after exposing one group to the stimulus, the researcher can show that the effect, or change, occurred after exposure to the presumed cause; that the stimulus caused the effect can be shown by comparing the scores on the dependent variable for those who received the stimulus to those in the control group who were not exposed to the stimulus.
3. There must be concomitant variation—that is, changes in the independent variable are related to consistent, systematic changes on the dependent variable. At the simplest level, demonstrating that an effect occurs in the presence of the stimulus but not in its absence does this. However, more rigorous tests are also possible by systematically manipulating the amount of stimulus and then show-

ing that increasing amounts produce increasingly higher or lower scores on measures of the dependent variable.
4. It must be possible to rule out other possible explanations for the presumed effect. In experiments, this comes from control: control over the independent variable or stimulus and control for possible confounds.

To show a cause-and-effect relationship, experiments must be planned so that the impact of the independent or stimulus variable can be recognized. The most efficient and effective experiments will maximize the variance of the main variables of interest and minimize or control for other possibilities.

Controlling the Manipulation

This is often the hardest part of an experiment because, in creating the manipulation, it is so easy to introduce into it elements that may be unrelated to the underlying concept or construct of interest. Moreover, to maximize the variance that might be attributed to that manipulation, the manipulation must be robust. That is, the stimulus must clearly possess the attribute of the concept in sufficient amount to produce an effect even on subjects who may not be very susceptible to it. At the same time, the manipulation cannot be so strong as to be unrealistic.

Controlling for Confounds

To ensure that whatever effects do occur can be attributed to the stimulus variable, researchers typically minimize variation due to factors other than the manipulation by incorporating one or more basic control techniques into their experiments. These include eliminating confounds, blocking them, and holding them constant.

Elimination. In the real world independent and dependent variables never occur in isolation from all other possible variables. Removing all of those other variables from an experiment is rarely possible. However, it is almost always possible to eliminate some confounds such as distractions caused by the presence of family and friends or the ringing of a telephone by conducting experiments in a laboratory where those problems would not exist.

Blocking. Rather than eliminate a confound, this technique actually adds a confounding variable as another independent variable to compensate for the fact that manipulations are seldom pure; instead, they are most often confounded by some other concept. By adding a confound such as sex, for example, to a study of persuasion, the researcher can examine the independent effects of gender and persuasive method as well as the interaction between the gender of the persuader and the effect of exposure to the persuasive message. However, controlling for a confound through blocking requires using a fairly large sample in order to have enough subjects for data-analysis purposes in each subdivision. In the example, four groups would be needed: male/persuasion, male/no persuasion, female/persuasion and female/no persuasion.

Holding Constant. With this technique, confounding variables are turned into a **constant** measure whose value is the same or made to be the same for all subjects.

Technically, all forms of control do this. But we can also take specific steps to turn confounds into constants.

In experiments using several stimuli, **counterbalancing,** or rotating the order in which subjects are exposed to the various stimuli, will hold treatment effects constant.

Effects on the dependent variable that may be due to confounds related to the subjects themselves can be held constant by in effect using each subject as his or her own control. Comparing each subject's pretest and posttest scores to each other effectively eliminates any variable other than the stimulus if it is reasonable to assume that nothing about the subject has changed during the course of the experiment. However, if something specific to subjects might affect scores on the dependent variable, those confounds may be held constant by limiting the range of variation of a confound. If we suspected age might make a difference, we could, for example, eliminate it from consideration by working with only one age group such as children or senior citizens.

But it isn't always feasible or desirable to limit the range on a confound. In any case, some confounds may be impossible to recognize. Ideally we would get around this problem by selecting two large probability samples of subjects from the same population and then assigning one sample to the experimental treatment and one to the control group. But except for some field experiments, getting two large random samples of subjects to participate will be impossible.

Therefore, for laboratory experiments and for some field experiments, the most important form of control comes from **randomization.** Randomization means using probability techniques to assign subjects to treatment conditions so that each subject has an equal and known chance of receiving a stimulus or of being in the control group. The technique can be as simple as counting subjects and then assigning those with an even number to the experimental treatment and those with an odd number to a control group.

As a result of randomization, groups should be equivalent on both known and unknown confounds before the experiment begins. Therefore, any effect of the treatment cannot be due to differences between subjects in the experimental and control groups. But as with probability sampling, effectiveness depends on sample size. The larger the number of subjects who can be assigned through randomization, the more likely all possible confounds will be represented in, and therefore constant for, both groups.

However, randomization may be impractical or impossible. In such cases, most of the benefits of randomization can be realized through **matching.** With this procedure, subjects who are very similar are paired; one member of each matched pair is then assigned to the experimental condition, the other to the control. Matching may sometimes be necessary when working with individuals, but it generally is most appropriate when the subjects are groups rather than individuals as would be the case with many studies designed to test the effects of teaching methods or workplace procedures.

Neither randomization nor matching is likely to work perfectly with small groups of subjects, but **statistical** control can also help eliminate many of the confounds related to differences between subjects in the experimental and control groups. Any confound that can be measured can be held constant by using appropriate statistics in data analysis. In experiments, the most common of these is the analysis of variance (ANOVA), which is discussed more fully in the Data Analysis section of this chapter. Other techniques such as partial correlations and regression analysis are more often used in survey research, so they are discussed in Chapter 7.

The Experimental Setting

In planning an experiment, researchers always strive for as much control as possible. Part of this control comes from the setting where the research will be conducted. Consider two approaches a public relations agency might take to determine the effect of public service announcements intended to encourage people to recycle newspapers:

- Two groups of people watch a "new" television program in the auditorium at a shopping mall. During commercial breaks in the program, one group sees the public service announcements mixed in with product commercials; the other gets only the product commercials. At the end of the hour, both groups fill out questionnaires asking their opinions about the program and about their willingness to buy the products advertised, to recycle newspapers, and to engage in other activities unconnected with anything they have just seen.
- The public service announcements run in prime time on commercial television stations in one mid-sized Midwestern city but do not run in another very similar Midwestern city. While the PSAs are being aired and for a few weeks afterward, the public relations firm collects data on newspaper recycling in both cities and compares the numbers to recycling figures from before the advertising campaign began.

The first approach is the **laboratory** experiment. Subjects come to a setting chosen by the researcher where they are exposed to the stimulus and react to it. In this setting, the researcher can be sure the subjects actually saw the announcements and that any effect came after the "cause." Because the setting is the same for both the experimental group and the control group, many of the common confounds are eliminated. If the desired effect does not occur in the laboratory setting, it is unlikely to occur under home-viewing conditions. But if it does occur in the laboratory, the researcher cannot be sure that the effects will occur outside the lab.

The laboratory viewing experience is quite artificial; few people watch television programs with strangers; most do not or cannot give their undivided attention to any program. Therefore, some of the effect most likely came from the viewing situation. Some part may also have come from the **Hawthorne** or the **guinea pig effect.** Knowing their reactions are being studied, subjects often change their behavior simply because they know they are being observed. Many will also try to behave like "good little guinea pigs" by trying to appear more open-minded or intelligent than they really are. In succumbing to **research demand,** they may give whatever they perceive to be the socially desirable answers or they may simply try to please the researcher.

Those kinds of behavior can inflate the magnitude of any effect, but it can also mask a real effect. If subjects in the control group know they are in the control group, they may try to compensate for what they believe they are missing. Even if they do not know, as probably would be the case with the lab version of the recycling experiment, they may try to appear open-minded, intelligent, and good. They may give socially desirable answers or simply try to please a researcher.

These problems are minimized in the second version of the experiment. That kind of **field experiment** takes the stimulus to the subjects, who remain in their own, natural setting. In some situations, as with the PSAs, the researcher maintains control over the manipulation. In other field experiments, the researcher may simply capitalize on a naturally occurring manipulation.

The advantage of the field experiment is that the researcher can know for sure that any observed effects did and will occur in the real world, not just in a laboratory. But even in cases such as the PSA experiment where the researcher can easily check to see whether any observed effects came after the presumed cause, the researcher will not have the same kind of control over confounds that would be possible in a laboratory. Therefore, there will be less certainty that the stimulus really was responsible for the effect.

Ethical Concerns

Experiments raise more ethical concerns than other kinds of research with the exception of observational research.

Because experiments require exposing some people to the stimulus while others serve as the control, some degree of deception is almost always present. If the researcher conducting the laboratory PSA experiment told subjects the true nature of the experiment, they would almost certainly exhibit the guinea pig effect. The same effect might occur even in the field experiment version. Moreover, manipulation necessarily exposes people to something they otherwise might not be exposed to.

Even the most realistic and apparently benign manipulation may have unforeseen consequences for at least some subjects. In the PSA experiment, the recycling message would seem harmless. But in the laboratory version the television program in which it was embedded might turn out to be distressing to some subjects who ordinarily would not watch that kind of program. That would not occur in the real-world experiment because people would be watching only those programs they normally watch. But one can imagine the possibility that some people who see the ads might decide to help decrease the amount of newsprint sent to landfills by canceling their newspaper subscriptions.

Telling the subjects in the field experiment that they are or are not getting a public service announcement would be impractical, but getting consent from subjects for laboratory experiments or for field experiments in closed, naturalistic settings such as businesses, schools, or other organizations is always necessary. For consent to be informed, as it must be, subjects must be told in advance the nature and purpose of the research and warned of any foreseeable risks. Because not all risks are foreseeable, they should also be told there may be other unforeseeable risks and their likely magnitude.

Risks should never be greater than the benefits to individuals or to society that may accrue from gathering information by using people as the subjects for a research project. Where subjects cannot be told in advance of the true nature of a study or of all possible effects, the standard procedure is to **debrief** them at the conclusion of the study. The debriefing should inform subjects of the purpose of the study, tell them about the manipulation, alert them to any deception involved in it such as using made-up information or nonexistent experts as sources in messages, and, if the situation warrants, offer help or point subjects to resources for coping with problems associated with participation in the study.

Experimental Designs

The common understanding of an experiment as a kind of research in which a scientist "does something" to "find out what happens" because of doing that "something" nicely captures the essence of the method. But, of course, there are many kinds of things the scientist might do and many ways of doing them. There are also many ways the study might be designed.

In connection with experimental research, **design** may refer to the statistical techniques used to analyze data. Used that way, a study might be described as a same-sample t-test or as an analysis of variance design. These designs are discussed in the Data Analysis section.

In this section **design** refers to the total plan for assigning subjects to groups and then collecting the data. Depending on how well they control the common threats to validity, these designs are commonly classified as true experiments, quasi-experiments, or pre-experiments. To identify these designs, researchers generally use the nomenclature and notation system developed by Donald Campbell and Julian Stanley (1963).

In the Campbell and Stanley notation system:

- **R** = Random sample or random assignment of subjects to the experimental and control groups
- **M** = Matching used to assign subjects to experimental and control groups
- **X** = The manipulation, treatment, or administration of the independent or causal variable. For studies involving multiple independent variables or various levels or kinds of manipulations, subscripts may be added: X_1, X_2 to indicate two versions or treatments of the same independent variable and X_{1-2} to indicate two different independent variables administered at the same time as part of the same manipulation.
- **O** = Observation, measurement of dependent variable and known confounds. Subscripts may be added to indicate temporal sequence: O_1 would be the first observation, O_4 would be the fourth observation.

Symbols are read from left to right. For example, a notation of **R O_1 X O_2** means that subjects are randomly assigned to a group, and measurements for the dependent variable and confounds are taken both before and after subjects are exposed to the independent, stimulus variable.

True Experiments

These are the ideal. True experiments are the model for all research because they control for the common threats to internal validity: selection, history, maturation, testing, instrumentation, regression to the mean, and mortality.

The Classic Experiment. This is the basic design. All other designs are variations of it. With this design, researchers typically use randomization to assign subjects to experimental and control groups. Then at the same points in time, they collect data from the experimental and control groups on the dependent variable and on known confounds both before and after exposing the experimental group to the stimulus.

$$R \quad O_1 \quad X \quad O_2$$
$$R \quad O_3 \quad \quad O_4$$

This design can be used any time the purpose of a study is to determine whether a stimulus has an effect. It is appropriate for detecting change in opinion, knowledge, or behaviors such as buying a product, voting, or giving money to a cause as the result of exposure to an advertisement or public service announcement.

The only serious drawback to this design is that administering the pretest could artificially inflate or deflate posttest scores. This potential problem can be avoided by using either the posttest-only control group or Solomon four-group design.

Posttest-only Control Group Design. This design handles the problem of pretest-posttest interference by eliminating the pretests:

$$\begin{array}{ccc} R & X & O_1 \\ R & & O_2 \end{array}$$

The design controls very well for all of the major threats to internal validity; however, in some situations the absence of a pretest can make it difficult to tell whether part of any apparent effect may have resulted from problems with instrumentation and/or selection. Therefore, this design is most appropriate when the manipulation is fairly simple and when there is no reason to believe initial differences will arise between the two groups on measures of the dependent variable.

The posttest-only control group design is commonly used for studies of learning when both groups are equally familiar or unfamiliar with the topic as they might be in testing the usability of documentation for a new computer program. It is also used to test advertising effectiveness for products that are generally well known or that are truly novel.

Solomon Four-group Design. This design is really a combination of the classic experiment and the posttest-only experiment.

$$\begin{array}{cccc} R & O_1 & X & O_2 \\ R & O_3 & & O_4 \\ R & & X & O_5 \\ R & & & O_6 \end{array}$$

The advantage of this design is that it provides the data you will need to check for equivalence among groups and for instrumentation problems while also controlling for test-retest interference. Therefore, it is the design of choice for studying opinion change because it is so difficult to know in advance whether members of both the experimental and control group hold very similar underlying attitudes and beliefs and because subjects in this kind of study can be quite sensitive to research demand. For similar reasons, it is the best choice for many studies designed to detect knowledge or behavioral effects.

The main problem with the Solomon four-group design is a pragmatic one. Because it involves using four groups, it requires more subjects and perhaps more laboratory space and research assistance than is true for other true experimental designs. Therefore, it is usually more expensive.

True Experiment Variants. Note that the basic designs, as they were described in the previous sections, say little about the time interval between administration of the pre- and posttests. Neither did the notation indicate how many independent variables or levels or categories of the independent variable might exist.

Box 6.1. *Calculating the magnitude of effects*

To see how well an experiment controls for threats to internal validity, you can calculate how much effect each such threat had on your results. You can also calculate how much of any apparent effect may be attributed to your manipulation.

This example uses the Solomon four-group design because it offers the best control for threats to internal validity.

To do the calculations, simply plug the appropriate mean (average) pretest and posttest scores (calculated from the individual scores for all members of a group) into the equations. For example, if the individual pretest scores for subjects in Group A were 8, 11, 9, and 12, the mean pretest score for group A would be 10. Therefore, you would use 10 as the value wherever there is an O_1 in an equation.

	Pretest	Stimulus	Posttest
Group A	O_1	X	O_2
Group B	O_3		O_4
Group C		X	O_5
Group D			O_6

Calculating the Effects of Threats to Internal Validity

History—Maturation = (posttest for Group D: no pretesting, no experimental stimulus) − (average of all pretests)
= $O_6 - (O_1 + O_3 / 2)$ = Effect

Testing—Reactivity = (posttest for Group B: pretesting, no experimental stimulus) − (posttest for Group D: no pretesting, no experimental stimulus)
= $O_4 - O_6$ = Effect

Testing—Pretest Sensitivity = [(posttest for Group A: pretesting, experimental stimulus) − (posttest for Group C: no pretesting, experimental stimulus)] − [(effect of testing-reactivity)]
= $[(O_2 - O_5)] - [(O_4 - O_6)]$ = Effect

Selection (randomization) = (pretest for Group B − pretest for Group A)
= $(O_3 - O_1)$ = Effect

Calculating the Effect of the Manipulation X

1. Effect of X = [(effect of posttest-pretest for Group A) − (effect of posttest-pretest for Group B)] − (effect of Testing—Pretest Sensitivity)
    = $[(O_2 - O_1) - (O_4 - O_3)]$ − [Effect of Testing-Pretest Sensitivity]
    = $[(O_2 - O_1) - (O_4 - O_3)] - [(O_2 - O_5) - (O_4 - O_6)]$
2. Effect of X = effect of posttest for Group C − posttest for Group D
    = $(O_5 - O_6)$

Note: If the results from calculation 1 and calculation 2 are different, you can average these two estimates of the effect.

The time interval between administering the pre- and posttest may be quite short or very long depending on the purpose of the study and available resources. Studies using a very short interval will measure only short-term effects. For longer-term effects, a second set of observations at some later time may be added or the interval between the pretest and the posttest may be increased. Increasing the interval will decrease pretest-posttest interference, but increasing the interval or adding a second set of posttest observations will add to the cost of data collection. With very long intervals or with studies extended to include multiple posttests, some subjects may drop out of the study, making it prone to mortality effects. There may also be more confounds from history.

A **factorial design** will accommodate the need to study two or more independent variables at the same time. In this kind of study, each independent variable is called a **factor**. A **2-factor** design has 2 independent variables; a **3-factor** design has 3, and a **27-factor** design, even though impossibly complex and expensive, would have 27 independent variables. Each factor may also have several levels or categories. If each of two factors had two levels, the design would be a 2 x 2 factorial design; if each had three levels, it would be a 2 x 3; if each of three factors had four levels, it would be a 3 x 4 factorial design, and so on.

To illustrate the concept of factors and levels, suppose a researcher wants to study the effect of delivery channel and writing style on employee compliance with new directives. Channel and style would be the factors. Each might have two or more levels or categories; for channel, voice mail and e-mail, for example, and for style, informal and formal language. For the study, subjects would be randomly placed into four groups: Group I, voice mail, formal style; Group II, voice mail, informal style; Group III, e-mail, formal style; Group IV, e-mail, informal. With this 2 x 2 design, the results from subsequent measures of employee compliance could then be used to determine the effect of channel and style as well as the joint effect or interaction between channel and style.

Of course if only one independent variable were of interest, there would be only one factor, so the basic designs are appropriate. But there still might be several levels or categories of that factor. These levels can be tested against each other with or without a formal control group. If, in the previous example, our only interest were in comparing voice mail and e-mail, we could simply replace the control group with one of the delivery options:

$$R \quad O_1 \quad X_1 \quad O_2$$
$$R \quad O_3 \quad X_2 \quad O_4$$

In this version, X_1 could represent voice mail, and X_2 could stand for e-mail. If we wanted to compare both delivery options to the current practice of using traditional paper memos, the paper memos would represent no change or intervention, so we could treat paper memos as the control, or we could add them as an X_3.

Quasi-experiments

Because quasi-experimental designs lack at least one of the features of the true experiment, they do not control for all of the possible threats to internal validity as well as true experiments control for them. Their value lies in the fact that they provide good control for most of the threats while accommodating many situations where true experiments would not be feasible.

In their classic works on experimental design, Campbell and Stanley identify sixteen different quasi-experimental designs. Some of these are just variations on the true experiment. Superficially at least, others such as the counterbalanced design appear quite different. Here we discuss just those designs that are most commonly used.

Equivalent Group Pretest-posttest Design. In the true experiment, randomization is used to assign subjects to experimental and control groups. But in the real world that may not be possible. Instead, it may be necessary to create equivalent test and control groups through matching. Thus, the basic experiment would be diagrammed as:

$$M \quad O_1 \quad X \quad O_2$$
$$M \quad O_3 \quad \quad O_4$$

With matching, you assign subjects to groups so that subjects in the test and control groups are the same—they match—on known characteristics. Most often these characteristics are observable ones such as gender, race, and age. Although matching on known, observable characteristics provides little guarantee that the groups will be equivalent on other characteristics, this problem can often be overcome by including appropriate measures on the pretest.

If matching is successful, this design will control for threats to internal validity as well as a true experiment. However, using it instead of a true experiment is most appropriate whenever it would be unethical or even impossible to use randomization to assign subjects to the test and control group.

Because there could be real problems with using randomization to assign some workers in a factory or some students in a school to a test condition while others were assigned to the control group, it would be better to work with in-tact groups of subjects. To do this, you could recruit two very similar factories or two very similar schools for your study. You could then use one of these matched pairs to be the experimental group and one to be the control group.

Nonequivalent Control Group Design. Although using matching to create equivalent groups will often provide the same kind of control that you can get from a true experiment, sometimes it may be necessary or desirable to work with nonequivalent groups. A researcher might, for example, like to use the matched-pair design to study the effect of computer-assisted composition instruction, but equivalent schools might not be willing to participate in the study. But the researcher might also choose two very different schools such as one from an upper-middle-class suburban area and one in a lower-income inner-city area in order to account for the possible effects of school resources and/or family socioeconomic status. In either case the design would be:

$$\frac{O_1 \quad X \quad O_2}{O_3 \quad X \quad O_4}$$

With that design, school resources and/or socioeconomic status become additional independent variables. But our researcher might really want to know whether introducing computers into the classroom will improve the performance of inner-city students so that their test scores become more like those of students in suburban schools. In that case the design would be:

$$\frac{O_1 \qquad X \qquad O_2}{O_3 \qquad \qquad O_4}$$

The students in the suburban school would, in effect, serve as the control.

Although the comparative element in both designs enhances internal validity, neither design controls as well as the true experiment or the equivalent-group quasi-experiment for selection or for various interactive effects between selection or instrumentation and other threats such as testing.

Separate-sample Pretest-posttest Designs. With these designs, separately drawn probability samples of subjects are used as test and control groups. These samples may come from the same population, similar ones, or ones that may be quite different. Some of the manipulations may actually occur; others, indicated by (X) in the diagrams, are simply inferred. Therefore, there are many possibilities such as:

$$\begin{array}{cccc} R & O_1 & & (X) \\ R & & X & O_2 \end{array}$$

and:

$$\begin{array}{cccc} R & O_1 & X & O_2 \\ R & & (X) & O_3 \end{array}$$

and even:

$$\begin{array}{ccccc} R & O_1 & & (X) & \\ R & & O_2 & (X) & \\ R & & & X & O_3 \end{array}$$

Because the subjects in each group are usually a large random sample, it is often possible to extract equivalent subsamples for data-analysis purposes or to handle selection problems by employing statistical controls to compensate for any differences among groups. But most of these designs do not control well for the effects of history, maturation, or mortality. Instrumentation problems may also arise.

Nevertheless, a separate-sample pretest-posttest design can be useful whenever circumstances make it necessary or desirable to collect data at different times or in different places. Therefore, these designs are commonly used in field experiments. They are also useful for piecing together results from several surveys to create an approximation of a true experiment.

Time Series. This design adds a series of observations before and after administration of the stimulus:

$$O_1 \qquad O_2 \qquad O_3 \qquad X \qquad O_4 \qquad O_5 \qquad O_6$$

Although this design does not control well for history or for some instrumentation problems, it effectively controls for the possibility that scores on dependent variables or on important confounds may not be stable over time. This makes it very useful for examining both short-term and long-term effects of any intervention such as implementation of a new policy, the demise of a rival publication, or change in a publication's format or mode of delivery.

A variant that ignores the X to focus on changes in O over time is the basis for **tracking** studies and for **forecasting.** For advertising purposes, a downturn in consumption of a product might signal it is time to begin a new advertising campaign. For a nonprofit organization, a change in contribution levels would provide evidence needed to decide whether to expand activities or cut back on expenses.

Counterbalanced Designs. This quasi-experimental design is sometimes called a **Latin square** or a **repeated measure design.** It is appropriate for factorial studies in which there are several levels or categories of one or more independent variables.

What sets this design apart is that, instead of assigning subjects to different treatments, all subjects or groups of subjects are exposed to multiple treatments, but the order of exposure to the treatments is rotated.

X_1O	X_2O	X_3O	X_4O
X_2O	X_3O	X_4O	X_1O
X_3O	X_4O	X_1O	X_2O
X_4O	X_1O	X_2O	X_3O

In this diagram, the lines indicate that each group is separate; probability techniques were not used to assign subjects to a treatment order. Neither was there any attempt to match the groups. Nevertheless, the design can readily detect effects because each individual subject or group in effect acts as its own control. To convert the design as illustrated to a two-factorial one, a second X could simply be inserted before each observation: $X_1X_4O\ X_2X_3O\ X_3X_2O\ X_4X_1O$.

Counterbalanced designs are often used whenever there is reason to believe that the order in which several manipulations are presented might have some bearing on subjects' scores on the dependent variable. A typical example would be a study of the effectiveness of four different persuasive messages delivered by four different kinds of sources.

Pre-experiments

These designs are in many ways oversimplifications of the true experiment. The two simplest versions, the **one-shot case study:**

$$X \qquad O_1$$

and the **one-group pretest-posttest:**

$$O_1 \qquad X \qquad O_2$$

lack a control group.

Although neither version provides much control over common factors affecting validity, either one can appropriately be used in preliminary studies and in those where the interest is more in people's reaction to a stimulus than in showing a true cause-and-effect relationship. For example, researchers might use one of these designs with subjects similar to those in a target public to gauge people's likely reaction to an ad, a television program, or a persuasive campaign.

These designs are also appropriate for useability studies because, in them, the interest is in finding out whether people can follow a set of directions or are having problems with a specific message rather than in gauging any real change in behavior or opinions.

A third version is the **static group comparison:**

$$\underline{X \qquad O_1}$$
$$O_2$$

Like the nonequivalent control-group design, this one is most useful in situations where the interest is in determining whether a particular intervention can make two very different groups more equal on a target dependent variable. But because this design does not include a pretest and its control comes from a nonequivalent group, it does not control for common threats to internal validity as well as the posttest-only true experiment or the nonequivalent control-group quasi-experiment. Therefore, it is best reserved for situations where equivalent groups are unavailable or undesirable and administering a pretest would be inappropriate or impossible.

Quality Control

At their best, experiments make it possible to draw valid conclusions about cause-and-effect relationships. But drawing valid conclusions will be possible only if the researcher takes care to ensure the quality of the study—its reliability and internal and external validity. Although the main threats to reliability and internal and external validity are somewhat different, the three are interrelated.

Reliability

Without reliability, there can be no validity. Reliability will affect internal validity, which, in turn, affects external validity. Therefore, the first step in quality control is to make sure the study is reliable. With experiments, reliability problems are primarily ones of **instrumentation.** But there are two kinds of instrumentation problems: ones relating to manipulation and measurement and ones relating to the actual conduct of the experiment.

For most measures of the dependent and control variables, quality control requires pretesting by using one or more of the techniques for assessing the reliability of measures that were described in Chapter 5. The same techniques can be used with the independent variable if the manipulation involves a stimulus that is in a fixed form, such as several versions of a printed or videotaped message.

However, if the manipulation requires a live performance, any variation in how a human delivers the stimulus or conducts the control will destroy reliability. So will any differences in coding decisions among observers or changes over time by a single coder. With humans as the instrument, ensuring reliability means carefully training the performers or coders and then checking their work for consistency.

Other threats to reliability may creep into an experiment by administering the stimulus under different conditions. These are most easily dealt with in laboratory experiments by making sure that all groups use the same or very similar facilities, at the same time of day, have the same amount of time to complete the study, and are equally shielded from possible distractions. Although eliminating all of the factors that could threaten the reliability of a field experiment is virtually impossible, you should try to identify as many of them as possible and then take them into account in interpreting and reporting findings.

Internal Validity

Any problems with reliability will affect the internal validity of a study because at least part of the response to the stimulus may actually be the result of unreliable measures, testing conditions, or variations in performance by researchers. However, ensuring reliability is never quite enough to guarantee that a study is measuring what the researcher intends.

Creating a manipulation that addresses the concept underlying the independent variable and is uncontaminated by other factors is, perhaps, the most challenging part of quality control. Assessing the internal validity of any measure may be done by using the techniques outlined in Chapter 5. However, for the stimulus variable other **manipulation checks** are often necessary. One of the most useful of these is the **audit.** With an audit procedure, the manipulation and controls are submitted to panels of experts and subjects similar to those who will participate in the actual experiment. Comments from these panelists are then used to identify and overcome problems with the manipulation.

Other common problems with internal validity stem from the researchers, from the way the work is done, and from the subjects themselves.

As noted earlier, the behavior of researchers can threaten the reliability of a study. However, it can also more directly affect internal validity by introducing the possibility that changes in a dependent variable are the result of responses to extraneous characteristics of the research administrator rather than to the stimulus variable. Subjects may, for example, respond with more enthusiasm to a researcher who seems warm and genuinely interested in them than to one whose personality seems less attractive. In other cases, researchers may give off subtle cues as to the intended purpose or desired outcome of a study, and subjects will very likely respond to those cues.

To overcome or eliminate researcher attribute effects, subjects could be matched with researchers, the same researcher could be used for all groups, or a wide variety of researchers could be used to minimize the effect any one researcher might cause. To control for the unintentional giving off of cues, researchers could also hire assistants who do not know the research goals or the hypothesized results.

But even with those controls, some part of the effect may come just from being part of an experiment. Here, the most common problems are pretest-posttest interference and the guinea pig and Hawthorne effects. However, test anxiety on the part of some subjects may also present a problem.

If learning from taking a test could affect scores on the posttest, researchers generally use the Solomon four-group design. Eliminating or minimizing the Hawthorne effect often requires working with subjects who are unaware they are part of a study by, for example, conducting a field study instead of a laboratory one. If conducting research in the field using naturally occurring manipulations is impossible, researchers may, at some cost in terms of ethics, resort to misleading subjects about the true purpose of an investigation.

Truly test-phobic subjects should be eliminated from a study. However, the more serious threat to internal validity comes about when the experimental and control groups know about each other and are influenced by that knowledge. In such cases, interaction between groups may mean that each becomes aware of the stimulus others are receiving and uses that information to shape their responses on the posttest. In other situations, members of one or more groups may try to overcompensate for what they perceive to be a disadvantage resulting from receiving or not receiving the stimulus.

Overcoming these **contamination problems** requires taking steps to make sure that experimental and control groups remain separate by using strategies such as conducting all sessions simultaneously. Using counterbalanced or equivalent group designs can be especially useful in studies conducted in workplace or school settings where group members know each other, interact on a daily basis, and may also fear being left behind.

Although contamination may be considered a problem associated with subjects, the most common threats to internal validity stemming purely from the subjects themselves involve selection biases, regression effects, and maturation or mortality problems.

Members of subject and control groups must be equivalent in all important ways before the study begins. Otherwise, part of the posttest scores will be the result of potentially unmeasured and uncontrolled-for differences between members of the two groups. The best way to eliminate this problem would be to use two large random samples of subjects drawn from the same population. However, that is feasible only for certain field studies. For laboratory experiments, the best assurance of equivalence among groups comes from using the largest possible number of subjects and then assigning them to groups by means of probability techniques. Employing statistical techniques to check for equivalence among groups and as a control for any differences among groups that cannot be eliminated or controlled for by other means also helps.

Subjects selected on the basis of very extreme scores on one or more measures will almost always exhibit less-extreme scores with subsequent retests. They will "regress to the mean" score for the larger population of which they are a part. **Regression to the mean** is a natural phenomenon, but it is very easy to mistake it for the effect of the manipulation. The problem can be minimized by excluding from the study those subjects whose extreme scores make them most susceptible to the problem. Where studying subjects with extreme characteristics is the point of the study, quality control requires using only subjects with similarly extreme initial characteristics and then using randomization or matching techniques to assign them to groups.

Other threats such as maturation, history, or mortality are also natural phenomena, but they may also affect some subjects differently than they do others. Effects caused by any of these threats to validity may also be mistakenly attributed to the independent variable when, in fact, a researcher may be measuring how old or tired subjects are becoming, what else is going on in their lives, or ways in which those who drop out of a study differ from those who remain in it.

Minimizing or accounting for all of these effects requires using pretest and posttest measures, often in conjunction with the Solomon four-group design. Keeping the study as short as possible will also help. For studies that necessarily must extend over a period of weeks, months, or even years, maturation problems can sometimes be minimized by using subjects, such as young adults, who are less susceptible to maturation effects than children or senior citizens would be. Mortality effects can be minimized, though probably not eliminated, by building incentives for continuation into the design or by offering bonus compensation to subjects who stay involved through each phase of the research. Where nothing else works, employing statistical controls can help eliminate or account for effects from history, maturation, or morality.

External Validity

The requirement that research results be generalizable to some larger population is the true weakness of most experiments. Generalizability requires working with reasonably large probability samples of subjects. But with some exceptions, getting that large probability sample of subjects to participate in an experiment is virtually impossible.

Whenever researchers must work with self-selecting or convenience samples of subjects, as is almost always the case in laboratory experiments and sometimes in field experiments, external validity will suffer.

Even in those situations where experimenting with large probability samples is possible, external validity may be threatened by pretest-posttest interactions, interactions among various treatments, history, mortality, or reactive arrangements that make the testing situation somewhat unreal or unusual. In short, one can never be quite sure that results found in a laboratory experiment will carry over into more natural settings. Even with field experiments, subject selection, events external to the experiment, and reactive arrangements may make the results somewhat suspect.

Although researchers may initially sacrifice external validity in order to achieve the reliability and internal validity that are the true strengths of an experiment, over time they can lay claim to external validity. But to make a strong case for the generalizability of their findings, they must first remove or control for as many threats as possible. They must then conduct additional experiments to show that replications produce the same findings and that those findings also hold true in different settings and with different kinds of subjects.

Data Analysis

Although researchers may sometimes tuck in some qualitative information to help explain or clarify their findings, they almost always analyze experimental data using quantitative methods. Truly qualitative analysis is usually limited to studies using experimental design to choose settings or subjects for focus groups, observation research, and some variants of surveys and content analyses. Therefore we reserve discussions of qualitative data analysis for subsequent chapters. Here we treat only quantitative analysis.

Experiments are designed to determine whether the independent or stimulus variable makes a difference in scores on some dependent or outcome variable. Although pretest and posttest scores on the dependent variable—and on most if not all control variables—will almost always vary among subjects, those variations among subjects do not really matter. In social science research, aggregate data, not the results from individual subjects, are what matter. In an experiment, researchers are looking for similarity and differences in mean scores:

- similarity in the mean score on the dependent variable and on control variables between the experimental and control groups before the experiment begins

136 Basic Research Methods

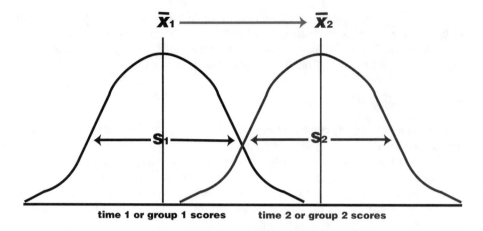

t-test: compares mean 1 to mean 2
anova: compares variance within groups to variance between groups

Figure 6.1. Tests of difference

- differences in mean scores on the dependent variable between the pretest and posttest for the experimental group(s)
- differences in the mean scores on the posttest between the experimental and control groups

Therefore, data analysis begins with a comparison of test and control-groups' pretest scores on the dependent and control variables. It then moves on to a search for change between pretest and posttest scores for the experimental group(s). This is followed by a check to see whether any significant difference occurs in posttest scores between the control group and the experimental group(s).

Fundamentally, this is a search for differences in mean scores, so quantitative analysis relies primarily on statistical tests of difference. The most widely used tests of difference are the t-test and analysis of variance. Both are designed for use with ordinal or higher levels of measurement.

The basic t-test compares mean scores within the same sample; an alternate version is appropriate for comparing mean scores from separate samples.

Although the t-test is relatively easy to calculate, it has some drawbacks. There is no direct way to interpret the results of a t-test. Manually calculating the statistic will require using a t distribution table, which can be found in any statistic book; with computerized data analysis, the printout will provide the t-value and the accompanying level of statistical significance. But even knowing the t-value and accompanying level of significance will tell you very little about the magnitude of any effect. Because the t-value depends so greatly on sample size, calculations such as those shown in Box 6.2 better indicate the magnitude of any effect.

But perhaps more important, t-tests can handle only one variable. Scores on that variable can come from just two groups or two points in time. Therefore, you may need to recode your data in order to use a t-test. If, for example, you had collected data on your subjects' level of education by asking the last grade they had completed, you would need to collapse that information into two categories such as "high school or

Box 6.2   *Conducting a t-test*

A t-test is often used with experiments to determine whether or not the mean on the dependent variable in the experimental group is significantly different from that of the mean in the control group.

Formula

$$t = \text{value} = \frac{\bar{x}_1 - \bar{x}_2}{S_{\bar{x}_1 - \bar{x}_2}}$$

where,

$\bar{x}_1$ = the mean on the dependent variable for group one
$\bar{x}_2$ = the mean on the dependent variable for group two

$$S_{\bar{x}_1 - \bar{x}_2} = \sqrt{\frac{SS_1 + SS_2}{n_1 + n_2 - 2} \times \left(\frac{1}{n_1} + \frac{1}{n_2}\right)} = \text{the standard error for the groups}$$

SS_1 = the sum of squares for group one
SS_2 = the sum of squares for group two
n_1 = the sample size for group one
n_2 = the sample size for group two

degrees of freedom (df) = $n_1 + n_2 - 2$

In order to determine whether there is a significant difference between the means, you compare the calculated t-value at the appropriate degrees of freedom and desired level of significance to the critical value in a t-distribution table. If the calculated t-value is greater than the t-value in the table (i.e. the critical value), the means for the experimental and control group are statistically different.

Example

This example uses SPSS with data from a 1996 survey of religion, media and politics. Questions on this survey are similar to those on the 2000 survey reproduced in Appendix C. While it is possible to do a t-test by hand, researchers often rely on computer data analysis programs such as SPSS or SAS to run such a test.

Research Question: Do individuals who view themselves as active in politics use more sources of political information about the presidential election than those who do not view themselves as active in politics?

Measures:

    total number of sources = the summation of the number of sources individuals indicated that they used for political information about the presidential election (inc. talk radio programs, National Public Radio, national television news, local television news, etc.).

    active in politics: "In general, how active are you in politics?" Response options of very, somewhat, not very and not at all recoded as two categories: active (very + somewhat) and inactive (not very + not at all).

(continued)

Output:

Group Statistics

	N	Mean	Std. Deviation	Std. Error Mean
Active	1	193	4.9741	1.5660	.1127
Totsource	2	179	5.5754	1.7349	.1297

Independent Samples Test

t-test for Equality of Means

Totsource	t	df	Sig. (2-tailed)	Mean Diff.	St. Error Diff.	95% Confidence interval of mean Lower	Upper
Equal variances assumed	−3.513	370	.000	−.6013	.1712	−.9379	−.2648
Equal variances not assumed	−3.500	358.8	.001	−.6013	.1718	−.9392	−.2634

Conclusion: The t-value in the table is −3.513 at 370 degrees of freedom. This value is significant below the .05 level. Therefore, the two means are significantly different. Individuals who perceive themselves as active in politics do use more sources for information about the presidential election than those who do not perceive themselves as active.

---

less" and "at least some college" in order to use a t-test to determine the effect of education on your subjects' scores on a dependent variable. Directions for recoding are in Box 6.3.

Because it is almost always possible to recode data into two defensible categories, the more serious problem with the t-test is that you can't add control variables to see whether they had any effect on scores on the manipulated variable. Neither is there any way to handle several factors. The only way to analyze data from a time series or Solomon four-group design would be to perform multiple t-tests.

Therefore, for all but the simplest analyses, researchers usually use the **analysis of variance (ANOVA).** Like the t-test, an ANOVA can detect differences in mean scores even when data come from a relatively small number of subjects.

But unlike the t-test, analysis of variance lets you examine the simultaneous effect of several independent variables or several factors on a dependent variable or make comparisons among several points in time or among several groups. Although you may still want to recode some of your data, you would not need to recode it into just two groups. Instead of using just "high school or less" and "at least some college," you could recode data on the last year of school into categories such as "college degree," "some college," "high school graduate," and "less than high school."

More sophisticated variants of analysis of variance, such as analysis of covariance (ANCOVA) and multiple analysis of variance (MANOVA) or covariance (MANCOVA), let you add control variables or test for the effect of repeated measures.

Box 6.3. *Recoding variables*

When researchers operationalize their concepts, they often allow for more precise measurement than they find they need or can use. Therefore, they may decide to recode the data they collected into fewer categories. These recoding decisions are usually made either because of the particular research question or because of the distribution of the data.

For example, variables are often recoded:

- if there are more than 15 to 20 values on a variable that you would like to display in an easily readable table.
- if you are interested in how a particular group compares to another. For example, if you wanted to know how those with less than a high school education compare to those with more than a high school education it wouldn't be necessary to have the finer distinctions offered by more defined categories.
- if there are very few individuals in a particular category, it might make sense to combine these individuals with those in a neighboring category—especially, if you would like to use a data analysis strategy such as crosstabs. Many categories will make it difficult to read such a table and make it more likely that there will be cells with few or no cases.

General rules for recoding variables:

- The new categories on the variable should be logically defensible and preserve the general shape of the distribution of cases on the variable.
- The new categories should be mutually exclusive and exhaustive, so that every case can be classified in one, and only one, category.

Example:

Data in this example come from a 1998 evaluation study of a college newspaper. Question: "Are you classified as a freshman, sophomore, junior or senior or are you in graduate school?"

To use data on class standing in a t-test, you could recode the five response options so you have just two categories: undergraduate and graduate students.

As you can see from the following frequencies, the number and percent of subjects in the new "undergraduate" category equals the sum of the number (59 + 49 + 50 + 76) and percent (21.7 + 18.0 + 18.4 + 27.9) of subjects in each of the categories that were combined. In this example, the number and percent of subjects in the "graduate" category remain unchanged.

Original Categories

Class Standing	Number	Valid Percent	
Freshman	59	21.7	
Sophomore	49	18.0	
Junior	50	18.4	
Senior	76	27.9	
Graduate	38	14.0	
(Total)	272	100.0	*(continued)*

	New Categories	
Class Standing	*Number*	*Valid Percent*
Undergraduate	234	86.0
Graduate	38	14.0
(Total)	272	100.0

The actual procedure for recoding will depend on the computer program you are using. You can find directions for reading computer output for frequencies in Box 7.1 in the chapter on surveys.

But for all its flexibility, an ANOVA has several problems. Manually calculating an analysis of variance requires you to consult an F table to determine whether any differences between groups are statistically significant; computerized data analysis will produce a printout that includes the F ratio and its accompanying level of statistical significance. But as with the t-value for a t-test, the value of F and the level of significance provide little information about the magnitude of any difference in mean scores. Even more important, calculating an ANOVA or having a computer do it for you will not tell you which group or groups differed from other groups. To find that out, you will have to perform more statistical tests. The most common of these are the **Tukey, Scheffé, and Least Significant Difference (LSD)**.

Main Points

- In an experiment the researcher manipulates the independent variable (the stimulus) and measures the effect of this manipulation on the dependent variable.
- The main advantage of an experimental design is that it provides the strongest evidence for showing cause-and-effect relationships. Carefully designed experiments can rule out rival explanations for an apparent relationship between the independent and dependent variable.
- To minimize the variation in the dependent variable due to other factors besides the manipulation of the independent variable, researchers often use one or more basic control techniques: eliminating confounds, blocking, and holding confounds constant.
- Researchers use randomization to assign subjects to treatment groups in an experiment in order to ensure that the groups will be equivalent on all known and unknown confounds before an experiment begins. When it is not possible to use probability techniques, researchers often use matching to assign subjects to groups.
- When they plan and conduct experiments, researchers strive for as much control as possible. Both the setting in which the research is conducted and the specific experimental design that is used can affect the degree of control possible.
- Both the Hawthorne effect (subjects changing their behaviors simply because they are being studied) and research demand must be guarded against because they can affect the outcome of an experiment.

Box 6.4. *Conducting an analysis of variance*

An analysis of variance (ANOVA) is basically an extension of the t-test. This statistic can be used to test for significant differences between two or more groups and to test for the simultaneous effects of more than one independent variable on a dependent variable.

Formula

For an ANOVA, sums of squares are calculated between groups (of subjects), within groups (of subjects) and in total (the sum of the between and within figures). The sums of squares between groups and within groups are each divided by their respective degrees of freedom to obtain the mean squares between and the mean squares within.

The $MS_b = SS_b / df$ $df = K - 1$, where K = the number of groups
The $MS_w = SS_w / df$ $df = N - K$, where N = the total sample size

The F ratio is then calculated using these mean squares as the formula below indicates:

$$F = \frac{MS_b}{MS_w}$$

The calculated F ratio at the appropriate degrees of freedom and level of desired significance is then compared to the critical value in an F-distribution table. If the calculated value is equal to or greater than the value in the table, the results are considered to be statistically significant.

Note: In contrast to the t-distribution table, an F-distribution table uses two degrees of freedom (the two calculated above) instead of one. One of these degrees of freedom is listed across the top of the table and one is listed down the side.

Example

This example uses SPSS with data from questions on a 1996 survey of religion, media and politics that are similar to the questions on the 2000 survey reproduced in Appendix C.
Research Question: Did individuals who had a high level of interest in the election use a greater number of sources of political information about the election?
   Measures:
      total number of sources = the summation of the number of sources individuals indicated that they used for political information about the presidential election (inc. talk radio programs, National Public Radio, national television news, local television news, etc.).
      election interest = "How interested were you in the presidential election?" Four response options of very, somewhat, not very and not at all.

(continued)

Output:

ANOVA

Totsource	Sum of Squares	df	Mean Square	F	Sig.
Between Groups	62.106	3	20.702	7.839	.000
Within Groups	979.903	371	2.641		
Total	1041.909				

Conclusion: The calculated F ratio in the table (7.839) is significant at or below the .000 level. Therefore, individuals with high levels of interest in the election used more sources of political information about the election than those with low levels of interest in the election.

- A laboratory experiment is conducted in a controlled setting. A field experiment is conducted in a natural setting.
- The main advantage of a laboratory experiment is control over potential confounds. The main disadvantage is that the setting is often quite artificial. Because of the artificiality of the setting, you can't be sure whether an observed effect would occur outside the lab.
- Deception is common in laboratory experiments, but deception raises ethical concerns. Researchers must be especially careful that the risks of participating never outweigh the benefits. Researchers who use deception must always debrief subjects at the conclusion of the experiment.
- The main advantage of a field experiment is that a researcher can be sure that any observed effects did and will occur in the real world. The main disadvantage is that the researcher has less control in a natural setting and can thus be less certain that the stimulus really was responsible for the effect.
- Experimental designs are commonly classified as true experiments, quasi-experiments, or pre-experiments depending on the degree to which the design controls for common threats to validity.
- True experiments have three characteristics: manipulation of the independent variable, a posttest measure of the dependent variable, and use of randomization to assign subjects to test and control groups. By possessing these characteristics, true experimental designs effectively guard against each of the main sources of internal validity.
- True experimental designs include the classic experiment, posttest-only control-group design, and the Solomon four-group design.
- Factorial designs allow researchers to assess the effects of two or more independent variables with each having two or more treatment levels.
- A quasi-experimental design lacks at least one of the key features of the true experiment. Therefore, they do not control for all possible confounds affecting the internal validity of a study.
- Common quasi-experimental designs include the equivalent group posttest, nonequivalent control group, and separate-sample pretest-posttest designs, the time series, and the counterbalanced design.

- Pre-experimental designs are oversimplifications of the true experiments. Because they do not control for many of the threats to internal validity, they are most appropriate for preliminary studies and for studies where the interest is more in people's reactions to a stimulus than in showing true cause-and-effect relationships.
- Examples of pre-experimental designs include the one-shot case study, the one-group pretest-posttest design, and the static group comparison design.
- In experiments, reliability problems are primarily due to instrumentation problems related to manipulation and measurement and to inconsistencies in the way the work is done.
- The main weakness of most experiments is the inability to generalize the findings to the population of interest. Even though experiments tend to be low in external validity, external validity can be established through replication.
- Experimental data are most often analyzed using a t-test or an analysis of variance (ANOVA).

Terms to Know

control group
manipulation
cause and effect
matching
blocking confound
holding confounds constant
randomization
statistical control
counterbalancing
pretest
posttest
laboratory experiment
field experiment
Hawthorne effect
debriefing
manipulation check
experimental design
true experiment
quasi-experiment
pre-experiment

classic experimental design
posttest-only control group design
confound
Solomon four-group design
factorial experimental design
counterbalanced design
Latin square or repeated measure design
equivalent group pretest-posttest
time series
nonequivalent control group
separate-sample pretest-posttest
research demand
one-shot case study
one-group pretest-posttest
static group comparison
audit
t-test
analysis of variance (ANOVA)

Questions for Discussion and Review

1. Explain why and how the true experimental design provides the strongest evidence for cause-and-effect relationships. In your answer, pay specific attention to the four criteria for showing a cause-and-effect relationship.
2. What ethical concerns do experiments often raise?
3. Differentiate among random sampling, randomization, and matching. Which of these techniques is most commonly used in experimental research? Why? In what situations is each of these techniques most appropriate? Least appropriate? Justify your answers.

4. What are the relative strengths and weaknesses of laboratory and field experiments? How do they compare on reliability and internal and external validity? How can you improve the reliability and internal validity of a field experiment? How can you improve the external validity of a laboratory experiment?
5. How do true experimental designs, quasi-experimental designs, and pre-experimental designs differ from one another? When would it be most appropriate to use each kind? When would it be least appropriate? Why? Justify your answers.
6. Which threats to internal validity are likely to be present in each of the true experimental, quasi-experimental, and pre-experimental designs discussed in this chapter? Fully explain and justify your answers.
7. Using one of the many search engines available, search the Web for information concerning the Hawthorne effect (hint: Use the search terms Hawthorne effect and experiments). Who discovered the Hawthorne effect? Describe the nature and findings of the experiment(s) that he conducted.
8. Find two studies in a communication journal that use experimental design. For each study, (a) identify the experimental design the researcher(s) used. (b) Explain how well the experiment controls for potential confounds, and (c) tell which, if any, of the three basic control techniques (elimination, blocking, and holding constant) were used. Describe these procedures in detail.
9. In an effort to determine which of two sets of coupon offers will be more effective in bringing in customers, the owners of a local sandwich shop send one set of coupon offers to people on their list of current customers and place the other set of coupon offers in the Sunday paper. The shop owners then count the number of coupons redeemed during a two-week period following the distribution of the two sets. For this study, (a) identify the experimental design the owners used, and (b) evaluate the owners' experiment for internal and external validity. Be sure to tell which threats are controlled for and which ones are left uncontrolled. (c) Propose an improved experimental design to more effectively obtain the information the owners seek.
10. To evaluate the effectiveness of a three-week workshop intended to decrease the likelihood of break-ups, a researcher identifies couples with extremely low scores on a communication-ability scale. Those who agree to participate in the study attend the workshop, after which their communication skills are reevaluated using the same communication-ability scale. (a) Identify the experimental design the researcher used, and (b) evaluate the experiment for internal and external validity. Tell which threats are controlled for and which ones are left uncontrolled. (c) Propose an improved experimental design to determine the effect of the communication workshop.
11. For each of the following situations, (a) tell whether you would conduct a laboratory or field experiment, and identify the specific true experimental, quasi-experimental, or pre-experimental design you would use. Justify your answer. Then for each study (b) identify the independent and dependent variables, (c) explain how you would manipulate or measure your independent variable, (d) provide a conceptual and operational definition for "effect," and tell how you would measure it, and (e) explain how many subjects you will need, how you will recruit them, and how you would assign them to groups. Finally (f) evaluate your research plan for ethical problems, and (g) assess the internal and external validity of your experimental design by explaining how well it addresses the common threats to internal validity.

A study to:
   a. test the effect on aggression toward women of listening to songs that degrade them.
   b. evaluate the effectiveness of online versus hard-copy documentation for secretaries at a major company who will be using a new word-processing program.
   c. determine the short-term and long-term effects of a radio advertising campaign.
   d. determine the effect of a product recall by a major company on consumer attitudes toward the company.
   e. determine whether it is more effective to announce changes in company policies to its workers via e-mail or via interoffice memo.
   f. decide which of three advertising campaigns is likely to be most effective.
12. Explain when you would use a t-test and when you would use analysis of variance (ANOVA) to analyze experimental data.
13. Go to the website for the General Social Survey (http://www.icpsr.umich.edu/gss). Using the extraction program, extract the variables TVHOURS, SEX, PARTFULL, and HEALTH and enter them into a data-analysis program such as SPSS or SAS.
   a. Use the codebook to examine the variables and their meaning. Then develop at least three hypotheses in which **TVHOURS** is the dependent variable and the following variables are independent variables: **SEX, PARTFULL** and **HEALTH.**
   b. Now using the t-test procedure, test each of the three hypotheses you created in part a by comparing mean scores on the dependent variable for the different categories of the independent variables. Tell which, if any, of the three hypotheses appear to be supported by the data.
   c. Do an analysis of variance (ANOVA) to determine the simultaneous effect of the independent variables on the dependent variable identified in part a.
14. Volunteer for an experiment by contacting the psychology department at your college and asking about opportunities for participating in laboratory experiments. Write about your experience.

Readings for Discussion

Byers, D. L. Jr., Hilgenberg, C. and Rhodes, D. M. (1999). Telemedicine for patient education. *The American Journal for Patient Education,* 13(3):52–61. The authors use the one-shot posttest-only design with replications to evaluate a method for delivering a Lamaze class to people living in rural areas.

Campbell, D. T. (1988). Reforms as experiments. In D. T. Campbell, ed., *Methodology and epistemology: Selected papers.* Chicago: University of Chicago Press, pp. 261–289. In this paper, written in the context of the Great Society programs of the 1960s, the master of experimental design makes a case for using the principles of experimental research to evaluate new programs for effectiveness.

Cantor, J. and Omdahl, B. L. (1999). Children's acceptance of safety guidelines after exposure to televised dramas depicting accidents. *The Western Journal of Communication,* 63(1):57–71. The authors point out some of the problems of recruiting and working with children as research subjects.

Donohew, L., Lorch, E. P. and Palmgreen, P. (1998). Applications of a theoretical model of information exposure to health interventions. *Human Communication Research, 24*(3):454–468. This is an example of time series analysis.

McQuarrie, E. F. (1998). Have laboratory experiments become detached from advertiser goals? A meta-analysis. *Journal of Advertising Research, 38*(6):15–26. The author examines 443 published experimental studies to see how well they match his reality criteria.

Nelson, D. C., Almanza, B. A. and Jaffe, W. F. (1996). The effect of point-of-sales nutrition information on the entrée selection of patrons of a university foodservice operation. *Journal of Nutrition in Recipe and Menu Development, 2*(4):29–50. The authors report on a field experiment.

Reinard, J. C. and Arsenault, D. J. (2000). The impact of forms of strategic and nonstrategic *voir dire* questions on jury verdicts. *Communication Monographs, 67*(2):158–187. The authors use a posttest-only design with counterbalancing and a no-question option as controls.

Smith, M. D. and Hand, C. (1987). The pornography/aggression linkage: Results from a field study. *Deviant Behavior, 8*(4):389–400. This is one of very few studies examining the real-world effects of sexually explicit material.

Taraban, R., Maki, W. S. and Rynearson, K. (1999). Measuring study time distributions: Implications for designing computer-based courses. *Behavior Research Methods, Instruments and Computers, 31*(2):263–269. The authors use a counterbalanced design and multiple measures.

References and Resources

Bausell, R. B. (1994). *Conducting meaningful experiments: 40 steps to becoming a scientist.* Thousand Oaks, CA: Sage. This easy-to-read book provides basic information and useful tips for planning and carrying out experimental research.

Campbell, D. T. and Stanley, J. C. (1963). *Experimental and quasi-experimental designs and research.* Skokie, IL: Rand McNally. This little book is the classic work written by the masters of experimental research.

Cook, T. D. and Campbell, D. T. (1979). *Quasi-experimentation: Design and analysis for field studies.* Skokie, IL: Rand McNally. Another classic, this one has excellent chapters on time series designs, validity, and data analysis.

Culbertson, H. M. (1989). Statistical designs for experimental research. In G. H. Stempel III and B. H. Westley, eds., *Research methods in mass communication,* 2nd ed., pp. 221–246. Englewood Cliffs, NJ: Prentice Hall. This chapter in a standard graduate-level textbook provides a good overview of quantitative data analysis.

Harris, R. J. (1994). *ANOVA: An analysis of variance primer.* Itasca, IL: F. E. Peacock Publishers. This book provides detailed information about using the t-test, basic, and more advanced forms of analysis of variance. It also covers the chi-square goodness-of-fit test, regression, and log linear transformations.

Haskins, J. B. (1981). A precise notational system for planning and analysis. *Evaluation Review, 5*(1):33–50. This is the developer's explanation of a diagramming system that is sometimes used instead of the Campbell and Stanley system.

Rosenthal, R. (1976). *Experimenter effects in behavioral research,* 2nd ed. New York: Irvington. This classic provides a thorough examination of one of the most troubling threats to reliability and validity.

Westley, B. H. (1989). The controlled experiment. In G. H. Stempel III and B. H. Westley, eds., *Research methods in mass communication,* 2nd ed., pp. 200–220. Englewood Cliffs, NJ: Prentice Hall. This textbook chapter provides a short but fairly sophisticated introduction to experimental design.

7

Surveys

WITH A TRUE survey, researchers gather information in the form of self-reports from a large probability sample of subjects by using a questionnaire that consists of a fixed set of questions arranged in a fixed order.

Typically there are many subjects: no fewer than 100 for simple exploratory studies to as many as several thousand for a national political opinion poll. In a true survey, the goal is to study either an entire population or to choose enough subjects through probability sampling so that you can use statistics to generalize findings to the population of interest.

The questions on the questionnaire may limit respondents to a few, fixed options; they may allow them to answer in their own words, or they may be a mixture of both **fixed-response** and **open-ended** questions. The key here is that all subjects respond to the same questions in the same order and have the exact same response options as all other subjects.

Because of these features, survey data are usually analyzed quantitatively. Statistics make it possible to describe the distribution of characteristics among respondents and then make inferences, within the limits of sampling error, to the entire population. Statistics also make it possible to test hypotheses by providing meaningful information about the strength and statistical significance of relationships among concepts.

A single survey is, however, a snapshot of how things are at one point in time. Standing alone, a survey cannot provide the control for time order, history, or maturation effects necessary to provide truly convincing evidence of cause-and-effect relationships. Moreover, all of the data consists of self-reports. At least some of the answers from some subjects may be suspect, especially if the questions are highly sensitive or have a socially desirable answer or if they call for people to remember the past or say what they will do at some future time.

Even if one could assume that all subjects could and would answer all questions honestly and accurately, a true survey is not well suited for delving deeply into people's thought processes to learn why they may have answered as they did. As a result, it can be very easy to misconstrue answers or make assumptions about motivations or about reasons for relationships among variables that may not be warranted.

Today, however, the most troubling limitation is the decline in response rates. Busy lifestyles, increased telemarketing, and the advent of cell phones and caller identification make it much more difficult to reach subjects and then convince people to answer survey questions than it was even five years ago.

But when a survey is done properly, the data are very reliable. The measurement form of internal validity will also be high. Even though self-reports have their limits, people remain the experts on what they as individuals know, think, feel, and do and why they believe and behave as they do. And, in spite of declining response rates, a true survey has better external validity than any other kind of human-subject research. Survey researchers can collect data nationwide or from several different regions almost as easily and cheaply as they can from a single location. Therefore, the survey remains the easiest, most cost-effective way to collect large amounts of high-quality data. Surveys also raise fewer ethical problems than most other kinds of human-subject research.

Communication Applications

Because of their obvious advantages, surveys remain the workhorse in applied communication research. No other method is used more often.

- Politicians rely on surveys to find out what people know and like or dislike about them, whether their candidacies are feasible, and what issues or issue positions to emphasize, deemphasize, or ignore. Once elected, they use surveys to keep in touch with their constituencies and make policy decisions that will address their wants and needs.
- Journalists turn to survey data when they report on trends or investigate social conditions. They conduct their own surveys to gather and then disseminate new information about the opinions, behaviors, and needs of people in the area they cover.
- Media organizations use surveys to connect with their audiences and to help them make decisions about content or modes of delivery such as a change in radio format or a switch from evening to morning publication.
- Businesses and nonprofit organizations use survey techniques to build a database about their customers, clients, and contributors; to monitor their interests and concerns; and to evaluate the effectiveness of public relations, advertising, and marketing efforts.

Much of the work planning, conducting, or evaluating survey research falls to those working in public relations, marketing, or advertising, but communication professionals also rely heavily on survey data collected by national or regional polling firms or other research providers.

One of the best-known firms conducting advertising effectiveness research is Starch INRA Hooper. Their posttest recall research, often called the **Starch test,** involves face-to-face interviews with approximately 300 subjects who are shown periodicals and then questioned at length to determine whether they saw certain ads and, if so, what they remember about each one.

In broadcasting, Arbitron and Nielsen are the leading providers of audience data and station ratings. This information is based in part on self-reports from a very large probability sample of households.

Surveys are also the basis for **segmentation** studies that help organizations target their products, services, or messages to appropriate publics. In **psychographic research,** these segments are developed from survey data on people's demographic characteristics, lifestyle, attitudes, and values. Many of the data used for these segmentation studies come from companies such as Simmons Market Research Bureau (SMRB) and Mediamark Research, Inc. (MRI), who give their clients access to extensive national survey data on people's demographic characteristics, their self-images, media habits, and product usage.

Ethical Concerns

Although survey research is generally low risk, it is not risk free. Except for those studies gathering sensitive information and those conducted in institutional settings, the greatest risk comes from sending data collectors into the field to collect data through face-to-face interviews.

In field collection through face-to-face interviews, potential danger always lurks. Therefore, to protect interviewers, data collectors should be sent out in pairs whenever possible. They should also be equipped with cell phones or pagers so they can quickly and easily call for help if they should find themselves in a threatening situation. To protect respondents, interviewers should be subjected to background checks and bonded. Their work should be carefully monitored.

But aside from those risks, the biggest ethical concern today is the use of survey techniques by those with ulterior motives. It is relatively easy to word questions or arrange them into a questionnaire in ways that unfairly maximize the chances of finding what a researcher or client wants to find and minimize the chances of uncovering countervailing information.

It is also easy to disguise telemarketing efforts or **push polls** as survey research. In push polling, subjects are asked a series of questions about their issue positions or feelings about a candidate. The interviewer then attempts to "push" people toward a desired position by asking follow-up questions designed to plant doubts in their minds about their original positions. Sometimes they do this by asking people to respond to statements that are unabashedly false or that are based on rumors.

In spite of what a client may want in the way of results, ethical survey researchers will insist on following the same standards that apply to all kinds of social science research. Standards developed by the American Association for Public Opinion Research (AAPOR), the major professional organization for survey researchers, require pollsters to obtain informed consent from survey subjects and to take appropriate steps to ensure that participation is voluntary. Standards also require respecting subjects' right to privacy and informing them of any known risks.

Subjects must be given the information they will need in order to give informed consent before they begin answering questions. Therefore, this information should be presented as part of an introductory statement. This statement may be sent to prospective subjects in a letter inviting them to participate in a survey, it may be read to them when they are first contacted by phone or in person by whoever will administer the survey, or it may be printed prominently on the first page of a questionnaire that subjects will fill out without researcher assistance. According to AAPOR standards, the introductory statement should include information on:

- the general purpose of the study
- who is conducting the survey
- who is sponsoring the survey

- how and by whom the information will be used
- what steps, if any, will be taken to ensure subjects' privacy
- any known risks associated with participation in the survey, especially in situations where anonymity or confidentiality cannot be ensured
- a statement that participation is voluntary; subjects may refuse to participate without fear of repercussion. They may also refuse to answer any questions or drop out before completing the survey without fear of repercussion.

In most situations survey researchers do not ask their subjects to sign a written consent form. Consent is implied if, after being given the necessary information about the survey, subjects begin answering the researcher's questions. However, consent forms signed by parents or guardians are necessary when gathering information from minors or other vulnerable populations. They are also a good idea if answering questions may incur some risk.

Because surveys conducted by phone or face to face at people's homes are, by their very nature, somewhat intrusive, it is important to contact subjects only during times that are likely to be reasonably convenient. That means avoiding initiating contact before about 9 A.M., after about 9 P.M. or during the normal dinner hours. It is also a good idea to include, as part of the introductory statement, some estimate of how much time the survey will take.

To ensure that participation is voluntary, researchers must accept and respect a subject's right to refuse to participate in the study as well as the right to refuse to answer any questions the researcher may ask. Researchers may politely attempt to convince a subject to participate in the study or to answer a question by explaining the importance of obtaining responses from everyone chosen in the sampling procedure, but ethically they cannot badger, harass, or threaten those who refuse to participate or to answer individual questions.

The right to be and remain anonymous or to have answers used confidentially in ways that make it impossible to link individual subjects with their answers to questions is particularly important in studies conducted among employees, students, or members of an organization. It is also very important in studies seeking to uncover information about illegal, unethical, or socially undesirable behaviors or opinions. In any of these situations, subjects may justifiably believe that any information they reveal could be used against them.

In situations where anonymity or confidentiality cannot be guaranteed, subjects must be clearly informed of any possible consequences of linking their identity to their responses. As part of this, subjects also need to know to whom, under what circumstances, and for what purposes information linking them to their responses may be revealed.

True Survey Designs

The typical one-shot true survey is modeled after the one-shot case study pre-experiment. However, the independent variable is measured by responses to one or more questions. Data for the dependent variables and for many possible confounds come from responses to other questions.

A survey, then, can be conceptualized as a statistical experiment. But because the independent variable is measured by means of a self-report instead of through manipulation by the researcher, you can never be quite sure whether subjects were or were not exposed to the independent variable. And because the model is a pre-experimental design, a single survey cannot adequately control for the effects of history or maturation. Neither can it provide the evidence of time order necessary to show a causal relationship.

To some extent these problems can be overcome by employing statistical techniques or by conducting additional surveys in order to create a research design that matches or approximates that of a true experiment or quasi-experiment.

Cross-cultural Designs

Instead of conducting a single survey, you can conduct a single survey in several locations or with several populations. You can also conduct several surveys simultaneously, each in a different area or with a different population. You might, for example, draw a single random sample stratified by location or by population and then use the stratification in data analysis to create a design approximating the posttest-only true experiment. Using that same basic experimental design as the model, you might also use matching to draw separate samples from similar regions or populations. Alternatively, you could draw separate samples from very different areas or populations to approximate the nonequivalent posttest-only experimental design.

The strength of these approaches is that they improve validity by adding a comparative element. Instead of having just one snapshot, you would now have several snapshots taken simultaneously in several settings or of several different groups. This comparative element reduces problems associated with subject selection and history.

But the weakness is that these designs still do not control well for maturation. Neither do they solve the time-order problem. Therefore, they are most useful when the research problem calls for comparing subjects on one or more dependent variables rather than drawing conclusions about cause and effect.

Longitudinal Designs

To study change over time or to draw causal inferences, researchers can conduct a series of surveys over an extended period of time.

For **trend** studies, researchers usually conduct multiple surveys over a period of months or years, using a new sample from the same or a very similar population each time. The model here is the separate-sample pretest-posttest design.

Trend studies provide better control over history and maturation than can be achieved by using a single survey or by conducting several surveys simultaneously. The weakness, however, is that the trend study provides stronger evidence of societal change than of change in individuals.

To study change in individuals, you can conduct a **panel** study by drawing a single sample and then surveying those same subjects at several different times. In its simplest form, a panel study will approximate a one-group pretest-posttest pre-experimental design; however, conducting some of the surveys before and some after a naturally occurring event will turn the panel design into a time-series quasi-experiment. Adding additional panels drawn from the same populations can turn the panel survey into an approximation of the true pretest-posttest experimental design; selecting panels from similar or quite different populations will produce pretest-posttest quasi- or pre-experimental designs.

With panel designs, the initial sample will need to be larger than for any of the other survey designs because there will always be some mortality. If the intervals between surveys are too short, pretest-posttest interference can also be a serious problem. However, the panel design provides excellent control for history and maturation effects. Therefore, it is better than the trend design for examining how people may change over time. In cases where some of the surveys occur before and some after a naturally occurring event, the panel design is also better for showing true causal relations.

Although both trend and panel designs provide better control for history than the one-shot survey can, external events related to life experience and life stage will always differ among subjects. Many of these individual differences among people have been traced to what was going on in the world during critical periods in their lives. The formative experiences of those growing up during World War II were different from those of their children coming of age in the 1960s whose experiences, in turn, differ from those of their children.

To account for these differences, researchers sometimes perform a **cohort analysis** in conjunction with trend surveys. If, for example, you have access to newspaper readership surveys from the end of World War II to the present, you can use the data to follow each cohort, or group of people born at approximately the same time, to determine whether their newspaper use stayed the same or changed over time. To do this, you would compare reading by those who were between ages 16 and 25 in 1950 to reading by those 26 to 35 in 1960, reading by those between 26 and 35 in 1960 to reading by 36- to 45-year-olds in 1970, and so on, eliminating cohorts as they die off and adding new ones as they come of age.

Quasi-surveys

In a true survey researchers use a fixed set of questions in a fixed order to gather information from a large random sample of subjects, but many variations on that basic technique are possible. The most common modifications include use of other kinds of samples and the introduction of some flexibility into the actual collection of data.

Nonprobability Sample Surveys

If generalizability is not the goal, collecting data from a nonprobability sample of subjects can be cheaper, quicker, and more convenient than surveying a probability sample.

For tracking networks of communication or influence and for locating hard-to-find subjects, snowball sampling is often the only kind of sampling that will work. If only some people's opinions are of interest, a purposive sample of, for example, the top 20 contributors to an organization could provide all the information a client might need.

A convenience sample of 10 to 50 subjects usually works quite well for pretesting measures. It may also be good enough for gathering preliminary data or for getting input on programming or some sense of a public's interests and concerns. Alternatively, you might use a very small probability sample. Even though the large sample error makes it difficult to generalize from samples under 100, very small probability samples tend to introduce more diversity at low cost than would be true for a convenience sample.

If more input is desired, self-selection through responses to a questionnaire placed in print media or posted on a website works quite nicely. Although results can never be generalized to an entire population regardless of how many people respond, this kind of quasi-survey effectively locates pockets of people with strong opinions or ones who may pose a threat to organizational goals.

Interview Surveys

Relying on a fixed questionnaire made up of mostly fixed-response items is usually the only realistic approach to gathering information from large numbers of subjects. At the same time, relying solely or even primarily on fixed questions imposes some severe limitations. The data collected will be much better for answering questions of "who," "what," "where," and "when" than for understanding the "why" or "how." Theory can provide some guide as to why relationships may exist, but explanations based on correlational data will always be open to question. Many people may think, feel, or behave the way theory predicts, but some will not. Even some whose responses do match the theory may have given answers predicted by theory for other reasons.

Therefore, you can usually obtain better information about people's thought processes, reasoning, and motivations by working with smaller samples and abandoning the fixed-format questionnaire in favor of conducting the kind of in-depth interview routinely used by journalists. By probing or asking follow-up questions, you can obtain information you might otherwise miss. This information may come in the form of detail beyond what you would get from answers to fixed-response questions or additional information your subjects may unexpectedly volunteer. But some of it will also come from the words your subjects use and the way they string those words together. Therefore, you will need to record the interview in order to preserve your subjects' natural language. This natural language is the evidence you will use for data analysis.

Data-collection Techniques

Interview surveys are almost always conducted in person. But researchers can use a variety of techniques to collect data from true surveys and other kinds of quasi-surveys. Some of these techniques are most appropriate when the researcher collects data; others are intended for use with self-administered questionnaires. The choice of whether to conduct a researcher-administered survey or rely on self-administration and then which of the techniques to use depends on the nature of the subjects and the kind of information sought as well as on time and budgetary constraints.

Researcher-administered Techniques

These techniques tend to be more expensive, but they have the advantage of giving the researcher greater control over who is actually responding to a questionnaire. Data collection also is generally quicker than for most self-administered surveys.

Telephone. This is undoubtedly the most widely used method for gathering survey data in the United States. Almost everyone has or has access to a phone; with modern technology, random-digit dialing can capture unlisted numbers. Questionnaires can be computerized so interviewers cannot easily skip questions or make errors in recording answers. Responses can be directly imported into data-analysis programs. This eliminates the cost of data entry and makes the findings available very soon after interviews have been completed.

Even in cases where random-digit dialing and computer-assisted interviewing software cannot be used, it is still possible to reach almost anyone and, with proper training of interviewers, get reliable, high-quality data relatively quickly and cheaply.

Although unexpected and probably unwanted phone calls are somewhat intrusive, the method provides few true safety risks for either the data collector or the subject. With proper safeguards, threats to subjects' privacy should be minimal to nonexistent.

The major limitations with this method come from the kinds of questions that can be asked and with response rates.

Because subjects must be able to remember what they are hearing, most will find it very difficult to answer complex questions or choose among responses if they are given too many options. They will, for example, find it hard to rank more than four or five items in a list. And because there is no direct contact between interviewer and subject, you cannot use questions that require showing examples such as a page layout or a product.

As response rates decline, external validity is becoming increasingly problematic. Response rates today typically range from 40 percent to about 70 percent of those actually contacted. The rates vary by geographic region. Within regions, they are usually lowest for racial and ethnic minorities and for those with lower incomes and education. They are generally highest for specialized populations such as members of an organization who have some personal or professional reason for cooperating.

Face to Face. As an alternative to the phone interview, researchers sometimes choose to do personal interviews. With the exception of the mall-intercept interview, this strategy is generally the most expensive option. It is the only one that works well for the in-depth interview survey.

If interviews are conducted at subjects' homes or places of business, the data collectors' travel time will have to be added to the costs of locating subjects, convincing them to participate, and then interviewing them. If subjects are invited to be interviewed at some other location, participation rates will suffer. Subjects will also most likely expect some compensation, thus offsetting any savings in data collectors' travel time.

Even though problems with logistics and the expense of conducting face-to-face interviews at subjects' homes or places of business are obvious disadvantages, the personal interview can be used equally well with fixed questionnaires and less-structured survey interviews.

Because of their more personal nature, researchers can delve into topics in more depth and use more different kinds of questions. Subjects can be shown items and asked to respond to them. They can also cope with more complex questions and rank somewhat longer lists of items than subjects can when they are interviewed by phone.

With face-to-face contact, data collectors can see their subjects so they can more readily ensure that they are interviewing the appropriate people. But because of this face-to-face contact, there is also greater danger of subject-researcher interactive effects. Not only may subjects respond differently depending on the gender, race, or personality attributes of the interviewer, but the researchers themselves may be more prone to bias in recording subjects' responses. In any case, interviewing may be less reliable because it is much easier to deviate from a questionnaire or protocol in a face-to-face setting than it is when talking on the telephone.

Some of these problems are less severe with mall-intercept interviews. In the mall-intercept interview, you can gather in-depth information by using many different kinds of questions. Data-collection costs are generally lower, interactive effects less common, and response rates higher than for other kinds of face-to-face interviewing. However, the higher response rates come at the expense of external validity. Mall intercepts are essentially a convenience sample or, at best, a random sample of those who happen to be at a particular location at a particular time.

Self-administered Techniques

Where once these might have been considered second-class options, the combination of increasing labor costs and declining response rates for researcher-administered techniques now make self-administration an attractive option.

Group Administration. This technique combines some of the features of the researcher-administered survey with those of truly self-administered ones. In its most common form, a researcher or research assistant distributes questionnaires and gives general instructions for completing them to subjects who are assembled in one location. Subjects then complete the questionnaire without further input from the researcher.

This data-collection strategy is frequently used to evaluate events, meetings, workshops, and instructional programs. The course evaluations given at many colleges at the end of each semester is a typical example. Although the group-administered survey is often used to obtain information from people who have assembled primarily for some other purpose, it is also a common component of focus group research, which we discuss more fully in Chapter 9.

Group administration is usually a truly low-cost kind of survey research even taking into account the fact that data almost always must be manually entered into a computer program. It is generally quite fast. Because people are assembled together, response is immediate. Completion rates are usually high even with fairly lengthy questionnaires.

But when researchers or research assistants are present during data collection, both their presence and the group nature of the activity can be a disadvantage. Comments from the person who distributes the questionnaires can affect reliability and validity. If the data collector's behavior or comments lead subjects to believe the researcher expects or wants a particular outcome, they may tailor their answers accordingly. Moreover, the presence of others may intimidate some people. And occasionally subjects will collaborate so that responses become more those of the group or some subset of it than of individual subjects.

Mail. Of all data-collection options, this is one of the least expensive but also the slowest. With it, the researcher sends a questionnaire via mail to respondents who write their answers on the form and than mail it back to the researcher.

A mail survey is more effective than other techniques for gathering extensive information from subjects who are generally predisposed to cooperate but who might not have all the desired information at their fingertips. But most people simply will not take the time to answer lengthy mail surveys; those who do answer may have a hard time following directions. Therefore, questionnaires usually must be shorter and the questions much simpler than would be necessary with researcher-administered techniques.

Even though response rates of 30 to 50 percent for a general population and sometimes much higher for specialized ones now approach and may even surpass response rates for telephone surveys, external validity is more suspect. There is no way to ensure that the person to whom the survey was addressed is the one who actually responds.

Disk by Mail. With this technique, subjects receive the questionnaire on a computer disk. As a variant of the traditional mail survey, it has all of the mail survey's strengths and weaknesses plus a few of its own.

The main advantages over the paper-and-pencil format is that the disk can usually be formatted so that responses do not need to be entered manually into a data-analysis program. Having people answer on disk also eliminates problems associated with trying to decipher their handwriting.

However, not everyone has a computer, so the technique can be used only if the researcher can be reasonably sure subjects have access to a computer and will be comfortable using it to answer questions. But even in that situation, differences in computer hardware, operating systems, word-processing programs, and support for color and graphics can affect people's willingness and/or ability to respond.

Internet. This is a variant of the mail and disk-by-mail techniques that has become increasingly popular in the last ten years. Instead of mailing the questionnaire, researchers post it on the Internet.

To obtain a large number of responses from a convenience or self-selecting sample, the questionnaire can be posted on an organization's website or sent to a chat room. For a true Internet survey, access to the questionnaire can be controlled by sending the site's address to a census or probability sample via mail or e-mail.

Creating a questionnaire for this kind of survey can be more expensive than for other methods, but this technique usually eliminates the cost associated with manually entering data into a computer program, so overall costs may actually be lower than for some other techniques. The chief advantage, however, is the ability to use links to many kinds of textual, visual, or audio materials that could not possibly be incorporated into other types of surveys.

Because of the convenience factor and because the Internet is still somewhat of a novelty, response rates are typically higher than for mail and telephone surveys. But like disk by mail, the Internet survey isn't well suited for reaching a general population. Although Internet access has increased markedly, many people still do not have computers; of those who do have them, some do not have Internet access while others may not be comfortable responding online. And, as is true with mail and disk-by-mail surveys, there is less assurance than with researcher-administered techniques that the person recruited into a sample is the one who actually responds.

Preparing the Survey Questionnaire

Two types of data-collection instruments are used with survey research: the protocol and the questionnaire. A **protocol** consists of lists of topics to be covered, possible open-ended questions, and directions that help the interviewer be consistent across subjects in probing for information and responding to situations that may arise during the interview. The **questionnaire** consists of a fixed set of questions intended to be asked in the same way of all subjects. An example of a questionnaire is in Appendix C.

Because of their flexibility, protocols are the instrument of choice for gathering qualitative data through in-depth interviews with small numbers of subjects. They are also commonly used with focus groups and observational research. We discuss these applications in Chapters 9 and 10. In this section we discuss the questionnaire used with true surveys and most quasi-surveys because it is so well suited for gathering quantitative data from large numbers of people.

A well-designed questionnaire contributes immeasurably to the reliability and validity of survey research. But in order to ensure reliability and validity, you must be certain the instrument for data collection does not bias the responses or make it difficult for people to administer or complete the survey. Therefore, in preparing the questionnaire, you must pay attention to the length of your survey, the way you word and arrange the questions you want subjects to answer, the instructions and other helps you give them, and the layout and visual appeal of your questionnaire.

Questionnaire Length

As a general rule, surveys should be as short as possible. Survey length is directly related to cost and response rate. Long surveys cost more than short ones. In most cases they also have lower response rates.

To keep surveys to a manageable length, all questions should be directly related to the purpose of the study. Extraneous questions should be eliminated. Still, if there is any question about whether certain questions are needed, it's usually advisable to include them. In the long run, the danger that some subjects will refuse to respond to a survey or drop out before completing it because of a few extra questions is often less of a problem than discovering too late that you should have included a few more questions.

Therefore, there can be no hard-and-fast rule as to how long is too long. Length depends on the:

- purpose of the study
- research budget
- subjects' motivation and demographic characteristics
- time and place of data collection
- data-collection technique
- type of questions

In general, questionnaires administered face to face or in group settings where the group was assembled specifically to participate in a research project can be longer than ones to be administered in other ways. Telephone surveys can usually be longer than mall-intercept ones. Among self-administered surveys, traditional mail surveys must generally be shorter than ones sent on disk or posted on the Internet. But the only way to know for sure whether a questionnaire is too long is to pretest it, first by reading it aloud, and then by trying it out on a small group of people similar to those who will be your subjects.

Types of Questions

With researcher-administered questionnaires, some questions about demographic characteristics are almost never asked. Inquiring about a person's gender, for example, could be quite offensive. That information is usually simply recorded on the basis of voice or appearance. In some settings, the same may be true for race or age. But with those exceptions, almost any of the measures discussed in Chapter 5 can be adapted for use as survey questions.

Most questions on every survey will be included to elicit the information necessary for achieving the research project's objectives. However, a few types of questions are specific to survey research. They facilitate data collection and analysis.

Filter questions separate those who are eligible to answer a question or set of questions from those who should not answer. It would, for example, make little sense to ask people who have never read a particular magazine to evaluate articles in that magazine. Therefore, a question about readership, placed before the evaluation ones, can be used as a filter. The **screener** used to identify appropriate respondents is another example of a filter. **Contingency** questions are a variant of the filter. Subjects' answers to these questions determine which questions they will be asked to answer next.

The **probe** is an unscripted question used to get a respondent to answer the question with which it is used or to answer it more fully.

A **sleeper** is, in effect, a trick question inserted into a questionnaire. Its purpose is to identify those people who are merely guessing in response to factual questions or providing opinions on subjects about which they have no real knowledge. To do this, these questions call for people to respond to questions whose content is fictitious. An example would be asking respondents to tell how well they believe "Senator Bob Barnes" is performing his duties when, in fact, there is no senator or other elected person named Bob Barnes.

To identify people who may not be paying attention, researchers may also insert **check** questions. These ask the same basic question twice at different points during the interview but in slightly different forms. In most cases, one version will be worded positively, the other negatively:

> I always get the information I need in time to do my job.   SA   A   N   D   SD
> I often don't get the information I need when I need it.   SA   A   N   D   SD

People who know their own minds and are paying attention should provide consistent answers. Others can be eliminated from the sample, or their responses to other questions can be compared to those from people who do provide consistent responses.

Question Wording

But even if people know their own minds and are paying attention, you may still not get meaningful data. Even subtle differences in the words researchers use can produce big differences in findings. Consider these two questions, used in national surveys by respected organizations:

> The Gallup Poll asks respondents: Do you approve or disapprove of the way (name of president) is handling his job as president?
> The Harris organization asks: How would you rate the job (name of president) is doing? Would you say he is doing an excellent, pretty good, only fair, or a poor job?

Both questions can be counted on to give a true assessment of public support for a president, yet the Harris poll routinely turns up much less support for a president than does the Gallup poll. And you can get results that fall between those from the Gallup and Harris polls by asking:

> Suppose you were to grade President (name of president) A, B, C, D, or F for the way he is handling his job as president. What grade would you give him?

The difference lies in the way the questions are worded and in the number and kind of response options.

As those examples suggest, there is no perfect, right, or wrong way to word questions. But there is one guarantee: People will respond to your words. They will answer the question you ask. Or, in a worst-case scenario, they will answer the one they think you asked.

Although different wordings are often acceptable, some ways of asking questions are guaranteed to create problems. To avoid asking questions that will produce biased, misleading, or uninterpretable answers:

- Match your language to your subjects' level of understanding. You can't expect your subjects to understand big words, technical, or specialized terms such as "concomitant," "habeas corpus," "outcome-based education," or "double-entry bookkeeping." Never use a big word when a little one will do. If you must ask about a technical or specialized concept, include a definition or substitute the definition for the word.
- Be sure your questions are clear. Respondents aren't mind readers. If you want to find out whether people like the content of your newsletter, don't just ask how well they like the newsletter. Some will interpret that as an invitation to comment on frequency of publication, quality of writing, or lots of other things that may not be important. Ask them to tell you how satisfied they are with the kinds of topics the newsletter covers or ask how much they like or want stories on various topics the newsletter covers or that it could cover. Give your subjects a frame of reference—a clue about the kind of information you want as an answer. Just be sure you don't signal a "correct" answer.
- Keep questions short and simple. Expecting people to respond to long, involved scenarios is asking for trouble. Interviewers will have trouble reading the item; subjects will respond either to the first or to the last part rather than to the statement as a whole.
- Don't ask for more detail than people can reasonably be expected to give. People simply can't tell you how many hours they watched television during the past year or whether their income from all sources is $35,000 or $36,000. Instead of asking about behavior over the past year, ask about yesterday, today, or a typical day. Offer ranges as response options whenever possible. If you really need detailed information on people's behavior, ask your subjects to keep a diary.
- Avoid asking leading questions. These suggest, either explicitly or implicitly, a certain response. "Don't you agree there is too much sex on television?" clearly signals that you expect a "yes" answer. Many more will answer "yes" than if you had asked the more neutral "Do you agree or disagree that there is too much sex on television?"
- Avoid loaded words. These also produce leading questions, but they do it by using ideological or value-laden words such as "patriotic," "honest," or "radical" to describe a person, organization, issue, or issue position. By using this kind of language, you are again suggesting how subjects should respond.

162   Basic Research Methods

- Avoid double-barreled questions. These ask people to give one answer to a combination of questions. If you ask a question such as "Do you think taxes are too high and the government should cut back on military spending?" as a fixed-response item, people who think taxes are too high but military spending is too low (or vice versa) won't know how to answer; even if people do answer, you can't really count on the responses to tell you how people feel about both taxes and military spending. Allowing open-ended responses won't work much better. Many people will answer one part and skip the other, leaving you with a lot of missing data. Whenever you see an "and" in a question, look at the question carefully. Divide double-barreled questions and multiple-part ones into two or more questions.
- Avoid double-bind questions. These are questions of the proverbial "Have you stopped beating your wife?" type. Regardless of whether a person answers "yes" or "no," the response implies an affirmative answer to the unasked questions of whether the person ever has beaten his wife. But of course, the subject may never have beaten his wife. This kind of problem can be avoided by using filter and/or contingency questions to eliminate the double bind.
- Avoid embarrassing subjects unnecessarily. If your survey is about spousal abuse, you obviously can't avoid asking questions about wife- (or husband-) beating, but you can soften things a bit: "Some people have done things that have injured their partner. Have you ever done that kind of thing?" To avoid embarrassment, it can also be a good idea to separate some questions such as those inquiring about marital status and number of children so the answers to the questions don't seem so closely connected.
- Avoid using negative terms in the question. You will confuse people and get a lot of answers that don't reflect subjects' true feelings if you ask them to agree or disagree with a statement like "The company should not publish a newsletter." If you want to know whether the company should publish that newsletter, use the negative as one of the response options: "Do you think the company should or should not publish a newsletter?"

Question Order

The order in which you ask questions can greatly influence people's willingness to complete the questionnaire as well as their responses to individual questions. Question order can also facilitate collecting and recording responses or make a nightmare of those tasks and of subsequent data analysis.

But arranging the questions appropriately is as much art as science. Although there often is no perfect order, here are a few things to keep in mind:

- Begin with interesting, easy, and nonthreatening questions. Especially for phone surveys, getting people to start answering questions is the hard part; once people begin responding, they will usually keep answering questions, even sensitive ones that they might initially have refused to answer.
- Put the really important questions near the beginning of the questionnaire. Some people will inevitably drop out before the survey is complete, so this will help ensure getting as many responses as possible for the main questions.
- Group questions by topic or type. Doing this, and also introducing each group of questions with a statement that provides a clue about the subject for a set of ques-

tions, helps people focus and provide meaningful answers. It also makes things easier for data collectors who must read the questions and record answers.
- Arrange questions and groups of questions in a logical order. This helps people understand the questions; in researcher-administered surveys, it also makes it easier for data collectors.
- Insert filter and contingency questions before they are needed to separate those who should answer a question or set of questions from those who should not answer them. This saves time and helps guard against confusing respondents. It also prevents getting a lot of meaningless answers.
- Proceed from general questions to specific ones both within the survey as a whole and within groups of questions. Doing this helps prevent subjects from using the specific questions as cues for answering the more general ones.
- Place open-ended questions before fixed-response questions on the same topic. This is particularly important for questions of salience. If you first ask how important several things are and then ask what is most important, many more people will select items from the set of fixed-response questions than would select them if they could not get cues from those items. But note: Open-ended probes after fixed-response questions are appropriate for finding out why people gave a particular response.
- Put the demographics at the end. This information is too personal to ask at the beginning of a survey. In general it is also less important than the answers to questions about knowledge, opinions, or behaviors. However, to ensure getting at least some of this information from those who do not complete the survey, some researchers like to ask one or two demographic questions as part of a screener.

Helps

In most cases, both data collectors and research subjects will need some help if they are to complete a survey properly. Giving them this help contributes directly to reliability by helping ensure that everyone completes the survey in the same way. It also increases external validity by making it easier for people to complete the survey. Therefore, a good questionnaire will include instructions for completing the survey. All but the very shortest will also have some transitional elements to introduce sets of questions and move from one set of questions to the next.

Instructions. There are two types of instructions. **Basic instructions** tell people how to indicate their answers to individual questions or sets of questions. **Procedural instructions** give directions for moving through the questionnaire.

Basic instructions are usually quite simple; however, they should be very specific. They must, for example, tell people very clearly whether they should place an X in a box or circle an item and whether they must select only one response or whether they can choose several. If response options for questions require subjects to choose a number from a range of numbers, basic instructions must also tell people what high and low numbers mean. If they must rank items, they need to be told clearly, ideally in several ways, whether they should use "1" or "10" for the item they rank as best or most important.

In researcher-administered surveys, procedural instructions give special instructions to the interviewer. These include, but are not limited to, directions for which response options should or should not be read, when it is appropriate to probe for further information, and what information may be recorded without asking a question.

In both researcher-administered and self-administered surveys, procedural instructions also help ensure that respondents will answer only those questions intended for them. Regardless of how a subject answers a question, this kind of instruction should make it clear which question the subject should answer next and which questions should be skipped. These instructions must be absolutely accurate. If some people are told to go to "Question 6," there should be a Question 6 and that must be the question they are supposed to answer. If some people are told they may skip the next question or set of questions because of their response to the last question, you must make absolutely certain they are not being told to skip any question they really should answer.

Transitions. These provide subjects with a frame of reference and/or a justification for answering a question or set of questions. Good transitional elements contribute to the response rate by helping people understand the research purpose and by encouraging them to move through the survey. But poorly done ones have the potential for biasing the results or causing people to drop out of a survey before completing the questionnaire. Therefore, transitional instructions must be short and simple.

In researcher-administered surveys, a transition consisting of a simple conversational statement such as "Now I'm going to ask you a series of questions about . . ." generally works best. In self-administered surveys, topical headings such as "media use" or "political activity" usually provide all the information subjects need.

Questionnaire Layout

Once all of the questions and instructions for a survey have been developed and arranged in a logical order, they must be physically assembled into a questionnaire that is visually appealing and easy to use. Questionnaires that are crowded, messy, or otherwise unattractive or hard to read may decrease the response rate. With both self-administered and researcher-administered surveys, they can also invite problems with recording or interpreting responses from those who complete and return the questionnaire.

To make questionnaires visually appealing and easy to use:

- Choose an appropriate type font and size. Other possibilities are also appropriate, but for questionnaires created with word-processing programs, 12-point Times Roman is a good choice.
- Use typographical elements to distinguish between questions and instructions for subjects and those for data collectors. For questionnaires sent on disk, posted on the Internet, or entered into a CATI system, different colored backgrounds are frequently used to separate procedural instructions and transitions from the questions themselves. A common convention for paper questionnaires that will be used with researcher-administered surveys is to use standard upper- and lower-case text for questions and response options, all uppercase lettering for instructions to interviewers and response options such as "no opinion" that will not be offered to subjects, and italics for transitional elements.
- Avoid crowded pages or screens. Leave extra space between questions and use adequate margins.
- Make sure there is a clearly defined space to record each answer.
- Leave enough space to answer open-ended questions but not so much as to look intimidating.
- Never split questions and/or response options across pages or screens.

- Arrange fixed-response options in a logical, easy-to-use form.

    How would you rate our customer service?
    _____excellent
    _____good
    _____average
    _____fair
    _____poor

    This format is easier on both respondents and interviewers than:
    How would you rate our customer service?
    _____excellent   _____good
    _____average   _____fair   _____poor

For researcher-administered surveys, you can precode items by placing the numbers you will enter into the computer directly on the questionnaire next to each response option for fixed-response questions. This saves time whenever you must manually enter data into a computer. However, precoding is more problematic for self-administered surveys. Unobtrusively placed, the numbers rarely present a problem, but there is always the possibility that some respondents will interpret a high or low number as signaling the desired response.

Quality Control

True surveys are widely used in communication research because, when done properly, they possess excellent reliability and validity. Variants using nonprobability or very small probability samples sacrifice external validity, but they can also score well on reliability and on the measurement form of internal validity. Nevertheless, reliability and validity can suffer unless researchers plan and then carry out their surveys carefully.

Reliability

For a study to be reliable, it must be replicable. But a study can be replicated only if the researchers specify all the procedures for data collection and analysis and then follow them exactly. In the true survey, with its reliance on a fixed questionnaire, replicability should be excellent.

However, any deviation from the set procedures will cause reliability problems. Those deviations can easily happen simply because humans are the data-collection instruments and people make mistakes. Interviewers may misread a question, skip questions that should have been asked, fail to record or misrecord some answers, or react inappropriately to respondents' answers or their queries about the survey.

In telephone surveys, reliability problems related to skipping questions or failing to record responses can be virtually eliminated by using computer-assisted telephone interviewing. This kind of problem can also be eliminated in most disk-by-mail and Internet surveys by setting up the questionnaire so that skipping questions or misrecording responses becomes almost impossible. For other kinds of surveys, however, the best protection comes from hiring good interviewers, providing them with thorough training, and then monitoring their work to make sure they do the interviews properly.

Internal Validity

Although only panel surveys provide the kind of control for time order that is necessary for showing a cause-and-effect relationship, most internal validity problems, as well as some reliability ones, stem from problems with question wording and/or question order. But, except in survey variants using truly open-ended journalistic interviewing techniques, these problems can be minimized by borrowing questions that have been previously tested for reliability and validity or that seem to have worked well in similar studies. When new questions must be developed, problems can be identified using the techniques discussed in Chapter 5.

More specifically survey-related threats to internal validity stem from data-collection procedures. These include problems associated with interviewers who misread questions, who react inappropriately to subjects' responses, or who try to help them understand or answer questions. Any of these actions can change the meaning of a question or introduce bias into subjects' responses. The remedy here is to train interviewers thoroughly in proper survey-administration techniques.

But even if interviewers behave properly, many people will respond, fail to respond, or tailor their responses according to their perception of the interviewer. Therefore, it is important to hire interviewers who can quickly and easily establish rapport with and gain acceptance from a wide variety of people. For face-to-face interviews, matching interviewers and interviewees on gender, race, or ethnicity can also be a good idea.

Other survey-related threats to internal validity include problems associated with the sensitization that can come from telling too much about the survey or its sponsorship in the survey introduction. If people know too little about a survey or its sponsorship, they may conclude the research is not legitimate and refuse to cooperate. This, of course, will decrease external validity. But upon learning its purpose or sponsorship, they may tailor their answers accordingly. This will affect internal validity.

Problems of this kind can be detected only through pretesting the questionnaire. If that pretesting indicates people are reacting to the purpose of the study or its sponsorship, the introduction may need to be reworded so that the survey's true purpose is somewhat veiled. To maintain that veil, some items unrelated to the survey's true purpose may need to be included in the questionnaire.

If subjects may be suspicious of a sponsor's true motives or may be tailoring their answers in response to their knowledge of or feeling about the research sponsor, the problem can be alleviated by enlisting a well-known research firm or a researcher from an established university to do the actual work. With this strategy, the emphasis in the introduction would be on who is doing the work rather than on who is sponsoring it. This strategy is especially useful in situations where respondents may be afraid to answer sensitive questions truthfully unless they have the assurance that the information is being gathered by a respected outsider who can and will protect their privacy.

External Validity

Many external validity problems stem from sampling. They can be alleviated by following procedures outlined in Chapter 4 to ensure drawing a probability sample that is large enough for making the desired inferences through statistical analysis. But in survey research, the major threat comes from low response rates and low completion rates.

General strategies for improving response rate include:

- Using callbacks for researcher-administered surveys and follow-up reminders for mail and Internet ones. Trying to contact respondents several times will both increase sample size and enhance representativeness. Similarly, sending reminders to those who do not return mail surveys or complete Internet ones promptly can increase the response rate by 20 percent or more.
- Offering token gifts or monetary rewards for participating. This is somewhat controversial because of ethical considerations. But it works. In any population some people will respond only to rewards. Offering a reward will get some of them to participate; almost all of those who do not need rewards will still respond.

With telephone surveys, a few respondents who cannot be contacted because their answering machine is always turned on can be persuaded to participate by leaving a brief message and a call-back number on their voice mail systems.

For face-to-face surveys and self-administered mail and Internet ones, personalizing introductory mailings and cover letters or survey introductions is very helpful.

For mail and disk-by-mail surveys, using a colorful stamp instead of metered mail and hand-addressing the mailing envelope—anything to make the mailing look personal instead of like bulk, junk mail—will also increase the response rate. And of course, providing a self-addressed, stamped return envelope is mandatory.

But getting people to begin answering survey questions is only half the battle. Generalizability will also suffer if too many people drop out before completely answering the questions. This mortality problem is greatest if a questionnaire is too long or if very sensitive questions are asked too early or are worded so as to seem too demanding. Therefore, to reduce mortality it is important to pretest the questionnaire to make sure there are no problems with length, question wording, or question order.

For researcher-administered surveys, it is also important to hire and then train interviewers so that they are able to establish rapport with respondents and keep them talking. Because interviewer demographics may also affect response and completion rates, matching interviewer characteristics to those of the subjects they will be contacting can also help.

Data Analysis

The data gathered from true surveys are almost always analyzed quantitatively using statistical techniques; data from variants using nonprobability samples or small nonprobability ones are more often analyzed using qualitative techniques. However, just as some information such as that from open-ended questions may be analyzed and reported qualitatively even in studies that are primarily quantitative, qualitative studies may make use of some numbers.

Before you begin your analysis, you may need to code answers to open-ended questions so that each kind of answer is given a number that you can enter into a computer. This is an exercise in content analysis, which is discussed in the next chapter.

After coding the responses, you can use the assigned numbers for a statistical analysis. You can also use them with the "select" function in your data-analysis program to locate certain responses that you might want to quote in your findings.

If you are doing a purely qualitative analysis, you still may need to do some coding to help you locate quotes or sort responses by type as a first step to finding patterns in your data. But just as some quotes from open-ended questions can help people understand findings from a quantitative analysis, some quantitative information can enhance a qualitative analysis. It will, for example, be more informative to say "Ten of the 17 people in this study said X" than to say "Most people said X."

Quantitative Analysis

Before you begin your analysis, you may need to create some new, composite measures by following the procedures given in Chapter 5. But once you have all the measures you will need in order to answer your research questions or test your hypotheses, you are ready to begin analyzing the data.

Quantitative analysis of survey data proceeds from an examination of responses to individual questions to a search for statistically significant relationships among answers to two or more questions. Therefore, you begin by examining the **univariate statistics** and then proceed to bivariate ones and, if necessary, to the **multivariate statistics**.

Univariate Statistics. These statistics are based on the frequency distribution. This distribution shows how many respondents and what percentage of all respondents gave particular answers to each survey question. For a true survey, information from the frequency distribution can also be used in conjunction with information on sample size and sampling error to make inferences about the distribution of responses to each question that would most likely be found if everyone in the population had been interviewed.

Used alone, frequencies provide descriptive information that is sufficient for answering the questions many survey research projects are designed to answer, but the accompanying measures of central tendency are also very useful. Of the three measures of central tendency, only the mode makes sense for variables measured at the nominal level. However, the mean or median can simplify or enhance the reporting of findings from variables measured at the ordinal level or higher.

Frequencies for a series of ordinal measures asking, for example, how often people use each of several different information sources can help answer questions about how many people use each source. But making comparisons across a number of sources will be easier and the findings will be clearer if you use the mean, or average, scores. Using the median is also appropriate, especially in cases where the mean may be unduly affected by a few very high or very low scores. You can determine this by examining the range of scores.

Bivariate Statistics. Testing hypotheses and answering many research questions requires determining whether any meaningful, consistent relationship exists between the way people answer a question (designated the independent variable) and the way they answer a second question that provides the measure for a dependent variable.

For a quasi-survey of a nonprobability sample of respondents, researchers sometimes use the tests of difference discussed in Chapter 6 because they work well with both nonprobability and small probability samples. With them, the level of significance tells whether a difference exists between groups, but tests of difference do not let you generalize your findings to some larger population. Therefore, they are rarely used with true surveys.

For true surveys, researchers use measures of association to examine bivariate relationships. These statistics detect whether any systematic relationship exists between scores on one variable and scores on a second variable. Unlike the t-test and analysis of variance, they are designed for use with data gathered from a large probability sample of subjects.

Box 7.1. *Interpreting a computer printout for frequencies*

A frequency printout shows the distribution of subjects' responses to an individual question. This information, used in conjunction with measures of central tendency, and measures of dispersion, helps you describe your data set. Frequencies are also used to calculate sampling error.

Example:

The frequencies are from a 1998 survey of college students' use of a campus newspaper and other media.

Question: "Do you use the Internet for e-mail a lot, some, a little, or not at all?"

Computer Printout:

Use the Internet for e-mail

		Frequency	Percent	Valid Percent	Cumulative Percent
Valid	None	21	7.6	8.1	8.1
	Little	14	5.1	5.4	13.6
	Some	48	17.3	18.6	32.2
	A lot	175	63.2	67.8	100.0
	Total	258	93.1	100.0	
Missing	System Missing	19	6.9		
	Total	19	6.9		
Total		277	100.0		

Interpreting the printout:

- The label at the top of the table identifies the variable.
- The first column labels the data by first dividing it into "valid," "missing," and "total" categories and then subdividing "valid" and "missing" into categories that are more meaningful for data analysis.
- Valid data are from subjects who answered the question. The categories "none," "little," "some," and "a lot" indicate their possible answers to the question.
- In most cases, "missing data" is the result of subjects not answering a question, either because they refused to answer it or were not eligible to answer it. But sometimes researchers may instruct the computer to ignore certain responses, or declare it missing. When this is done, "missing" may be subdivided into "system missing" and "declared (or user) missing" categories.
- The column labeled "frequency" provides a raw count of the number of respondents who gave a particular response. For example, 48 respondents indicated that they had used the Internet some for e-mail; 258 respondents (the "total") answered the e-mail questions. But there were really 277 respondents (the final total); data from 19 are missing.

(continued)

- In the percent column the raw numbers have been converted to percentages of the total number of respondents. Any missing data are included in this calculation. For example, 17.3% of the 277 total respondents indicated that they had used the Internet some for e-mail.
- The valid percent column is a recalculation of the percentages without the missing data. This column is often more meaningful because it reports the percentages of those who answered the question. For example, of the 258 respondents who answered this question, 18.6% indicated that they used the Internet some for e-mail. If every respondent is asked a particular question and there are no missing data, the valid percent column and the percent column will be identical on the printout.
- The last column reports the cumulative percentage. For example, 32.2% of the respondents who answered this question indicated that they had used the Internet for e-mail none, little, or some of the time.

Box 7.2. *Commonly used measures of association*

These measures indicate how strongly two variables are related to each other or to what extent they occur together in the data. The level of measurement of the variables in question determines which measure is most appropriate. Except in the case of chi-square, measures of association for nominal variables indicate the strength of the relationship. They do not indicate the direction of the relationship. Measures of association for ordinal, interval, and ratio variables indicate the strength and direction of the relationship.

Chi-square: This statistical test of significance indicates whether a systematic relationship exists between two variables by comparing expected values in a table with the actual values in a table. The chi-square statistic is designed for **nominal level data,** but it can be used with higher levels.

Formula:

$$\text{Chi-square} (\chi^2) = \sum \frac{(f_o - f_e)^2}{f_e}$$

df = $(r - 1)(c - 1)$ where r = the number of rows and c = the number of columns in the table

where,
f_o = the observed frequency in a cell of a cross-tabulation table, and
f_e = the expected frequency in a cell if the two variables are not associated with each other. The expected frequency (f_e) for a cell equals the product of the row and column totals for that cell divided by the total sample size.

- In general, a large chi-square indicates that a relationship of some sort exists between the two variables, and a small chi-square implies the absence of a relationship. To determine whether a particular chi-square is large enough, you

(continued)

need to look in a chi-square distribution table and compare the computed value to the critical value at the appropriate degrees of freedom and desired level of significance.
- The value of chi-square does not give you any information about the strength or nature of the relationship. It just tells you whether the observed distribution of scores is what you might get by chance.

Phi: This statistical test of significance tells the strength and statistical significance of the association between two **nominal level variables.** Phi is appropriate only for $2 \times k$ tables. This means that one of the variables should be a simple dichotomy (e.g., male/female).

Formula:

$$\text{phi } (\varphi) = \frac{\sqrt{\chi^2}}{n}$$

where,

n = sample size

- Values range from 0 for no relationship to 1 for a perfect association.

Cramer's V: This statistic handles problems associated with the number of categories in the rows and columns of a table. It also gives some indication of the strength of the relationship between the two variables. Cramer's V is suitable for **nominal data.**

Formula:

$$\text{Cramer's V} = \frac{\sqrt{\varphi}}{n(k-1)}$$

where,

n = sample size
k = the smaller of the number of rows and columns in the table.

- Values range from 0 for no relationship to 1 for a perfect association.

Kendall's tau$_b$: This statistic provides information on the statistical significance and strength of an association between two variables. It is most appropriate for a *square table* (same number of categories in the rows and columns). A variation, Kendall's tau$_c$, is more appropriate for rectangular tables (ones with a different number of categories in the rows and columns). Both Kendall's tau$_b$ and Kendall's tau$_c$ are appropriate for use with **ordinal level data.**

Formula:

$$\text{Kendall's } \tau_b = \frac{C - D}{\sqrt{[n(n-1)/2 - T_x][n(n-1)/2 - T_y]}}$$

where

C = the total number of concordant pairs of observations. A pair of observations is concordant if the member that ranks higher on one variable also ranks higher on the other variable.

(continued)

D = the total number of discordant pairs of observations. A pair of observations is discordant if the member that ranks higher on one variable ranks lower on the other variable.
T_x = the number of pairs tied on x (the independent variable). A tie occurs when two observations fall into the same category of the variable.
T_y = the number of pairs tied on y (the dependent variable), and
n = the sample size

- Values range from –1 (a perfect negative association) to +1 (a perfect positive association). A value of zero indicates no relationship between the variables.

Gamma: This test of association is similar to Kendall's tau, but it does not make adjustments for ties or for table size, so it tends to yield higher values than Kendall's tau. This test is most appropriate for **ordinal level data.**

Formula:

$$\gamma = \frac{C - D}{C + D}$$

where,
C = the total number of concordant pairs of observations and
D = the total number of discordant pairs of observations

- Values range from –1 for a perfect negative relationship to +1 for a perfect positive relationship. A value of zero means there is no relationship.

Spearman's rho: This test of association indicates the strength and direction between two **ordinal level variables.** It is most appropriate for ranked data.

Formula:

$$\text{Spearman's rho } (\rho)\ (r_s) = 1 - \frac{6 \sum D^2}{n(n^2 - 1)}$$

where,
D = the difference between the two rank scores for each item/subject,
n = the total number of items/subjects

- Values range from –1 for a perfect negative relationship to +1 for a perfect positive relationship. A value of zero means there is no relationship.

Pearson correlation coefficient (r): This measure indicates the degree to which variation (or change) in one variable is related to variation (or change) in another variable. It is most appropriate for variables measured at the **interval or ratio level.**

Formula:

$$r = \frac{N \sum XY - \sum X \sum Y}{\sqrt{[N \sum X^2 - (\sum X)^2][N \sum Y^2 - (\sum Y)^2]}}$$

(continued)

> where
>
> X (independent variable) and Y (dependent variable) stand for the original scores
> N is the number of pairs of scores, and
>
> - Values range from –1 (a perfect negative association) to +1 (a perfect positive association). A value of zero indicates that the two variables are not at all related to each other.

With the exception of chi-square, the value for the bivariate statistics tells the strength of any relationship. Measures of association intended for use with ordinal or higher levels of measurement will tell whether a relationship is a linear positive one—high scores on one variable are associated with high scores on the second variable, and low scores with low scores—or a negative one—high scores on one variable go with low scores on the other variable.

Because nominal measures have no direction or amount, chi-square and Cramer's V will not tell whether any relationship between two variables is positive or negative, but they can sometimes pick up curvilinear relationships that other measures of association miss.

The level of statistical significance tells the likelihood that any systematic relationship between scores on two measures might not hold up if data were gathered from the entire population—that the findings are not "real"; they might be due to chance alone or to the vagaries of sampling.

Strength and significance are related to each other; however, sample size can have a marked effect on significance. With small samples, the statistics may indicate that a fairly strong relationship is not significant, but if the sample is large enough, even an extremely weak relationship may be statistically significant. For this reason, researchers sometimes accept a significance level of .1 instead of the more common .05 if they are working with very small samples. They may also accept a higher level of significance if the strength or direction of the relationship is supported by theory and if the consequences of being wrong about the reality of the relationship are not severe.

Although academic researchers sometimes rely on just the statistics and their accompanying level of significance, applied researchers often give greater weight to information from a cross-tabulation table, sometimes also called a contingency table. Each box in a crosstab table shows how many subjects who gave a particular answer to one question gave each possible answer to the second question. Including some of this information in reports helps clients visualize the findings.

For surveys of an entire population, the percentages given in the crosstab table will be enough to test your hypotheses or answer your research questions. Any differences in numbers and percentages may or may not be important for the purposes of a study, but they are real. But where sampling has been used to select subjects, the numbers and percentages in the contingency table can be difficult to interpret. In many cases, it will be impossible to tell whether any apparent relationship is statistically significant. Therefore, it will be necessary to look at the accompanying statistics to determine the strength and significance of any relationship between two variables.

Sometimes, however, apparently weak relationships result from having too many "boxes" in your crosstab table. The statistics used to detect correlations or associations

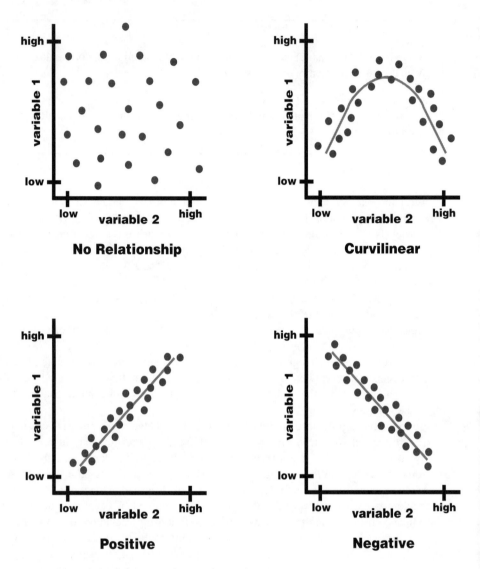

Figure 7.1. Types of relationships between two variables

between pairs of variables don't work very well if there are too many empty boxes (indicating that none of your respondents gave a particular pair of responses to the two questions) or if there are too few subjects who gave a certain pair of responses. Even if there are enough cases, usually at least ten, in each box, crosstab tables with a lot of boxes can be hard to interpret. These problems occur most often when you are working with questions that allow open-ended responses, but they can also occur even when your subjects could choose from among only a very few response options. Therefore, as you proceed through your bivariate analysis, you may find that you need to combine some response options by recoding them, resulting in fewer categories. Directions for recoding variables are in Chapter 6.

When recoding, it is a good idea to create a new variable instead of changing the scores on your original measure. Even though having fewer categories often works best in bivariate analysis, the statistics designed specifically for multivariate analysis work best with interval and ordinal measures that have many different values.

Multivariate Statistics. Relying solely on bivariate statistics can easily lead a researcher to reach overly simplistic, even misleading conclusions. Regardless of the strength and statistical significance of any relationship found through the use of bivariate statistics, a host of other factors besides the one labeled the independent variable most likely had some influence on people's scores on the dependent variable. By using

---

Box 7.3. *Interpreting a computer printout for a cross-tabulation table and measures of association*

Cross-tabulation tables are used to examine bivariate relationships. Therefore, they are useful for answering research questions or testing hypotheses.

Depending on the computer program you use and the options you choose, this kind of table will tell you the number and/or percentage of subjects with a particular score on one variable who had each possible score on another variable. Examining these figures helps you explain your findings. But used alone, they will not tell you how strong any apparent relationship between two variables is or whether it is real (i.e., statistically significant). That information comes from the measures of association. Again, depending on the computer program you use and/or the options you choose, you can have the computer calculate all measures of association or just the one you need to answer your research question or test your hypothesis.

Example:

To illustrate the use of cross-tabulation and measures of association, we are using data from a 1998 survey of college students' media use to answer this research question: Are graduate students and upperclassmen more likely than undergraduates to say international news coverage is an important determinant of whether they read a particular newspaper?

To answer this question, we need to examine the relationship between the independent variable, GRADE, and the dependent variable, INT-NEWS, each of which is measured by a single survey question.

GRADE: "Are you currently classified as a freshman, sophomore, junior, or senior, or are you a graduate student?" Recoded into three categories: "underclassmen" (freshmen + sophomores), "upperclassmen" (juniors + seniors), and "graduate."

INT-NEWS: "In deciding whether you will read a particular newspaper, is international news coverage very important to you, somewhat important, neither important nor unimportant, not very important, or not at all important?" Recoded into three categories: "important" (very important + somewhat important), "neutral" (neither important nor unimportant), and "unimportant" (not very important + not at all important). *(continued)*

Computer printout:

	GRADE			
INT-NEWS	Under-classmen	Upper-classmen	Graduate	Total
Unimportant count	19	15	3	37
% within INT-NEWS	51.4%	40.5%	8.1%	100.0%
% within GRADE	17.8%	11.9%	7.9%	13.7%
Neutral count	31	31	8	70
% within INT-NEWS	44.3%	44.3%	11.4%	100.0%
% within GRADE	29.0%	24.6%	21.1%	25.8%
Important count	57	80	27	164
% within INT-NEWS	34.8%	48.8%	16.5%	100.0%
% within GRADE	53.3%	63.5%	71.1%	60.5%
Total count	107	126	38	271
% within INT-NEWS	39.5%	46.5%	14.0%	100.0%
% within GRADE	100.0%	100.0%	100.0%	100.0%

	Value	df	Asymp. Sig. (2-sided)
Pearson chi-square	5.153	4	.272
Likelihood Ratio	5.225	4	.265
Linear-by-Linear Association	5.035	1	.025
N of Valid Cases	271		

		Value	Asymp. Std. Error[a]	Approx. T[b]	Approx. Sig.
Nominal by Nominal	phi	.138			.272
	Cramer's V	.098			.272
Ordinal by Ordinal	Kendall's tau-b	.126	.055	2.298	.022
	Kendall's tau-c	.109	.048	2.298	.022
	gamma	.219	.094	2.298	.022
	Spearman Correlation	.137	.059	2.271	.024[c]
Interval by Interval	Pearson's r	.137	.058	2.261	.025[c]
N of Valid Cases		271			

[a] Not assuming the null hypothesis.
[b] Using the asymptotic standard error assuming the null hypothesis.
[c] Based on normal approximation.

(continued)

Interpreting the cross-tabulation table:

- Variable names are printed in all capital letters. The column headings "underclassmen," "upperclassmen," and "graduate" identify the categories of the independent variable GRADE. Row labels "unimportant," "neutral," and "important" identify the categories for the dependent variable INT-NEWS.
- The column and row labeled "total" give the frequency distribution for each variable and are often referred to as the marginals. They give the number and percentage of all subjects who gave a particular response on one of the variables. The box, or "cell," in the lower right corner gives the total number of subjects (271) who answered both the GRADE and INT-NEWS questions and the percentage (100%) of all subjects who answered both questions.
- Other cells in the table give the number and percentage of respondents who gave each unique combination of responses such as "underclassmen" and "unimportant" or "upperclassmen" and "important"
- Each cell in this printout, including the cells in the "total" column and row, has three sets of figures. The labels included in the box with those for the variable INT-NEWS call these figures "count," "% within INT-NEWS," and "% within GRADE."
- The top figure in the "total" column tells how many subjects gave the response identified by the corresponding row label. In other boxes, this figure gives the "count" or number of subjects who gave a particular combination of responses. For example, 37 subjects said international news is "unimportant"; 107 subjects are "underclassmen." The top figure in the upper left box tells us that 19 subjects who are underclassmen said international news is unimportant to them; the figure in the lower right cell tells us that 27 graduate students said international news is important.
- The second set of figures in each cell, labeled "% within INT-NEWS," is the row percentage. It tells us the percentage of subjects who are in the category identified in that row who are in the category identified by the column heading. Looking first at the marginals, we see that, working across the table, row percentages will always add up to 100%; the column percentages will not. The row percentage tells us that the 37 subjects who said international news is important are all—or 100%—of the subjects who gave that answer. The column percentage tells us that the 107 underclassmen are 39.5% of the subjects who answered both questions. The second number in the upper left box tells us that 51.4% of the subjects who said international news is unimportant are underclassmen.
- The third set of figures in each cell, labeled "% within GRADE," is the column percentage. It tells us the percentage of subjects who are in the category identified in the column who are in the category identified by the row label. For this set of figures, the column percentages will add up to 100%; the row percentages will not. From the marginals, we see that the 107 subjects who are underclassmen are 100% of those in the sample who are underclassmen, but the 37 who said international news is unimportant are just 13.7% of all of the subjects in the sample. The third figure in the upper left box tells us that 17.8% of all underclassmen said international news is unimportant.

(continued)

Note: The raw numbers, the "count," can be hard to interpret. For most purposes, the percentages are less prone to misinterpretation. However, you must be careful to use the set of percentages that best answers your research question. In many disciplines it is customary to make the independent variable the column variable and put the dependent variable in the rows as we have done here. Therefore, for our purposes, the column percentages are most meaningful. Examining them will show whether a change on the independent variable produces systematic change on the dependent variable.

- Examining the column percentages will give a partial answer to our research question. The column percentages in the top row or cells tell us that 17.8% of the underclassmen but only 11.9% of the upperclassmen and 7.9% of the graduate students said international news is unimportant. The bottom row of cells tell us that 53.3% of the underclassmen, 64.5% of the upperclassmen, and 71.1% of the graduate students said international news coverage is an important determinant of whether they will read a particular newspaper.

Tentative Conclusion: There appears to be a systematic relationship between GRADE and INT-NEWS, but to find out how strong that relationship is and whether it is statistically significant, we need to examine the appropriate measures of association.

Interpreting measures of association:

- The first box below the cross-tabulation table gives the value, number of degrees of freedom (df), and level of statistical significance for three versions of the chi-square statistic. Of these, the first one, identified as "Pearson's chi-square," is the one most commonly used. It is the one described in Box 7.1.
- The second box gives information for other measures of association. For these measures, the number listed in the column labeled "Value" tells us the strength of the relationship. Except for phi and Cramer's V, it will also tell us whether the relationship is positive or negative. The number listed in the column headed "Approx. Sig.," tells us whether the relationship is statistically significant.
- The measure of association you should use depends on the level at which you have measured your variables. In our example, both GRADE and INT-NEWS are measured at the ordinal level. Therefore, we could use either gamma, Spearman's rho, or, because our cross-tabulation table has the same number of columns as rows, Kendall's tau$_b$.

Note: We might also want to check the chi-square. Although it is designed for nominal measures, it can be used with all levels of measurement. When used with ordinal or higher measures, it will sometimes pick up a curvilinear relationship that will be missed with measures of association such as Kendall's tau, gamma, Spearman's rho, and Pearson's r, which can detect only linear relationships.

(continued)

> - In our example, Kendall's tau$_b$ is most appropriate. The value of Kendall's tau$_b$ is .126. This tells us that there is a weak, positive relationship between GRADE and INT-NEWS. The chi-square statistic is not significant. This provides further evidence that we have a linear relationship instead of some kind of curvilinear one. The Kendall's tau$_b$ statistic is significant at the .022 level. That is well below the generally accepted level of .05. There are only 22 chances in 1000 that this relationship would not hold up if we were to survey all of the approximately 20,000 students enrolled at the college.
>
> Conclusion: In this case, the measure of association confirms the impression we gleaned from examining our cross-tabulation table. We now have the answer to our research question. Graduate students are more likely than upperclassmen, who, in turn, are more likely than underclassmen to say that international news coverage is an important factor in deciding whether they will read a particular newspaper.

multivariate statistics, you can control for other influences. You can also examine the simultaneous effects of several independent and/or control variables.

Upon reflecting on a finding that people who watch news regularly on national network television stations are more likely than occasional viewers to think the environment is an important political issue, you might, for example, suspect that at least some of the apparent influence of network news might be due to other factors such as people's own level of education.

One way to find out whether your hunch is correct would be to use the "select if" function in data-analysis programs to create a subsample of just those who are regular viewers of network news, use crosstabs and correlations to examine the relationship between issue salience and education level, then repeat the process with just the occasional viewers of network news. If the correlations show that the most educated viewers think the environment is an important issue regardless of how much news they watch, you would have justification for concluding that television viewing is not the only influence. Education level also matters. However, this process will not determine how much it matters.

The **partial correlation** statistic is the simplest way to eliminate the effect of any confounds. Partial correlation is based on Pearson's r, but it "partials out" the effect of other variables such as education level on the correlation between your independent and dependent variable. Calculating the partial correlation will, in effect, tell you how strong the correlation between television news viewing and the salience people attach to the environment as an issue would be if education level were not a factor. By comparing this value to that of Pearson's r, you can get a sense of the size and kind of effect the variables you consider controls or confounds have on the basic bivariate relationship that interests you.

Box 7.4. *Partial correlations*

The partial correlation statistic is based on Pearson's r (see Box 7.2). Like Pearson's r, it shows the strength and significance of a bivariate relationship, but partial correlation controls (or adjusts) for the effect of additional variables. The effect of these other variables is said to be "partialed out," hence the name. In practice it is rare to use more than two or three control variables.

As with Pearson's r, values range from –1 to +1, but the partial correlation is usually lower than the value for Pearson's r because the portion of the relationship attributable to other variables has been removed.

Partial correlation is designed for variables at the **ordinal or higher level.** However, it is possible to include nominal variables in the analysis by treating them as **dummy variables.** Directions for creating dummy variables are in Box 7.6.

Formula:

$$r_{12.3} = \frac{r_{12} - (r_{31})(r_{32})}{\sqrt{1-(r_{31})^2}\sqrt{1-(r_{32})^2}}$$

where,

$r_{12.3}$ = the correlation between variable X_1 and variable X_2 controlling for variable X_3,
r_{12} = the correlation between variable X_1 and variable X_2,
r_{31} = the correlation between variable X_1 and the control variable X_3, and
r_{32} = the correlation between variable X_2 and the control variable X_3

Example:

Question: Controlling for gender (SEX), what proportion of the variation in how often an individual uses the Internet to play games (PLAY) is explained by how often an individual uses the Internet (INTERNET)?

PLAY: "Do you use the Internet to play games, a lot, some, a little, or not at all?"

INTERNET: "Do you usually use the Internet almost every day, four or five days each week, two or three days each week, about one day each week, less than once a week, or don't you use the Internet?"

The correlation between INTERNET and PLAY is .171. The relationship is significant at the .01 level. The control variable SEX is correlated with both INTERNET (–.123; sig. = .05) and PLAY (–.131; sig. = .05).

Using the formula, we see that

$$r_{12.3} = \frac{.171-(-.123)(-.131)}{\sqrt{1-.(-.123)^2}\sqrt{1-(-.131)^2}} = \frac{.171-.016}{(.992)(.991)} = \frac{.155}{.983} = .158$$

The size of the correlation has been reduced somewhat from .171 to .158 from .171.

Note: When the partial correlation is squared, the result reflects the proportion of variation explained by the control variable and explained by the independent variable. Thus, about 2.5% of the variation ($.158^2 \times 100$) in using the Internet to play games is explained by INTERNET use after removing the effect of gender.

But the easiest way to determine the independent and joint influences of two or more variables on some other variable is to use **regression analysis.** A regression analysis is ANOVA's twin. But where an ANOVA determines whether variance between groups is greater than variance within a group, a regression equation uses the scores on one or more variables to see how well those scores predict scores on a dependent variable.

In our example, you could instruct the computer to enter responses to the issue salience question as the dependent variable and network news viewing and education level together as predictor variables. In this case, the value, or **beta,** and associated level of significance for each predictor variable would tell how well each one predicts the importance people attach to the environmental issues. The value and level of significance for the overall equation tell how well the combination of news viewing and education level predict issue salience.

Like ANOVA, regression analysis is really a very flexible tool. So you could instruct the computer to do the calculation by first entering into the regression equation the predictor variable that is the best predictor, then adding other variables in decreasing order of importance or you could have the computer enter them from least to most important. Or as an alternative, you could enter variables **stepwise,** in blocks or groups, in any order that makes logical or theoretical sense.

With any of these methods, the computer printout will tell you how well each independent or control variable and each set of independent and control variables predicts scores on the dependent variable.

But one drawback to a regression analysis is that any dependent variables that you enter into your regression equation should not be highly correlated with each other. If they are, you should combine them into a composite measure. Another problem is that regression is designed to work with ratio measures having many different values. You can usually use regression with ordinal or interval data without any problem. However, if you want to include a nominal measure as one of your independent variables, you will first have to turn it into a series of **dummy variables.**

Qualitative Analysis

If your only interest is to incorporate some qualitative material into an otherwise quantitative analysis, the task is rather simple. As you code open-ended questions, you can simply flag potentially useful responses in some way and set them aside. Later, you can reread all of the ones you flagged for each question to see whether they provide any insight into how you should interpret your statistics. At the same time, you can select the most interesting responses and set them aside to quote as illustrations for your findings.

But doing a truly qualitative analysis is much more complex. It is both a reporting project and an exercise in content analysis. As a reporting project, you will need to condense your material by categorizing answers according to their main idea and then subcategorizing them to indicate subdimensions such as direction and intensity. But in doing this, you can't just rely on the words; you will also have to look at the narrative structure in order to determine whether there are latent meanings. Finally, you will have to come up with explanations for your findings.

Because you will be working from documents—the text of your interviews—you will actually be doing a qualitative content analysis. Although we present the basics in this chapter, we provide more information on managing, coding, and analyzing

Box 7.5. *Multiple regression analysis*

Regression analysis is a technique for looking at the simultaneous effect of multiple independent variables on a dependent variable. It does this by determining how well subjects' scores on the independent variables predict their scores on the dependent variable. As with partial correlation, this technique is used **for ordinal or higher levels of measurement.** Nominal measures may be included using the **dummy variable** technique (see Box 7.6).

Regression analysis is very flexible. All variables may be examined simultaneously, or they may be explored individually or a group at a time in the order specified. Adding variables individually or as a group in effect treats these added variables as control variables. Regression analysis is only appropriate if

- there are at least 10 cases for each variable in the analysis. As a result, this technique is better suited for analyzing survey data than it is for analyzing laboratory experiments. For experiments, the ANOVA is most appropriate.
- the frequencies for the variables in your analysis approach a normal distribution. Check the frequency distributions of your variables.
- your independent variables are relatively independent of each other. Check the correlations between them; correlations greater than .5 will cause problems.

Formula:

$$Y = \alpha + \beta_1 X_1 + \beta_2 X_2 + \beta_3 X_3 + \ldots + \beta_k X_k$$

where,

$Y =$ the dependent variable
$\alpha =$ the conditional mean of Y when each of the independent variables $= 0$. This is the point at which the regression line crosses the vertical axis.
$X_1, X_2,$ etc. $=$ the independent variables
$\beta_1, \beta_2,$ etc. $=$ the partial regression coefficients or slopes

Example:

In a 1996 survey to assess the need for a master's program in public relations, subjects were asked how likely they were to enroll in such a program (ENROLL). Some said they were very likely to enroll, others said they were unlikely to enroll, while others were more ambivalent. In this situation, regression can help us determine which variables might help us explain this variation. Four variables were identified that might help us explain this variation.

IMP: "How important is a master's degree for people in your field?"

DEMAND: "How much demand do you believe there is in this area for a program like this?"

PROGCOST: "A total of 32 credits is necessary to complete the proposed master's degree. The anticipated cost is $375 per credit hour. In your opinion, is this cost excessively high, somewhat high, about right, or lower than expected?"

PERSBEN: "How much would this kind of degree benefit you personally?"

These independent (or predictor) variables were regressed on the dependent variable ENROLL.

(continued)

Computer printout for a regression analysis:

Listwise Deletion of Missing Data

Equation Number 1 Dependent Variable .. ENROLL likelihood of enrolling

Block Number 1. Method: Enter   IMP   PROGCOST   DEMAND   PERSBEN

Variable(s) Entered on Step Number
1.. PERBEN   degree personally beneficial
2.. PROGCOST   anticipated program cost
3.. IMP   importance of degree
4.. DEMAND   demand for degree

Multiple R	.57563
R Square	.33135
Adjusted R Square	.29201
Standard Error	.83709

Analysis of Variance

	DF	Sum of Squares	Mean Square
Regression	4	23.61182	5.90296
Residual	68	47.64845	.70071

F = 8.42422   Signif F = .0000

------------------------------Variables in Equation------------------------------

Variable	B	SE B	Beta	T	Sig T
IMP	−.131953	.187855	−.074943	−.702	.4848
PROGCOST	−.069171	.128790	−.054286	−.537	.5930
DEMAND	−.273742	.194235	−.160487	−1.409	.1633
PERSBEN	.860131	.152946	.643981	5.624	.0000
(Constant)	.980125	.746833		1.312	.1938

End Block Number 1 All requested variables entered

Interpreting a regression analysis:

- The column labeled B contains the regression coefficient for each independent variable. This coefficient is expressed in the units of measurement of the specific variable. Because variables are seldom measured in the same units, it is often more useful to examine the Beta, which is a standardized coefficient. Note: The coefficients (both B and Beta) estimate the effect of a particular independent variable on the dependent variable controlling for the effects of the other independent variables entered in the equation.
- The Betas (estimates) and the level of probability, given in the column labeled "Sig T," let you determine which of the independent variables predict values on the dependent variable. Because the Betas are standardized, you can tell which variable has the greatest effect on the dependent variable by comparing their values. The signs for the Beta tell you the direction of the relationship between each independent variable and the dependent variable.

(continued)

> - R^2 and Adjusted R^2 and their related statistics tell how much of the variance in the dependent variable can be explained by the independent variables. The Adjusted R^2 is a more conservative estimate of the amount of variance explained.
>
> Conclusion: In our example, R^2 tells us that our four independent variables (IMP, DEMAND, PROGCOST and PERSBEN) explain 33% of the variance in our dependent variable, ENROLL. Or more conservatively, Adjusted R^2 tells us that they explain 25% of the variance.
>
> The significant, positive Beta for PERSBEN tells us that subjects who believe the program will be personally beneficial to them are more likely to say they may enroll in the program than are those subjects who believe they will derive less benefit from enrolling. Because only the Beta for PERSBEN is statistically significant (sig. = .000), we can conclude that the other independent variables contribute little, if anything, to explaining the variance in the dependent variable.

textual material in Chapter 8. Therefore, you should read that chapter before you try to analyze information from survey interviews. You may also find it helpful to read the qualitative analysis sections in Chapters 9 and 10 on focus group and observational research.

Qualitative researchers do not usually think in terms of variables the way quantitative researchers do. Still, their analysis usually progresses from a univariate analysis to a bivariate or multivariate analysis.

With data from in-depth interviews as an example, your first step would be to transcribe tapes of your interviews. Although it is possible to work directly from tapes, most researchers find it much easier to work from transcriptions. Some researchers like to enter the text of their interviews into a computer program designed for qualitative analysis; others prefer to work from hard copy. Qualitative analysis programs allow you to enter keywords or codes directly into your database; these keywords or codes facilitate locating information and making comparisons across documents. But, as we explain in Chapter 8, you can do the same thing manually.

Once you have the text for all your interviews, you begin your analysis by examining them in order to identify the most common responses to key questions and other answers that are less common. After you have located those typical and atypical answers, you look for information that might explain similarities and differences in responses to individual questions and for information that might provide an alternate explanation. Some of this information may come from notes about the demographic characteristics of interviewees, but much of it will be found in the subjects' own words. After that, you look for explanations—answers to one question or information from other sources that explain answers to other questions or that put those answers in perspective.

In qualitative analysis, words are the evidence that are used in lieu of statistics to present the findings, so a part of your search through your interview texts will be devoted to locating the quotes that best illustrate typical and atypical responses and that best explain patterns of responses for each of your subjects and similarities and differences in responses and patterns of responses across interview subjects.

Box 7.6. *Creating a dummy variable*

A dummy variable is a categorical variable recoded to assign the values 0 and 1 for each category so that 1 indicates the presence of the category and 0 represents its absence. Using this procedure makes it possible to analyze nominal variables using statistics such as regression analysis, which are designed for use with higher-level measurement.

When creating dummy variables, the rule is to *create one fewer than the total number of categories used in the original measure.*

1. If a categorical variable has **only two categories** as in the case of *sex* (male, female), a dummy variable assigns 0 to one category and 1 to the other.
   Example: Male = 0
              Female = 1
   This dummy variable represents female/nonfemale (or presence and absence of femaleness) as opposed to male/female as it did in its original form.
   Note: The mean of the variable now equals the proportion of females in the sample.

2. If a categorical variable has **more than two categories,** more than one dummy variable must be created.
   Example: Race is often coded to include three categories: white, black, and other. To transform race into a dummy variable, we could recode the categories as follows:
   White: transformed into white/nonwhite with white being coded 1 and all others (black and others) coded 0
   Black: transformed into black/nonblack with black being coded 1 and all others (white and others) coded 0

We do not create a dummy variable for the other category because this group is already represented by the two dummy variables that have been created because people in this category are the only subjects who are coded as 0 on both dummy variables.

However, in performing this search for patterns and explanations, the key to a good analysis is to use all the evidence. This means:

- Examining the text or notes from every interview. Every subject's responses are equally important, even when working with large random samples. Each subject's response is even more important when working with samples that were drawn using nonprobability techniques.
- Examining all of each document. Information relevant to one question may sometimes be found as part of a subject's answer to another question.
- Using the same criteria to include or exclude answers that shed light on key research questions. Inconsistencies can easily skew the findings.
- Searching as diligently for atypical answers and countervailing evidence as for information that is typical and/or that supports a hypothesis. In the absence of tests of statistical significance, care and caution are the only safeguards against drawing erroneous conclusions.

Main Points

- For a true survey, researchers gather information from an entire population or from a large probability sample of subjects by using a questionnaire consisting of a fixed set of questions arranged in a fixed order.
- Although surveys raise fewer ethical concerns than most other kinds of human-subject research, ethical survey researchers follow American Association for Public Opinion Research (AAPOR) standards for conducting surveys and disseminating the findings. These standards require pollsters to obtain informed consent from survey respondents and to take appropriate steps to ensure that participation is voluntary. Standards also require respecting subjects' right to privacy and informing them of any known risks.
- Other ethical concerns involve sending data collectors into the field to collect data through face-to-face interviews and the use of survey techniques by those with ulterior motives.
- True surveys are the easiest, most cost-effective way to collect large amounts of high-quality data.
- When done properly, surveys provide data that are reliable. However, any deviation from set procedures will cause reliability problems.
- Because subjects for a true survey are either an entire population or a large probability sample, findings can be generalized to the population of interest. However, external validity will suffer if response rates are low.
- Strategies for improving response rates include using callbacks for researcher-administered surveys and follow-up reminders for mail and Internet ones and offering token gifts or monetary rewards for participating.
- Although the measurement form of internal validity should be very good, problems with question wording and/or question order can decrease reliability and internal validity.
- Surveys usually resemble a one-shot case study pre-experimental design. Therefore, they cannot provide the kind of control for time-order, history, or maturation effects needed as evidence for a cause-and-effect relationship.
- To approximate a true experimental design, researchers can conduct cross-cultural or longitudinal surveys.
- Cross-cultural surveys require conducting separate surveys simultaneously in several different locations or with several different populations. Doing this improves validity by adding a comparative element; however, cross-cultural designs cannot control for maturation or time order.
- To study change over time or to draw causal inferences, researchers conduct longitudinal surveys. They may also perform a cohort analysis.
- There are two main types of longitudinal survey designs: trend studies and panel studies. Trend studies involve conducting multiple surveys over a period of months or years using a new sample drawn from the same population each time. Panel studies involve drawing a single sample and then surveying those same subjects at several different points in time.
- For a cohort analysis, researchers collect survey data at several different times but analyze it by examining scores from a subset of subjects of similar age to see whether they change as a subset, or cohort, ages.
- With any of the survey designs, researchers can collect data in many ways. For a researcher-administered survey, they can collect data through face-to-face inter-

views or by phone. Self-administered options include mail, disk-by-mail, and Internet surveys as well as group-administered ones.
- A well-designed questionnaire contributes greatly to the reliability and validity of survey research. Questionnaires should be visually appealing, easy to use, and as short as possible. They must include instructions and other helps so people can complete them easily and accurately.
- Any of the measures discussed in Chapter 5 can be adapted as questions for a survey. Some kinds of questions are also specific to survey research. These include filter questions, screeners, contingency questions, probes, sleeper questions, and check questions.
- All questions should be related to the purpose of the survey, arranged in a logical, appropriate order, and worded so neither the questions nor the response options bias the results.
- Data gathered from true surveys are almost always analyzed quantitatively using statistical techniques. The statistical analysis usually begins with an examination of frequencies and univariate statistics such as the mean, median, and mode. Survey researchers usually use measures of association to examine bivariate relationships and regression analysis to examine multivariate relationships.
- Qualitative analysis is most often used with data gathered from quasi-surveys. In contrast to the true survey, quasi-surveys are used to study relatively small numbers of subjects, who may be chosen using nonprobability techniques. Instead of using a fixed questionnaire, researchers may conduct in-depth interviews.
- Qualitative analysis usually follows the following steps: (1) transcription of interviews, (2) identification of common responses to key questions and other answers that are less common, and (3) identification of information that helps explain similarities and differences in responses to individual questions.

Terms to Know

true survey
segmentation research
psychographic research
push poll
AAPOR standards
response rate
cross-cultural survey
longitudinal survey
trend study
panel study
cohort analysis
quasi-survey
researcher-administered survey
face-to-face interview
telephone survey
self-administered survey
mail survey
disk-by-mail survey
Internet survey

fixed-response question
open-ended question
screener
filter question
contingency question
probe
sleeper question
check question
leading question
double-barreled question
double-bind question
basic instruction
procedural instruction
quantitative analysis
univariate statistics
frequency
bivariate statistics
cross-tabulation table
measure of association

group-administered survey
protocol
questionnaire
qualitative analysis

multivariate statistics
partial correlation
regression

Questions for Discussion and Review

1. Discuss the advantages and disadvantages of surveys in relation to experiments.
2. Discuss the advantages and disadvantages of (a) true surveys and quasi-surveys; (b) a single survey, cross-sectional survey, and longitudinal survey; (c) trend studies and panel studies; (d) researcher-administered, self-administered, and group-administered surveys; and (e) face-to-face, telephone, mail, disk-by-mail, and Internet administration techniques. Tell when each might be appropriate and when it would be inappropriate.
3. Create a hypothesis that could best be tested by (a) conducting a cross-cultural survey; (b) using a panel design; (c) doing a cohort analysis.
4. Examine the survey questionnaire in Appendix C, and identify the following types of questions: filter, contingency, and probe. Also identify any basic and procedural instructions.
5. Prepare a report on the content of each of these Internet sites. In it, tell how you might use each to learn more about survey research. Be sure to include information on the organization that created and/or maintains each site.
   a. American Statistical Association Research Methods Section http://www.minority.unc.edu/~kalsbeek/asa/srms.html
   b. Telephone Sampling Questions and Answers http://srlweb.berkley.edu:4229/res/tsamp.html
   c. Internet Sites Related to Survey Research http://wrl.uic.edu/srllink/srllink.htm#Ethics
   d. Centre for Applied Social Surveys http://www.scpr.ac.uk/cass
   e. The Question Factory http://www.erols.com/bainbri/qf.htm
6. Go to the Code of Professional Ethics and Practices section of the American Association for Public Opinion Research (AAPOR) website (http://www.aapor.org) and examine the standards for reporting survey data. Then find a newspaper or magazine article that reports on the findings from a survey or a poll. Prepare a brief report telling how well the article meets the AAPOR standards for reporting survey research. Tell what information, if any, should be added to the article in order for it to conform to AAPOR standards. Explain how including that missing information would help readers understand the survey results.
7. Find two articles in communication journals in which the authors use quantitative data analysis to analyze the results of a survey. For each one, (a) describe its purpose and tell what research questions the researchers sought to answer or what hypotheses they tested; (b) describe the survey design and data-collection technique; (c) identify the population, sampling procedure, sample size, and response rate; (d) identify the concepts and the questions used to operationalize them; (e) identify the univariate, bivariate, and/or multivariate statistics used to analyze the results; (f) describe any steps the authors took to ensure or improve the reliability and internal and external validity of the work; (g) explain how

well the research and reporting appear to conform to AAPOR standards; and (h) tell what, if anything, might realistically have been done to improve the studies.

8. As a research director in the marketing and public relations division of the XYZ company, you are expected to help develop a new advertising campaign to improve public awareness of the company's products and services. But before doing that, you need to find out what people know about the company and how they feel about the company and its products and services. You decide that some kind of survey is the best method for collecting the necessary information. Design that study. You may assume anything reasonable about the nature of the company and its products or services. But you must (a) develop three research questions or hypotheses to guide your research; (b) tell what population you will study, how many subjects you will need, and how you will select them; (c) explain whether you will do a mail survey, disk-by-mail survey, or Internet survey or whether you will interview subjects face to face or by phone; and (d) tell how you will analyze your data. Justify your decisions.

9. Regardless of the kind of survey you designed for question 8, prepare an introductory cover letter and a questionnaire suitable for use in a mail survey. Your questionnaire need be no longer than two or three pages, but your questions must provide the information you will need to answer the research questions or test the hypotheses you created in question 8. Your questionnaire must also include several demographic questions, at least one open-ended question, and at least one filter question or contingency question. Also be sure to include instructions and transitions wherever they are needed. Pretest your questionnaire with at least three different people and make any revisions that are needed to address any problems or concerns raised during pretesting.

10. Go to the website for the General Social Survey (GSS) located at http://www.icpsr.umich.gss, access the data archive, and extract the variables NEWS, LOCVOTE, POLEFF13, and GOVERNOR. First, specify three hypotheses you could test using those variables. Then, using the data-analysis program of your choice:
    a. Run frequencies and measures of central tendency for each variable. Describe the frequency distribution for each variable, tell which measure of central tendency is most appropriate for each variable, and explain what the value for that statistic tells you.
    b. Use the cross-tabulation procedure and measures of association to test your hypotheses. Interpret your output. Tell which measure or measures of association are appropriate for testing each hypothesis. Using the information provided by those measures, describe in words the strength and nature of the relationship between your variables, tell whether the relationship is statistically significant, and explain what that means. Use data from the cross-tabulation table to help clarify and explain your findings.
    c. Choose one of your hypotheses, and recode one of the variables into fewer, but still meaningful, categories. Rerun the cross-tabulation and measures of association, and compare the new results to the results you obtained in part b.

11. Go to the GSS website at http://www.icpsr.umich.gss and examine the GSS variable RADIOHRS. Then find at least three independent variables that you believe will help explain the variation on RADIOHRS. Extract the variables and regress your three independent variables on RADIOHRS. Interpret the output.

Readings for Discussion

Backen, C. M., McLaughlin, M. M. and Garcia, S. S. (1999). Assessing the role of gender in college students' evaluation of faculty. *Communication Education,* 48(3):193–210. The authors use MANOVA to analyze survey data from a convenience sample of students.

Lindstrom, P. B. (1997). The Internet: Nielsen's longitudinal research on behavior changes in use of this counterintuitive medium. *Journal of Media Economics,* 10(2):35–40. This is a secondary analysis of longitudinal survey data.

Michelson, M. R. (1998). Explorations in public opinion—presidential power linkages: Congressional action on unpopular foreign agreements. *Political Communication,* 15(1):63–82. The author combines voting records with pre-existing survey data, which she analyzes separately according to whether the survey questions were good or bad.

Peiser, W. (2000). Cohort trends in media use in the United States. *Mass Communication and Society,* 3(2–3):185–206. The author uses pre-existing survey data to track media use in the United States and compare the trend to that found in his previous research in Germany.

Sudman, S. (1983). The network polls: A critical review. *Public Opinion Quarterly,* 42(3):490–496. This one is a bit dated, but the author makes some good points about surveys conducted by or for the mass media.

Supovitz, J. A. (1999). Surveying through cyberspace. *American Journal of Evaluation,* 20(2):251–263. The author offers eight lessons from the perspective of an evaluator.

Weir, T. (1999). Innovators or news hounds? A study of early adopters of the electronic newspaper. *Newspaper Research Journal,* 20(4):62–81. This study is based on separate surveys: one conducted by telephone and one posted on the Internet.

Wright, D. L., Aquiline, W. S. and Supple, A. J. (1998). A comparison of computer-assisted and paper-and-pencil self-administered questionnaires in a survey of smoking and alcohol and drug use. *Public Opinion Quarterly,* 62(3):331–353. The authors combine surveys to create an experimental design.

References and Resources

Blalock, H. M. (1964). *Causal inference in nonexperimental research.* New York: W. W. Norton. In this classic, the noted researcher explains a statistical technique for drawing causal inferences from data collected at a single point in time.

Call, R. A., Otto, L. B. and Spenner, K. I. (1982). *Tracking respondents: A multimethod approach.* Lexington, MA: Lexington Books. This work provides helpful advice for those who need to keep track of subjects for panel studies.

Davis, J. J. (1997). *Advertising research: Theory and practice.* Upper Saddle River, NJ: Prentice Hall, pp. 441–486. These pages provide good directions for conducting, using, and evaluating segmentation research to target communications.

Dillman, D. A. (2000). *Mail and internet surveys: The tailored design method.* New York: J. Wiley. This is the best available source of information on designing and conducting Internet surveys.

Fink. A., Bourqe, L. B., Fielder, E. P., Frey, J. H., Oishi, S. M. and Litwin, M. S. (1995). *The survey kit.* Newbury Park, CA: Sage. This set of books, with varying authors, covers all the basics. Titles include *A survey handbook; How to ask survey questions; How to conduct self-administered and mail surveys; How to conduct interviews by telephone and in person; How to design surveys; How to sample in surveys; How to measure survey reliability and validity; How to analyze survey data;* and *How to report on surveys.*

Fowler, F. (1993). *Survey research methods,* 2nd ed. Newbury Park, CA: Sage. This standard work covers all the basics.

Fowler, F. J. and Mangione, T. W. (1990). *Standardized survey interviewing: Minimizing interviewer-related error.* Newbury Park: Sage. This book covers methods for creating questions and for hiring, training, and supervising interviewers.

Kvale, S. (1996). *InterViews: An introduction to qualitative research interviewing.* Thousand Oaks, CA: Sage. Chapters on analyzing data are especially good.

Magnusson, D., Bergmann, L., Rudinger, G. and Torestad, B. (1991). *Problems and methods in longitudinal research.* Cambridge: Cambridge University Press. This is a good source of information on using surveys to collect data over extended time periods.

Patten, M. L. (1998). *Questionnaire research: A practical guide.* Los Angeles: Pyrczak Publishing. This workbook provides thorough instructions for writing questions, creating questionnaires, and then conducting a survey.

Rea, L. and Parker, R. (1997). *Designing and conducting survey research.* Chicago: American Marketing Association. This practical text covers all the basics. Examples and exercises focus on marketing research.

Rubin, H. J. and Rubin, I. S. (1995). *Qualitative interviewing: The art of hearing data.* Thousand Oaks, CA: Sage. This is a thorough overview of methods for collecting, analyzing, and interpreting data from in-depth interviews.

Scott, J. (2000). *Social network analysis: A handbook.* Thousand Oaks, CA: Sage. This book outlines the theoretical basis and key techniques for network analysis.

Stanford, S. (1989). Statistical designs in survey research. In G. H. Stempel III and B. H. Westley, eds., *Research methods in mass communication,* pp. 173–199. Englewood Cliffs, NJ: Prentice-Hall. The author gives a fairly sophisticated overview of quantitative data-analysis procedures.

Tanur, J. M., ed. (1992). *Questions about questions: Inquiries into the cognitive basis of surveys.* New York: Russell Sage Foundation. Chapters critically examine problematic features of self-report measures used in survey research and the influence of data-collection techniques on them.

Tourangeau, R., Rips, L. J. and Rasinski, K. (2000). *The psychology of survey response.* Cambridge: Cambridge University Press. This exercise in theory building also offers insight into how people understand and answer questions and on how modes of administration and data analysis can affect results.

Traugott, M. W. and Lavrakas, P. J. (2000). *The voter's guide to election polls,* 2nd ed. New York: Chatham House. Two expert survey researchers use an easy-to-read question-answer format to tell everything you ever wanted to know about political polling.

Trumbo, C. (1995, August). Longitudinal modeling of public issues. *Journalism and Mass Communication Monographs, 152*:1–53. This is a good source of information on analyzing longitudinal survey data.

Wilhoit, G. C. and Weaver, D. H. (1990). *Newsroom guide to polls and surveys.* Bloomington, IN: Indiana University Press. This little book is intended as a crash course in survey research for media professionals.

8

Content Analysis

CONTENT ANALYSIS REFERS to procedures for studying the content and meaning of messages. Conceptually, it can be considered a survey of documents, broadly defined.

Documents suitable for content analysis may be almost anything that is in a format stable enough to allow you to scrutinize it carefully. They may be books, magazines, newspapers, reports, minutes of meetings, letters, diaries, web pages and other Internet content, transcripts of conversations or interviews, movies, television or radio programs, music, photographs, art work, exhibits, or displays.

Like the traditional survey, a content analysis is systematic and objective. In its classic form, it is also quantitative. But unlike the survey, the data come from archival records—the documents pre-exist. With few exceptions, they were produced prior to the start of the research project and for unrelated reasons. And, unlike the survey researcher, content analysts both ask the questions of the documents and answer on their behalf by recording information about their content. To do this, they use coding sheets and a code book that are analogous to the questionnaire and accompanying instructions used with surveys.

Because of their similarity to a survey, content analyses have most of the strengths and weaknesses of survey research. Like surveys, a well-done content analysis will produce findings that are reliable and valid.

Creating and then following defined procedures makes content analysis very reliable. By following the same procedures, another researcher should get the exact same results from a content analysis even if idiosyncrasies in creating the coding schemes may sometimes raise questions about just what one is measuring.

With care in choosing documents and creating measures, content analysis will also score high on internal and external validity. Researchers will, of course, make some subjective choices in deciding which documents to analyze and in creating coding procedures. But this is no different from the decision making that goes into choosing subjects and creating measures for any other kind of research. As with surveys, developing and using defensible measures and procedures for selecting subjects for study minimizes the possibility that a researcher's personal idiosyncrasies will affect the findings.

But content analyses also have strengths and weaknesses that are method specific.

Content analysis is the only method that allows researchers to study the past. By using media accounts, diaries, or other archival records, they can get a sense of life in the 1700s as easily as they can learn about issues and problems that people living today are concerned about. And, as with survey research, their studies know no geographic bounds. They can as easily study documents from other countries as they can study ones produced in their hometown.

There are no history or maturation effects even if analysis is interrupted or proceeds over a long period of time. This makes content analysis very convenient. Because the documents are in a fixed format, the work can be done any place and at any time that suits the investigator. And because direct contact with other people is not a factor, reactivity effects are nonexistent. The method is safe. Ethical problems tend to be minimal.

But even with computer assistance, content analysis tends to be time consuming and labor intensive. This can make it expensive if researchers must be hired specifically for the project. Otherwise, the only out-of-pocket expenses are for producing the code sheets and code books and perhaps some costs for acquiring or photocopying documents.

Used alone, content analysis is not sufficient for making claims about motives, the meaning people will derive from messages or message effects. It is, however, an appropriate first step in studying them. Without firm knowledge that a message exists in a particular medium, it is almost impossible to provide convincing evidence that attention to that medium causes an associated behavior. But with evidence of message content and style in hand, the researcher is in a strong position to conduct follow-up studies to find out how and why a message came to be the way it is or what people do with or because of that message.

Communication Applications

Although not as widely used as survey research, content analysis has many uses in applied research.

- Advocacy agencies use content analysis to monitor media portrayals of issues and groups and then use that information to promote their interests.
- Legislators at all levels of government use their own content analyses and those conducted by others to set policy and justify new laws.
- Media organizations use content analyses to refute charges of bias, identify lapses in coverage, improve representativeness, and make other adjustments in news coverage or program content.
- Journalists use content analysis to report on societal trends and to examine the performance over time of businesses and government agencies.
- In advertising and marketing, content analysis is frequently used to monitor competitors' strategies and to detect changes in advertising practices over time.
- Businesses and nonprofit organizations use content analysis to evaluate the work of their public relations and public information staffs by examining the way media outlets use news releases and other information from their organizations. They also use it as an audit tool to evaluate their public image or monitor the environment.

Content analysis is the basis for historical and legal research. By studying records from the past, an organization facing problems can often determine what went wrong and what might or might not work to remedy the situation. By examining the legislative record and court decisions, an organization faced with a lawsuit or considering whether to file one of its own can decide how to proceed.

Ethical Concerns

Most documents of interest to media researchers are either government records that are open under state or federal law or they are books, media accounts, exhibits, or displays that are readily available to the general public. Ordinarily these kinds of publicly available documents may be subjected to content analysis without obtaining consent from their creators. Problems would occur only in those rare instances when researchers include so much of an original document in a final report that they exceed the limits of fair use under copyright law.

Privately held documents are a different matter. If documents are not generally accessible to the public, the researcher will need to negotiate access, perhaps pay for that access, and then obtain informed consent.

Ideally informed consent for privately held documents such as corporate records, correspondence, and diaries should come from the custodian of the documents and also from those who created them or are their subjects. However, in some cases such as examinations of correspondence or of customer or client records, obtaining informed consent from both those people who produced the documents and those who are their custodian's or their subjects may be impractical or even impossible. In other cases, the custodian's rights may override the interests of the creators or subjects of documents as would usually be the case in studying, with organizational approval, employee e-mail or phone conversations. In cases where informed consent from all parties is unnecessary or cannot be obtained, you need to take precautions to protect authors and subjects from unreasonable risk of harm or embarrassment. This may require eliminating names or other identifiers or creating pseudonyms.

Types of Content Analysis

Purists will insist that, in addition to being systematic and objective, a content analysis must be quantitative. That is, message content and characteristics must be converted to numbers that, in turn, can be submitted to a statistical analysis. But certain variants do not neatly fit the standard model. These include readability studies, which are quantitative but do not examine message content per se, and textual analyses that do analyze messages but do so in a purely qualitative way.

Readability Studies

In contrast to a traditional content analysis, which is commonly understood as a method for examining and describing the characteristics and content of messages, readability studies are designed to draw inferences from writing style about whether people can understand the message. This kind of research is most common in the field of education, where book publishers often insist that textbooks intended for use in elementary schools and high schools conform to grade-level standards, but it can be conducted on any textual material to determine whether it is suitable for an intended audience.

The most common techniques used in this kind of research are Rudolph Flesch's Reading Ease Formula or the Fog Index created by Robert Gunning. With both of these, readability is based on the number of short words and short sentences in sample passages. But in some cases the assumption that short sentences and short words will be easier to understand than long sentences and long words may be unwarranted. One

> **Box 8.1.** *Two methods for measuring readability*
>
> **Rudolf Flesch's Reading Ease Formula**
>
> In a passage of 100 words:
> 1. Determine the average sentence length and multiply it by 1.015.
> 2. Count the number of syllables and multiply by 0.846.
> 3. Add the numbers you calculated in Steps 1 and 2.
> 4. Subtract the number you calculated in Step 3 from 206.835. Use this table to interpret reading ease:
>
> | Score | Ease | Grade Level |
> |--------|-----------------|--------------------------|
> | 90–100 | Very easy | 4th Grade |
> | 80–89 | Easy | 5th Grade |
> | 70–79 | Fairly easy | 6th Grade |
> | 60–69 | Standard | 7th or 8th Grade |
> | 50–59 | Fairly difficult| Some high school |
> | 30–49 | Difficult | High school/some college |
> | 0–29 | Very difficult | College |
>
> **Robert Gunning's Fog Index**
>
> In a passage of 100 words:
> 1. Calculate the average sentence length.
> 2. Count the number of words of three syllables or more. However, do not include proper nouns, verb forms that become three syllables by adding -ed or -es (e.g., created), or words of three syllables or more that were created by combining one- or two-syllable words (e.g., bookkeeper).
> 3. Calculate the Fog Index by adding the average sentence length you calculated in Step 1 and the number of three-syllable words you counted for Step 2.
> 4. Multiply the number you obtained in Step 3 by 0.4. This number represents the years of education a reader needs in order to understand the passage.

way to alleviate this problem is to use the formula developed by Edgar Dale and Jeanne Chall that takes into account the proportion of common words in a passage. Another is to use the **Cloze** procedure, which involves recruiting subjects to read passages that have every k^{th} word in the text replaced by a blank. In the Cloze procedure, reading ease is based on the proportion of subjects who fill in the blanks correctly.

Textual Analysis

Textual analysis is any kind of qualitative examination of message content and characteristics, but more narrowly it refers to a type of analysis that combines aspects of social science research with features derived from the humanities. These methods are better at revealing the **latent,** or below-the-surface, meaning of messages than is the traditional quantitative content analysis. Quantitative content analysis is better suited to describing the **manifest,** or obvious and perhaps more literal, meaning of messages.

A **discourse analysis** focuses on the structure and function of messages and/or the ways in which people use language or interact with each other to make sense of their world. **Rhetorical criticism** uses standards of excellence to interpret and evaluate messages. Those standards are most often borrowed from work by Aristotle or Kenneth Burke. In the **critical studies** approach, the emphasis is on "reading between the lines" to uncover or deconstruct the latent, hidden, or "real" meaning of messages and then evaluate or criticize messengers' motives and/or message effects. Postmodernism, Marxist, and feminist scholarship are variants of this way of studying messages.

Textual analysis is more common in academic and theoretical work than in applied research, but the techniques do have practical applications. Discourse analysis, for example, can be a useful tool for uncovering detrimental communication patterns and then correcting them in dysfunctional families. Rhetorical analysis provides a way to discover why some speakers are persuasive and others are not. Critical studies offer insight into the implications of communication for culture and of the ways in which culture affects communication.

Choosing Documents

The process of choosing documents for analysis begins with identifying the relevant population. In content analysis, "population" refers to the set of all relevant documents. Specifying this population requires identifying:

- the institutional source for documents
- the time period of interest for the study
- the individual items that will actually be analyzed. These items for analysis are called the **sampling units.**

Depending on the purpose of the study, the sources of interest could be specific media such as newspapers, government agencies, businesses, or nonprofit agencies. The time period might be just a few days, an entire year, or a number of years. Items for analysis could be all issues of a publication, all articles or programs, all minutes, or customer files from each institutional source during the specified time period or just a subset of them. Therefore, the actual selection process is usually more complicated than it is for survey research. For example:

- Researchers hired by an advocacy group concerned about the media image of minorities might define the relevant population as "all stories mentioning a member of a minority group aired on network newscasts during the last twelve months." In this case, there would be purposive selection of the institutional sources. To locate the stories for analysis, the researcher could use the population of all newscasts aired on ABC, CBS, and NBC during the year and then examine all of the stories on those newscasts that mention minorities or just a subset of stories from all network newscasts. But the researchers might also select a sample of newscasts from each of the networks and then examine only the stories that ran on the selected dates.
- To prepare a report for the city council on issues local governments face today, researchers might decide the relevant information could be found in minutes of city council meetings. To choose the cities, they might draw a probability sample

stratified by population or choose to study minutes just from cities of a size similar to that of the one that commissioned the study. If there were just a few such cities, the researchers might examine minutes from all of the meetings in all of the cities. If there were many similarly sized cities, the researchers might randomly select cities for inclusion in the study or select just half a dozen that are most like the city commissioning the study. Regardless of how the researchers select the cities, they would then have to decide whether to look at minutes from just the past year or whether to study the minutes from several years and then how many years and which ones. They would also have to decide whether to study all minutes from the selected year(s) or just a subset of them and, if so, how to choose the subset.

Some steps in selecting the population of items for analysis may be combined or bypassed. But as the preceding examples suggest, choosing the items for a content analysis usually involves several steps. In each step, a researcher might choose a population or a sample; the selection could involve probability or nonprobability sampling. The decisions will be based on a number of factors including the purpose of the study and time and budgetary constraints. But it also depends on whether external validity is important.

As with all research methods, the ability to generalize findings depends on sampling procedures. Because most content analyses involve several samplings using both probability and nonprobability techniques, the ability to generalize and, if so, to which population or populations depends on the stage or stages at which each kind of sampling occurred.

If you select your document source for convenience, you will never be able to generalize to any other source; you can, however, generalize your findings to that source if you use probability techniques to select the items you actually analyze. But you can do that only for the year or years included in your sampling frame. If, for example, you randomly select items from your sources at five-year intervals, you can assess trends over time; however, you cannot draw firm conclusions about document content in years not analyzed.

In choosing a sampling procedure, the general principles for choosing subjects outlined in Chapter 4 apply. But there are also some problems that occur only when you sample documents. These problems arise from the use of indexes or databases as the sampling frame and from the existence of periodicity within document content.

Indexes and Databases

If your research project requires you to analyze documents on just one or a few subjects, locating those documents from within a much larger population of documents can be a time-consuming task. Therefore, it can be tempting to look for an index and then select for your study those items listed under relevant headings. Or you might look for a computerized, searchable database and let its search engine comb through the listing of documents and select ones that include certain key words.

Both indexes and computerized, searchable databases are widely available. Major newspapers, magazines, and journals publish annual indexes. In some cases, these are available online or on disk or CD-ROM. Some even publish abstracts or the full text of articles that you can search using a predetermined list of keywords or even keywords you provide. Of the available indexes, *Readers' Guide to Periodic Literature* is widely used to locate articles from a number of magazines. *Lexis/Nexis* lets you search by keyword for articles from major newspapers, news magazines, and network television newscasts.

Using an index or searching a database will almost certainly be faster than manually searching through a large population of documents. But some problems that may not be readily apparent may crop up.

Indexes and databases vary in quality. Some are much more comprehensive than others. Some newspapers index every article; others index only the articles in the first section. Both *Readers' Guide to Periodic Literature* and *Lexis/Nexis* leave out many media that you might be interested in. Computerized databases may also be available in different versions, each covering a somewhat different set of publications and/or documents.

With both indexes and databases, variations in indexing procedures and in selecting keywords can cause internal validity problems in longitudinal studies. Even when you are looking for documents from just one source during a single year, there may be problems of over- or underrepresentation. Index entries, keyword searches, and searches based just on an abstract can easily fail to identify documents that properly should have been included within a set of documents for analysis. Computerized full-text searches that identify items based on the occurrence of a word anywhere within a document will miss documents where a synonym was used rather than the selected keyword. More often they will produce a list with many items that are, at best, tangentially relevant because they cannot take into account the context in which words with multiple meanings are used.

Therefore, use indexes or computerized searches with caution. Deciding whether to rely on them must be on a case-by-case basis. Searching manually for relevant documents is definitely more time consuming, but it is advisable whenever you are uncertain of how published indexes or computerized search programs created their database or how their search engines function. Manual searching is also advisable if your analysis requires locating really rare items. And, of course, a manual search will be the only way to locate the documents you need if there is no index or computerized database to search.

Periodicity

If the purpose of your study is to draw generalizations about the content or other characteristics of documents from one or more sources for an entire year, you could analyze the population of all documents from the relevant source or sources for each of the years of interest. But in many cases that population will be so large that you will want to sample.

You could sample by drawing every k^{th} document from the set of all documents. But imagine doing that by going through every page of a newspaper for an entire year just so you could select stories. Or imagine doing it with all of the customer correspondence files from a major insurance company.

Although systematic random sampling will work, most researchers find it easier to sample dates and then analyze all of the documents for those dates. But whenever you sample dates, history creates a problem unique to content analysis.

Just as history—the time when a study is conducted—may have some bearing on survey findings, time may have some bearing in content analysis. But here, "time" does not refer to when you actually do the analysis. It refers to the period when the documents were created, published, or received.

With documents, content will almost certainly vary from year to year. In most cases, it will also vary by season, month, or day.

In newspapers and on television newscasts, feature stories are more common on weekends than on weekdays; education news will be heaviest at the beginning and end of the school year; stories about travel and outdoor activities will be more numerous in most areas during the summer than in winter. Special sections or segments such as ones devoted to religion, science, or business will be published or aired on some days but not others, and, of course, media content will be different on the days surrounding major holidays than it will be at other times.

The same kind of variations occur in documents produced or received by government, businesses, and other organizations. At some times of the year, fiscal matters will receive heavy emphasis; at other times the emphasis may be on project development or oversight. Seasonal variations in sales may affect the amount and kind of correspondence or phone calls a business receives. A casualty insurance company, for example, may get more claims and therefore more complaints related to losses from fire during very cold winter months and dry summer ones than in spring and fall.

You could choose a set of consecutive dates, but that strategy will never produce a sample that allows generalizing to any longer time period. Choosing consecutive dates is appropriate only for analyzing coverage after some event. It would, for example, be appropriate to study media coverage of a company for 30 days following a major product recall.

And with only 365 days (52 weeks or 12 months) to choose from, the usual procedures for creating a random sample won't work. Not only might your sample produce too many items from a particular day or season, but your procedure would violate the requirement that each element in your sample have the same, known chance of being included.

If your interest is in relatively rare content or if you need to study documents published in monthly publications, you will have to look through all issues to find them. There simply is no way to sample.

But through extensive work with newspapers and magazines, the research team of Daniel Riffe, Stephen Lacy, and Frederick Fico (1998) found that you can avoid periodicity problems and still sample dates if you need to study common content or other common characteristics of documents produced weekly or daily.

If your interest is in commonly occurring features in weekly publications, you can draw a representative sample by randomly selecting one issue from each month plus one additional issue selected randomly from the entire year. For daily publications you can draw a sample consisting of just two constructed weeks if you are interested in common features. You will need three or four constructed weeks for somewhat less-common content or features. For nightly television newscasts, you can also randomly select two days from each month.

To create a **constructed week,** draw a probability sample of dates stratified by day of the week. That is, for each constructed week, choose one Sunday, one Monday, one Tuesday, and so on randomly from all of the Sundays, Mondays, Tuesdays, and so on in the year.

Although these recommended sampling procedures are based on extensive studies with news media, there is little reason to think the same procedures would not work with other kinds of weekly or daily documents. But even when sampling television newscasts, news magazines, newspapers, or newsletters, it's a good idea to check your sample. In sampling newspapers, you might find you have too many days from a single month. And with such a small number of dates in your sample, it is quite possible to end up with too many holidays.

For studies covering a number of years, it is unclear whether it is better to use the same dates for each year or to draw a new sample of dates for each year. Both techniques can work, but either one can present a problem if, for example, it leads to including an Easter issue for one year but not for others.

Measurement in Content Analysis

Measurement in content analysis is the process of noting and recording the characteristics of documents that will be useful for answering the research questions or testing hypotheses. This process involves identifying the appropriate unit of analysis, developing the content and coding categories, and, for quantitative analyses, creating the enumeration system.

Units of Analysis

The sampling unit refers to the documents selected for study. The unit of analysis is what you actually count and assign to a category. The unit of analysis may be the same as the sampling unit, or it may be smaller than the sampling unit, but it can never be larger than the sampling unit. The sampling units might be letters or transcripts of phone calls. That sampling unit—the letter or phone call—could also be the unit of analysis. But the unit of analysis might also be a paragraph, sentence, or word.

The selection of a unit of analysis depends on the purpose of the study. If the purpose of a study of letters or phone calls were to find out why people write or call a company, it would be inefficient to select units smaller than the letter or call. However, if the purpose were to find out whether writers or callers make more negative or positive comments about the company, you could choose the letter or the call as your unit of analysis, but it would also be appropriate to choose a section such as a paragraph or sentence as your unit of analysis. With the document as a whole as the unit of analysis, you could record the total number of negative, neutral, and positive assertions. With the paragraph or sentence as the unit of analysis, you could code each paragraph or sentence as negative, neutral, or positive.

Category Construction

As the examples in the preceding section suggest, units of analysis are the basis for creating your measures. To do this, you first create concept categories for each kind of information you are seeking from each unit of analysis. Then you create response options or coding categories as your measurement device.

The concept categories are analogous to the questions researchers ask when doing survey research. The coding categories provide the "answers" to the "questions" you ask of each unit of analysis.

Your coding scheme may allow for both open-ended answers and fixed responses. Qualitative analysis typically uses the open-ended approach. Answers are found in the words of the document itself. Coding consists primarily of marking or flagging those answers in some way. For quantitative content analysis, open-ended questions are appropriate if the researcher cannot know ahead of time what the answers may be. For example, to record the names of businesses covered in the business section of a newspaper, you could create a short, numbered list of names and then add new names and numbers to it as your coding of items progresses.

For a quantitative content analysis, however, most coding categories will be fixed. That is, you will create a finite list of possible coding categories into which each item must fit. These coding options must provide a place for all possible answers. They must also be mutually exclusive. There should be only one logical place to code each item.

In general, it is better to keep the number of response categories relatively short. However, when in doubt, it is better to specify too many rather than too few options. It is always possible to combine rarely used options during data analysis. Adding categories once data collection begins is much more difficult.

Although coders can more easily become familiar with short lists of response options and use them consistently, most can cope with up to 20 nominal categories such as subjects or themes without much difficulty. They will, however, find it quite difficult to cope with more than five ordinal categories for judgments about quality, direction, or intensity.

Enumeration

In developing the coding scheme for a quantitative content analysis, researchers typically attach numbers to each fixed-response coding category or assign them to each option as it is added to an open-ended list. They use these numbers for statistical analysis.

Numbers attached to each nominal category have no real meaning. They can be assigned to coding categories in any order. However, coders will find it easier to do their work if you number nominal categories alphabetically or in some other logical order.

Because much of the data collected involves simply categorizing document content by type or subject, many of your measures will be at the nominal level. But higher levels of measurement are possible and, in many cases, advisable. With these higher levels, the way you assign numbers to categories is important. You must preassign numbers so they really do correspond to direction, amount, or intensity.

Code Sheets and Code Books

To facilitate data collection, researchers record data for quantitative analysis on standardized **code sheets.** Code sheets list all of the concept categories for which information is needed. They may also list all of the coding categories for each concept category, or they may just provide a place to record the number assigned to the coding category into which a unit of analysis fits.

Concept categories should be arranged for the convenience of the coders. In most cases, this means putting "bookkeeping" concept categories such as the date and type of document first. Concept categories that require coders to examine the entire document or reflect on its content before choosing a response option usually go at the end.

For any content analysis, recognizing that some documents belong in the population or sample will be easy, whereas others may be a judgment call. The same is true in coding. Therefore, developing a **code book** is a necessary part of every content analysis. A good code book makes coders' work easier. It also helps ensure consistency over time and among coders. An example of a code book and code sheets for quantitative and qualitative content analysis can be found in Appendix D.

A code book gives explicit instructions for deciding whether to include a document in the study. It also provides explanations for categories and response options and the rules for choosing among them. For latent content such as whether a message is favorable or unfavorable, decision rules must be especially clear. Therefore, the code book may also include examples to serve as a guide.

Using Computers for Content Analysis

Over the past 25 years or so, computer programs designed specifically for doing a content analysis have become increasingly available. Most of these programs do the work very well, but the work they do can be quite limited.

The simplest programs give you a list of all the words in a document or set of documents and tell you how many times each word was used. These programs tell you nothing about the actual messages, but they are very useful for drawing inferences about authorship or for comparing word choice between or among known authors or document sources.

Some programs have their own lists of words that have been categorized in some way to reflect shared meanings. These dictionary and thesaurus programs work by comparing words in your documents to words in these categorized lists to determine how common each kind of shared meaning is in your documents. These programs can be useful for drawing inferences about direction, intensity, or style. The problem, however, is that, in using them, you may find that the list of words in the dictionary is incomplete or the way the program categorizes words in its list is not entirely appropriate for your study.

Other commonly used programs simply code each document for the presence or absence of keywords. The best ones can treat sets of words such as a first and last name as a single unit. They will also let you use Boolean logic to determine, for example, the joint occurrence of two or more words within a document or to code for the appearance of a word except when another word is present. But even the best of these programs cannot recognize synonyms on their own. You will have to know all the different words authors may use to say the same thing and then enter those words as keywords. But regardless of how many keywords you use, these programs cannot take into account either the manifest or latent meaning attached to the keywords you select. Therefore, if some of your keywords have multiple meanings, real problems can arise.

Still other programs analyze documents for message structure. A few of these programs can analyze music, but most work only with text. Text-based ones and variants of them that calculate readability scores are widely used to help students learn to write better. As a content analysis tool, they can be used to draw inferences about writing quality or authorship.

Like message structure programs that analyze the way notes or words are linked to determine whether they conform to musical conventions or the rules of grammar, mapping programs determine how persons or groups mentioned within a document are related to each other. These programs can, for example, provide information on latent content such as the direction and intensity with which messages flow among participants in a debate or which of several nations dominates in media coverage of trade negotiations among them.

Using an appropriate program will let you bypass much of the work of coding documents and analyzing data. But whether it will really save time is another matter. Each document for analysis must be entered into the computerized content analysis program. If you can do this by scanning the documents, data entry can be fairly quick. But many documents are not scannable. They will have to be entered manually. And, in entering data either by hand or through scanning, you may encounter compatibility problems between the program you are using for data entry and the one you want to use for the content analysis.

Before you select or use a program, you should check to see how it really works and then whether you will encounter any insurmountable problems with incompatible hardware or software. In spite of their problems and limitations, computerized content analysis is an option worth considering. Almost all of the programs will do at least some things very well. Some will let you do things that would be very difficult to do without computer assistance. But in most cases, the real advantage of using a computer program for content analysis is the greater reliability with which a computer can do the work.

Preparing Documents for Qualitative Content Analysis

In the information we have presented so far, our emphasis has been on doing a quantitative content analysis. The directions in this section are for qualitative analysis, but they can be used to facilitate retrieving some qualitative information for use in quantitative content analyses.

Although the procedures for doing a qualitative content analysis are much the same as for a quantitative analysis, there are some differences.

As with quantitative analysis, you will need to prepare a code book or data-collector's guide. This guide often consists of just a list of the concepts or kinds of information you will need to look for in order to answer your research questions or test your hypotheses. But for each entry on the list, you may also need to write in some notes to help you recognize those concepts or other relevant information. You may also want to add some examples as you pretest your list or work your way through the actual coding.

That part is pretty much the same as for preparing the code book for a quantitative analysis. The real difference between collecting data for a quantitative analysis and for a qualitative one is the way you code your data.

For a qualitative analysis, you won't be turning the information in your documents into numbers because you won't be doing any kind of statistical analysis. And you probably won't need or want a coding sheet. But you almost certainly will need to develop some codes that are roughly analogous to the coding categories used in quantitative content analysis.

These codes are usually keywords that indicate the kind of information in a document. If you have a lot of documents, you may want to expand keyword codes to include symbols indicating such things as direction or intensity. You will probably also find it helpful to create some codes you can use to mark passages you may want to paraphrase or quote when you write up your findings.

But qualitative analysis usually follows the inductive model. Coding and data analysis often occur simultaneously, so you will probably find yourself developing or modifying codes as you go along. That means you may have to go back through your documents several times to code and recode them until you finally get all of them coded the way they will need to be coded for your actual data analysis.

If you are using a computer program designed for qualitative analysis, you can enter your keywords into a "keyword" section. The best programs also let you enter them directly into the text. This feature makes it much easier to compare documents and to locate the information or specific passages you want.

If you are not using a computer program, you code your documents by entering keywords or symbols into the margins of your documents next to the appropriate passages. To make it harder to miss documents with each kind of information during data analysis, you can also list those keywords at the top of each document. If the original documents do not have sufficient margins to let you write in those keywords, you can put them on a separate sheet of paper and attach that cover sheet to each document. You can also use different colors, each keyed to a code, to mark or highlight passages to make them easier to locate during data analysis.

Quality Control

To outsiders the results of a content analysis may appear to be subjective. That is a danger whenever the measurement instrument is a human researcher. It can even be a problem with computerized content analysis. But reliability and validity problems can be minimized if researchers take the proper steps to ensure the quality of their work.

Reliability

A lack of consistency in coding is the major threat to reliability. This problem is greatest when selecting documents for the study or choosing appropriate response options based on a document's latent content such as direction or intensity rather than on manifest content. But with humans as the data-collection instrument, problems may occur even with manifest content, such as a document's author, place of origin, or length.

Quite obviously coders might have different opinions about the intensity of a statement or whether it is favorable, neutral, or unfavorable. But one coder might also decide to include all letters signed with a person's full name or some variation of it; another might accept only the writer's full name. One might include the addresses, salutation, and closing in the measure of length; the other might measure just the body of the letter.

The likelihood that different coders will make different coding decisions or even that individual coders will change their minds over time increases greatly when the coding scheme requires making judgments or drawing inferences based on latent content.

Ensuring that all coders will make the same decisions and that individual coders will do their work consistently over time requires first developing clear boundaries for the study and coding categories, then creating a detailed code book and training coders to use it properly, and finally checking all work for reliability.

In the sampling stage, the requirement for clear boundaries requires creating a statement that fully defines and describes the documents of interest and then specific directions for deciding which documents should be included and excluded at each sampling stage. For the coding categories, the meaning of both the categories and of each response option needs to be spelled out. More detail will be needed for making judgments based on latent content than for coding manifest characteristics, but in both cases it is a good idea to provide a description of the category and each coding option. For complex coding schemes, adding examples of relevant and irrelevant items is also helpful.

All of this information must be written down in a code book so that data collectors can refer to it as necessary. Once the initial code book is prepared, it must be tested by using it to make sampling and coding decisions on a set of documents similar to those that will be included in the study. If, at this stage, coders encounter difficulties that cannot be addressed by clarifying the explanations or adding more examples to the code book, the sampling or coding schemes will have to be reworked and the entire code book revised and again tested. After actual data collection begins, any modifications to the scheme will almost inevitably require redoing at least some portion of the work that had already been done.

Once you have devised a workable scheme, everyone who will be involved in the coding will need to be trained to use the scheme correctly. But even the best-trained coders may not be entirely consistent in their work, especially if they work too long at one time or if the work is spread over a number of days or weeks. Therefore, all coding needs to be checked by using a reliability test such as Scott's pi. Directions for calculating Scott's pi are in Chapter 5.

To check for stability over time, at the end of a long coding session, each coder should recode a few items coded early in the session and then calculate the reliability coefficient based on the results of the early and late coding. When coding is split across days or weeks, each coder should recode a few items from the previous session and then calculate the reliability coefficient for decisions made on the two days. Where several coders are working on a project, intercoder reliability will also need to be calculated to make sure they continue to agree with each other as the work progresses.

At the end of the coding, the final inter- or intracoder reliability coefficient must be calculated by randomly selecting a small subset of items, recoding them, and comparing the results to the original coding. This reliability coefficient should be included in any report of findings. As a rule of thumb, most journals expect a minimum reliability coefficient of .9 if the calculation is based on simple agreement or if the calculation is for decisions about simple presence/absence of a word or symbol. With Scott's pi, which factors out chance agreement, coefficients of .7 are generally acceptable.

For truly qualitative studies, these procedures will need to be modified somewhat. It will still be necessary to create a code book, test the scheme, and train coders. But because there will be no numbers attached to the recording of data, reliability testing may take the form of an **audit.** With an audit procedure, an outsider who is familiar with the study purpose but unconnected to the actual project examines a subset of items and certifies coders' decisions as "reasonable" and "acceptable."

Internal Validity

In content analysis, most internal validity problems are measurement problems. For computerized content analysis, you must check the assumptions that went into creating the computer program to make sure it is doing what you think it is doing. For other kinds of content analysis, the key is to create good categories and category response options.

It is usually a good idea to borrow measures from previous studies whenever possible. This will help eliminate biases due to the self-interest of those conducting or sponsoring the research. In most cases measures can be assumed to measure what they are supposed to measure on the basis of face validity. But where time frames and study populations overlap or partially overlap, the availability of two sets of reasonably comparable data will also make it possible to assess measures for concurrent validity. In some cases, calling on outside experts to validate your measures can be a good idea, especially if you are analyzing historical or specialized documents or coding for latent content such as visual cues, humor, or satire.

External Validity

Whether generalizability is important will, of course, depend on the purpose of the study. Where it is important, using appropriate probability sampling techniques is mandatory.

If you randomly sample document sources but use nonprobability techniques to sample items for analysis, you will not be able to generalize your results to those randomly selected document sources. But even in cases where researchers take care to employ appropriate probability sampling techniques, two other factors may affect external validity: duplication of items and missing documents.

Organizations may have duplicate files or duplicate items within them. Television entertainment programs run and rerun; newspapers sometimes accidentally run the same item twice or repeat items in end-of-the-week roundups. Including duplicate items can inflate estimates from content analysis of the frequency, severity, or salience of some occurrence, but eliminating duplicates, particularly from studies of media content, may lead to underestimating coverage or impact. Therefore, there can be no universally applicable rule for how to handle the problem of duplicate coverage. The partial solution is to create explicit rules that reflect the purpose of the study and then take those rules into account in interpreting and reporting findings.

More threatening to external validity is the simple fact that not all documents will survive. Organizations accidentally or purposely lose, misplace, or destroy some records. Many old newspapers, movies, and radio and television programs have not been preserved; with others, some issues or programs may be missing, or their quality may be so bad as to make analysis almost impossible.

There is no way to tell whether missing documents are like or unlike those that survive. In any case, missing documents will lower the number of items included in a study. Therefore, quality control requires taking steps to recover missing documents whenever possible.

Museums, private collectors, and those involved in the making of early television shows or movies are your best bet for recovering apparently lost films, radio or television programs, and some newspapers. For newspapers that have been preserved, the quality of microfilm, microfiche, or microprint versions may vary. If the version obtained on interlibrary loan from one library has a missing issue or an unreadable page, contacting another library may turn up a copy with the missing issue or a more readable page. Missing organization records are the hardest to recover. Some may be available as government records, but for others, contacting long-time employees and former employees on the outside chance they may have some of the missing material is usually your only hope.

Data Analysis

Data from a content analysis can be analyzed using either quantitative or qualitative techniques or a combination of them. Reducing content to numbers can strip documents of their "life." Putting that life back in requires at least some reliance on evidence presented in narrative form. Therefore, making use of some qualitative information can very much improve a quantitative analysis. Used instead of quantitative analysis, it can provide insight beyond what can be obtained from statistical evidence.

Regardless of whether you are doing a quantitative content analysis, a qualitative one, or some combination of the two, the data-analysis process is essentially the same as for a survey. Therefore, at this point it may be helpful to review the data-analysis section in Chapter 7.

Quantitative Analysis

Although content analysis is usually considered a quantitative method, statistical tests are used less often than in experimental or survey research. Because many studies do not employ probability sampling techniques to select units for analysis and others are intended primarily to describe content characteristics of documents selected from a single source such as a newspaper, the emphasis is often on the univariate statistics—the frequencies that describe the distribution of content features of each set of items.

To analyze the relationship between two content features within a single source, researchers typically use bivariate measures of association such as chi-square or Pearson's r. Occasionally they use the t-test or analysis of variance to examine differences in content characteristics within a source or between or among sources.

Regardless of the kind of sampling you use, t-tests and ANOVAs can show whether any apparent differences among documents are real. Measures of association will give you some idea of the strength of any relationship between or among variables. However, you can use the accompanying tests of significance to generalize to a larger population only if you used probability sampling techniques to choose your units of analysis. And then you can use them only to make generalizations to the population of documents and to the years from which those units of analysis were selected.

Multivariate statistics are used even less often than bivariate ones. However, at times they are appropriate. Both analysis of variance and regression analysis can be useful for taking control variables or confounds into account. Regression analysis is also useful for making predictions. By creating dummy variables, you could, for example, see whether the topic of a newspaper story predicts its length or placement within the newspaper. By adding measures drawn from other sources, you can also assess the impact of external factors on message content or of messages on people. Adding data on circulation size or number of employees to a content analysis of items drawn from several media, for example, will give you some indication of the effect of resources on media content. Adding survey data, as is done in agenda-setting and cultivation studies, provides some evidence of media effects on public opinion.

Qualitative Analysis

The first step in a qualitative analysis is to examine and reexamine your documents until you are fully familiar with them, their content, and their characteristics. After that, you use keywords or symbols you have attached to your documents to help you figure out what you have found, check your assumptions, and then report your findings.

If you are using a computer program to help manage your data set, you can use it to retrieve all of the documents that you coded as containing each kind of information. Otherwise, the next step is to use the codes to help you sort through the set of all documents and create subsets, each containing the documents having a particular kind of information.

With either method, the next step is to determine the prevalence or frequency of concepts or kinds of information and other relevant content characteristics. A computer program can do this for you, or you can do it manually by going through each subset of documents and making notes about what you find.

After that, you can move on to a search for connections between content characteristics coded using different keywords and then to a search for reasons, explanations, or patterns that help explain the prevalence of message characteristics coded using the same keywords and connections between concepts or message characteristics coded using different keywords.

The best computer programs have features that will facilitate this search for patterns. Some will even help you test hypotheses. But at this point qualitative analysis becomes much more difficult than quantitative analysis. Because you can't use statistics to tell you the strength of a relationship or the probability that a finding is generalizable to some larger population, you will have to build a case for your findings in other ways.

Some of the evidence you will need in order to explain and then support your findings about message characteristics, patterns, and connections will be in the text of the documents themselves. In other cases, you may be able to find the reasons in external factors such as when the document was produced, who produced it, or why it was produced. But sometimes you will have to go beyond the message to find plausible explanations.

Where the quantitative researcher counts every document as equivalent to every other document, the qualitative researcher must use informed judgment to help decide how important each document and each passage within it is and how much it may contribute to finding an answer to a research question or count as evidence to support or refute a hypothesis. Being able to make these kinds of judgments contributes directly to the value of a qualitative analysis. But bringing individual judgments to bear in the analysis process requires you to know much more about your documents and about the historical and social context in which they were produced than you would need to know if you were doing a traditional quantitative content analysis.

Main Points

- Content analysis refers to procedures for systematically studying message characteristics and content.
- Used alone, content analysis cannot provide evidence of messengers' motives or of message effects. However, it is the only research method that allows the researcher to study the past.
- Ethical problems are minimal. Although publicly available documents can be subjected to content analysis without obtaining consent from their creators,

- researchers will need to negotiate access and obtain informed consent in order to use private documents that are not generally accessible to the public.
- In its classic form, content analysis is systematic, objective, and quantitative. Message content and characteristics are converted to numbers that can be subjected to statistical analyses.
- Readability studies and textual analysis are variants of the classic model. Readability studies use features of writing style to draw inferences about whether people can understand a message. They are quantitative but do not really examine message content. Textual analyses analyze messages qualitatively. Types of textual analysis include discourse analysis, rhetorical criticism, and critical studies.
- Quantitative content analysis is best suited to describing manifest content. Qualitative content analysis is better at revealing the latent meaning of messages.
- Choosing documents for content analysis usually requires multistage sampling. In each stage of the document-selection process, researchers may choose a population, a nonprobability sample, or a probability sample. Therefore the ability to generalize the findings from a content analysis depends on the stage or stages at which each kind of sampling occurred.
- Indexes and databases may help locate items for content analysis, but the methods used for indexing and selecting keywords can cause validity problems.
- Document content almost always varies from year to year. Variations may also occur by season, month, or day. Special sampling procedures minimize periodicity problems.
- In content analysis, measurement is the process of noting and recording the characteristics of documents that will be useful for answering research questions or testing hypotheses. The process involves identifying the appropriate unit of analysis, developing content and coding categories, and, for quantitative analyses, creating the enumeration system.
- The unit of analysis refers to what is actually counted and assigned to measurement categories.
- Concept categories in content analysis are analogous to the questions asked by researchers doing survey research. The coding categories provide the "answers" to the "questions" you ask of each unit of analysis. For a quantitative analysis most coding categories will be fixed. For a qualitative analysis, an open-ended approach is more common.
- Lack of consistency in choosing documents and in coding them is the major threat to reliability in content analysis. Reliability is increased by developing clear boundaries for the study and for coding categories, creating a detailed code book, training coders to use the code book properly, and checking all work for consistency in coding. Reliability coefficients such as Scott's pi should be calculated and included in any report of findings.
- Most internal validity problems are measurement problems. Computerized content analysis programs are very reliable, but to minimize validity problems, you should check the assumptions that went into creating the program to make sure it is doing what you think it is doing. For other kinds of content analysis, the key to internal validity is to create good concept and coding categories. Borrowing measures from other studies is also a good idea.
- Using appropriate probability sampling techniques is necessary for a study to be externally valid, but duplicate coverage and missing documents can also create problems.

- Data from content analysis can be analyzed both quantitatively and qualitatively. The data-analysis procedure is essentially the same as it is for surveys.

Terms to Know

readability study
Flesch's Reading Ease Formula
Gunning's Fog Index
Cloze procedure
latent content
manifest content
discourse analysis
rhetorical criticism
critical study
population
sampling unit
unit of analysis

index
database
periodicity
concept category
coding category
enumeration
code sheet
code book
Scott's pi
intercoder reliability
intracoder reliability
audit procedure

Questions for Discussion and Review

1. What are the main threats to reliability, internal validity, and external validity associated with content analysis? What steps can you take to minimize reliability and validity problems? In what ways are the problems and solutions the same as those for experiments and surveys? In what ways do they differ?
2. Why is selecting documents for content analysis often more difficult than selecting a sample in survey research?
3. Use the content analysis website maintained by Georgia State University (http://www.gsu.edu/~wwwcom/content.html) to find information for a report on computer software for content analysis or to prepare a bibliography of new publications and other useful works on content analysis.
4. What effect might using a computer program to select documents or analyze content have on the reliability and internal validity of a content analysis?
5. What are the advantages and disadvantages of quantitative content analysis? of qualitative textual analysis? Give an example of a situation where each might be appropriate and of where each would be inappropriate.
6. Use the Flesch Reading Ease Formula and the Gunning Fog Index to determine the readability of the main article on the front page, business page, and sports page of your local newspaper.
7. Search the communication research literature for a study that uses quantitative content analysis as its principal method of data collection and a second study that uses qualitative textual analysis. For each study (a) identify the research question the study addresses, (b) describe the target population, the sampling units and/or units for analysis and the procedures and rationales for selecting them, (c) identify the key concepts and procedures for measuring or recognizing them, (d) explain the data-analysis procedure(s), and (e) identify any techniques the researchers used to check the reliability and validity of their work. Finally, (f) tell which study's findings seem most credible. Justify your answer.

8. Use the *New York Times* index and a computerized searchable database such as *Lexis/Nexis* to identify articles about the homeless published in the *New York Times* last year. To what extent do you believe that each search has over- or underrepresented articles on this topic? Justify your answer. What would be the advantages and disadvantages of doing a manual search instead of relying on an index or database? What problems with periodicity might you encounter if you wanted to examine just a sample of stories on the homeless? How might you sample to minimize those problems?
9. Review the sample code sheet and code book in Appendix D.
    a. Give reasons why the researcher may have used four constructed weeks as the sample. Explain whether the results based on this sample are generalizable and, if so, to what population they can be generalized.
    b. Use the code book in Appendix D to identify relevant stories from the abstracts found on the Vanderbilt University Television News Archive website: http://tvnews.vanderbilt.edu/eveningnews.html. Note any problems you have and tell how you might modify the directions to alleviate the problems.
    c. Videotape a week's worth of network television news programs, and then use the code book in Appendix D to identify religion news stories and code them. Use Scott's pi to calculate your intracoder reliability and to compare your reliability to that of at least one of your classmates.
    d. Use the same coding scheme to code religion news from either a newspaper or a magazine. What parts of the scheme would have to be modified/changed for use with print media?
    e. Use the code sheet and code book in Appendix D as a guide in creating a code book to use for a study of a different type of news such as business or science.
10. For each of the following studies, tell whether you would do a quantitative or qualitative content analysis. Then (1) identify your target population, sampling unit, and unit of analysis, (2) explain how you would define key concepts such as "problem-solving strategy" and "role," and (3) identify the concept and coding categories you will need for a quantitative content analysis in order to measure or recognize your key concept as you have defined it. Justify your answers.
    a. problem-solving strategies in the movies
    b. changes in the role of men in television situation comedies
    c. the image of African Americans in cartoons
    d. the content of Internet sites on content analysis as a research methodology
    e. prosocial themes in music videos
    f. the effect of a training session intended to improve customer satisfaction with the way an insurance company processes claims
    g. media use of news releases from a state university
    h. bias in newspaper coverage of local high schools
11. Identify several kinds of information from sources other than your content analysis that you might add to your content analysis data for question 10h to help explain any differences you find in the way the newspaper has covered those local high schools.

Readings for Discussion

Andsager, J. L. and Powers, A. (1999). Social or economic concerns: How news and women's magazines framed breast cancer in the 1990s. *Journalism and Mass*

Communication Quarterly, 76(3):531–550. The authors use human coders and computerized content analysis for their time-series analysis of longitudinal data.

Bengston, D. N. and Fan, D. P. (1999). An innovative method for evaluating strategic goals in a public agency: Conservation leadership in the U.S. Forest Service. *Evaluation Review, 23*(1):77–100. The authors use computer-coded content analysis to evaluate attitudes expressed in online news media stories.

Bodle, J. V. (1996). Assessing news quality: A comparison between community and student newspapers. *Journalism and Mass Communication Quarterly, 73*(3): 672–686. The author uses Flesch readability scores as one measure of quality.

Brown, J. D. and Campbell, K. (1986). Race and gender in music videos: The same beat but a different drummer. *Journal of Communication, 36*(1):94–106. The authors analyze visuals.

Buddenbaum, J. M. (1990). Network news coverage of religion. In J. P. Ferré, ed. *Channels of belief: Religion and American commercial television.* Ames, IA: Iowa State University Press, pp. 57–78. Quantitative and qualitative data were collected using the content analysis scheme in Appendix D.

Collins, C. (1997). Viewer letters as audience research. *Journal of Broadcasting and Electronic Media, 41*(1):109–131. The author uses content analysis to draw inferences about audiences.

Esposito, S. A. (1998). Source utilization in legal journalism: Network TV news coverage of the Timothy McVeigh Oklahoma City bombing trial. *Communication and the Law, 20*(2):15–33. Instead of the more common document as the unit of analysis, the author uses the source.

Gerbner, G., Signorielli, N. and Morgan, M. (1995). Violence on television: The cultural indicators project. *Journal of Broadcasting and Electronic Media, 39*:278–283. This is one in a long series of articles on cultivation effects that the authors study by combining content analysis and survey data.

Jacob, W., Murdersbach, K. and van der Ploeg, H. M. (1996). Diagnostic classification through the study of linguistic dimensions. *The American Psychologist, 14*(1):8–17. Findings from this applied study are based on a structural analysis.

Jameson, D. A. (2000). Telling the investment story: A narrative analysis of shareholder reports. *The Journal of Business Communication, 37*(1):7–38. This is an example of a qualitative content analysis.

Lacy, S., Fico, F. and Simon, T. (1989). The relationships among economic, newsroom, and content variables: A path model. *Journal of Media Economics, 2*(1):51–66. This study illustrates one method for drawing inferences about temporal and causal relationships.

Lasswell, H. D. (1952). *The comparative study of symbols.* Stanford: Stanford University Press. In this short monograph, the pioneer researcher uses his work on political rhetoric, which was part of the larger Revolution and the Development of International Relations (RADIR) project, as a springboard for discussing the problems and methods of content analysis.

Musso, J., Weare, C. and Hale, M. (2000). Designing web-technologies for local government reform: Good management or good democracy? *Political Communication, 17*(1):1–20. The authors develop lessons for creating exemplary websites from their analysis of 270 existing sites.

Naccarato, J. L. and Neuendorf, K. A. (1998). Content analysis as a predictive methodology: Recall, readership and evaluation of business-to-business print advertising. *Journal of Advertising Research, 38*(3):19–34. The authors make creative use of information derived from a content analysis.

Neuzil, M. (1994). Gambling with databases: A comparison of electronic searches and printed indices. *Newspaper Research Journal,* 15(1):44–54. The author discusses strengths and weaknesses of common tools for locating documents.

Orbe, M. P. (1998). Construction of reality on MTV's "The Real World": An analysis of the restrictive coding of black masculinity. *Southern Communication Journal,* 64(1):32–47. This article illustrates the critical studies approach to qualitative textual analysis.

Rogers, R. and Marres, N. (2000). Landscaping climate change: A mapping technique for understanding science and technology debates on the World Wide Web. *Public Understanding of Science,* 9(2):141–163. The authors use computerized techniques to reveal relationships between and among websites of stakeholders in a debate.

Simonton, D. K. (1994). Computer content analysis of melodic structures. *Psychology of Music,* 22(1):31–43. This is an example of structural analysis of nontext documents.

Stevens, K. T., Stevens, K. C. and Stevens, W. P. (1992). Measuring the readability of business writing: The Cloze procedure versus readability formulas. *The Journal of Business Communication,* 29(4):367–382. Findings illustrate strengths and weaknesses of two common methods for determining the appropriateness of documents for an audience.

Wanta, W. and Gao, D. (1994). Young readers and the newspaper: Information recall and perceived enjoyment, readability and attractiveness. *Journalism and Mass Communication Quarterly,* 71(4):926–936. The authors employ an experimental design to compare results from computer-assisted content analysis with self-reports from a convenience sample of students.

Willey, S. (2000). The pitfalls of cyberspace and electronic database research. *Journalism and Mass Communication Educator,* 55(2):78–85. The author uses her own experience with *Lexis/Nexis* to explore problems associated with relying on computerized searches to locate items for content analysis.

References and Resources

Bauer, M. W. and Gaskell, G. (2000). *Qualitative researching with text, image and sound.* Thousand Oaks, CA: Sage. Part two of this book contains useful chapters on traditional content analysis and rhetorical analysis of audio and visual materials; Part three covers computer-assisted techniques.

Coupland, N. and Jaworski, A., eds. (1999). *The discourse reader.* New York: Routledge. This compendium of original research and writings covers the foundations, methods, and traditions of discourse analysis.

Flesch, R. (1974). *The art of readable writing.* New York: Harper and Row. This is the classic work on readability.

Gee, J. P. (1999). *An introduction to discourse analysis.* New York: Routledge. This book covers a variety of perspectives and theoretical approaches.

Hansen, A., Cottle, S. Negrine, R. and Newbold, C. (1998). *Mass communication research methods.* Wilmington Square, NY: New York University Press, pp. 130–224. These pages provide good information on techniques and problems associated with analyzing the moving images of film and television programs.

Holsti, O. R. (1969). *Content analysis for the social sciences and humanities.* Reading, MA: Addison-Wesley. The Association for Education in Journalism and Mass

Communication named this one of the twentieth century's top books on mass communication.

Kelle, U., Ed. (1995). *Computer-aided qualitative data analysis.* Thousand Oaks, CA: Sage. Chapter authors discuss issues and applications for theory building and hypothesis testing.

Krippendorf, K. (1980). *Content analysis: An introduction to its methodology.* Beverly Hills, CA: Sage. This book has excellent chapters on the conceptual foundations of content analysis, systems and standards for making inferences during the coding process, and some of the more sophisticated data-analysis techniques.

Riffe, D., Lacy, S. and Fico, F. G. (1998). *Analyzing media messages: Using quantitative content analysis in research.* Mahwah, NJ: Lawrence Erlbaum Associates. This practical guide is the best single source for information on research design, sampling, and computerized analysis.

Roberts, C. R., ed. (1997). *Text analysis for the social sciences: Methods for drawing statistical inference from texts and transcripts.* Mahwah, NJ: Lawrence Erlbaum Associates. Early chapters cover various theoretical and methodological perspectives; subsequent ones focus on use of computers in qualitative research.

Rybacki, K. and Rybacki, D. (1991). *Communication criticism: Approaches and genres.* Belmont, CA: Wadsworth. This undergraduate text has good chapters on five approaches and their application to public speaking, film, television, song, and humor.

Smith, M. J. (1988). *Contemporary communication research methods.* Belmont, CA: Wadsworth, pp. 235–277. This undergraduate text has very good chapters on the methods of discourse analysis and rhetorical criticism.

Stempel, G. H. III. (1989). Content analysis: Statistical designs for content analysis. In G. H. Stempel III and B. H. Westley, eds. *Research methods in mass communication,* 2nd ed., Englewood Cliffs, NJ: Prentice Hall, pp. 124–150. These pages provide a good overview of methods and issues in quantitative content analysis.

van Dijk, T. (1988). *News as discourse.* Hillsdale, NJ: Lawrence Erlbaum. This is a classic study using discourse analysis.

Weber, R. P. (1984). Computer-aided content analysis: A short primer. *Qualitative Sociology,* 7(1/2):126–147. This early work examines the applicability of computers for qualitative research.

Weber, R. P. (1990). *Basic content analysis,* 2nd ed. Newbury Park, CA: Sage. This work has good information on dictionary, thesaurus, and keyword lists as they are used in computer programs for content analysis.

Weitzman, E. A. and Miles, M. B. (1995). *Computer programs for qualitative data analysis: A software sourcebook.* Thousand Oaks, CA: Sage. Reviews of software programs, categorized by type, follow sections on applications and tips for choosing among types of programs and programs within types.

Zalaluk, B. L. and Samuels, S. J., eds. (1988). *Readability: Its past, present, future.* Newark, DE: International Reading Association. This work explains the techniques, strengths, and weaknesses of various methods of assessing readability.

9

Focus Groups

FOR THIS RESEARCH method, between six and twelve people who possess certain characteristics of interest to the researcher meet to provide qualitative data as a moderator leads them in a nondirective way through an open-ended, focused discussion of a specified topic.

In contrast to a survey, with its fixed set of questions, most of which allow for only certain response options, focus groups make it possible to probe beneath surface opinions to gain insight into people's real feelings and the reasons for their opinions and behaviors. This makes the method ideal for gathering ideas about what audiences, customers, clients, or voters want or do not want, how they react or may react to media content or communication campaigns, or what went right, went wrong, or could go wrong with a client's plans, programs, or communication efforts. The method is also useful in preliminary stages of research as a way to locate items for inclusion on a survey questionnaire or check the validity of alternate question wordings. It also is good for cross-checking interpretations of findings from research conducted using other methods.

The major weakness of the method is that the data come from just a small number of subjects selected through nonprobability techniques that are usually a cross between purposive sampling and self-selection. Therefore, findings are not truly generalizable. Indeed, given the vagaries of the sampling technique, a focus group may provide some information that is truly idiosyncratic while failing to provide other information that may be more common and perhaps much more crucial. This possibility makes it dangerous to rely solely on the results of a focus group to make final decisions.

Nevertheless, focus groups are popular because they typically are fairly inexpensive, especially when the richness of the data they provide and the fact that much of that information would be unavailable using other methods are taken into account. Between $1000 and $5000 per group will usually cover expenses for renting a facility for the group meetings, paying the moderator, developing any needed materials for the group session, recruiting subjects, and paying them for participation.

Recruiting appropriate subjects and finding and training the moderator can pose problems. Analyzing the results can be difficult and time-consuming. But focus groups can be set up fairly quickly—often in less than a week. Ethical problems are usually minimal.

Communication Applications

As a research method, focus groups had their origin in the studies of troop morale during World War II conducted by the noted sociologist Robert K. Merton and his colleagues, Marjorie Fiske and Patricia L. Kendall (1956). However, their use spread first to the fields of advertising and marketing and then into other areas. Even though less popular as a tool for theoretical researchers, focus groups rank alongside or just below the survey in popularity as a method for applied research.

- In marketing, focus groups are widely used to test products and product placement.
- Advertising researchers and public relations practitioners use them to pretest, monitor, and evaluate their campaigns.
- Public relations practitioners also use them as a tool for assessing the interests, desires, and needs of their publics.
- Candidates for public office routinely conduct focus groups to find out which issues will resonate with likely voters and how to package themselves and their messages in order to maximize their chances of being elected.
- Businesses, government agencies, and nonprofits use focus groups as a tool for assessing the wants and needs of clients and customers and for evaluating services to them. They also conduct focus groups with employees to involve them in the planning process or to find alternatives to current business practices or get help making decisions about product development and about advertising and marketing strategies.
- Program producers and film studios use focus groups to test public reaction to story concepts and conclusions.
- With the rising popularity of civic journalism, both journalists and the organizations for which they work have begun using focus groups extensively as a tool for finding out what issues and what kinds of information matter to people and how they want information packaged and delivered.

As the examples suggest, focus groups are well suited for use as an aid in the planning process. They can also be effective tools for monitoring and evaluating organizational efforts.

Ethical Concerns

The focus group is generally a low-risk method. However, it is not risk free. As a form of human-subject research, most ethical concerns center on protecting group members. However, because the method depends so heavily on the skill of the moderator, attention must also be paid to the integrity of the data-collection process.

From the standpoint of group members, participation must be voluntary. Those who are invited to be a part of a focus group must know that they will incur no penalty or repercussion if they refuse to participate or if they refuse to answer any uncomfortable questions posed to them during the session.

In most cases you can assume consent is voluntary if people show up for the focus group session. However, getting written consent from participants is a good idea if the sessions will focus on sensitive or controversial topics or if groups consist of employees or clients of the research sponsor. You will need written consent from parents or guardians if you will be working with children.

Regardless of whether written consent is necessary, participants or their parents or guardians must be given certain basic information when they are first asked to participate in a focus group. They must be told the general topic for discussion, any provisions for recording or videotaping the session, and whether they will be compensated for their time. They also need to know who is sponsoring the research, how the sponsor will use the information gathered, and whether there are any provisions for protecting their privacy.

All subjects also need to know where and when sessions will be held and how long they will last in order to decide for themselves whether the location is safe and the time is convenient. However, providing detailed information about the time and location for the session, facilities at the location, and transportation options is especially important if the participants are children, the elderly, or those who may be physically or mentally challenged.

Because it is almost impossible to ensure true anonymity to members of a focus group, assurance of confidentiality is very important. Without a strong guarantee that their comments will be used anonymously, their participation may not be truly voluntary and their comments may be less than candid. For consent to be truly voluntary, subjects should be told whether their full names or other identifying information will be known to other members of the group, released to the sponsoring organization, or used in any reports. Knowing this is particularly important for participants in groups made up of employees or clients of the sponsoring organization.

Before, during, and after the sessions, participants must be protected from both subtle pressures to conform and from possible embarrassment or outright humiliation should they offer comments that do not conform to the expectations or desires of the moderator, the sponsoring organization, or other group members. Because of the interactive nature of the data-collection and analysis process, those conducting the research must also be protected from internal or external pressures to guide the discussion or tailor findings to create particular results. To protect the integrity of the research, moderators must be free of biases that could affect the discussion. They must also be adept at dealing with all kinds of people and all kinds of situations.

Planning the Study

In order to get the most useful information from a focus group, certain basic issues must be addressed during the planning process. These include identifying and recruiting appropriate participants, creating the groups, finding and training a good moderator, developing the materials that will be used to elicit responses during the focus group session, and working out the logistics for conducting the sessions.

Participants

Typically focus group participants are chosen because they possess certain characteristics that indicate they may be able to provide information, ideas, or insight on the problem of interest to the researcher or research client. These basic characteristics may be demographic, attitudinal, or behavioral.

Locating Potential Participants. Once you identify the characteristics your subjects should possess, the easiest way to locate potential focus group members is to work from an existing list of people who possess the desired characteristics. Examples would include an organization's records of its employees, customers, or clients; published directories listing members of a profession or a special-purpose organization; and public records of voters and property owners.

With a list, you can simply call every k^{th} name and then attempt to convince those whose answers to a screening question suggest they possess the desired characteristics to participate in the study. But you probably won't be able to find a list if you want subjects with common demographic, attitudinal, or behavioral characteristics such as African American males, people who think television programs are too violent, or who routinely eat cereal for breakfast.

If no lists are available, you can often locate potential participants by advertising for volunteers, asking about willingness to participate in a focus group as part of a survey, or telephoning a random sample of households and then using screening questions to identify appropriate people. In many cases, however, it is easier to work through a research organization that specializes in conducting focus groups. These organizations can usually sort through their lists of people who are available to participate in focus groups to find those with the characteristics you desire.

Although locating a pool of potential participants can be difficult, it is only half the battle. Convincing people to participate and then making sure they show up for the focus group session can be equally time-consuming. Here, again, it may be easier to hire a research firm specializing in focus group studies to recruit participants than to try to do it yourself.

Recruiting Participants. The recruitment process usually begins with a phone call to potential participants two or three weeks before the sessions are to be held. In this call, researchers first use a screener to make sure the person contacted possesses the desired characteristics. Once they are sure they are talking to a person who is appropriate for the study, they provide the information necessary for informed consent while trying simultaneously to get the person to agree to participate in a focus group session.

If the person agrees to participate, the researcher then sends a follow-up letter confirming the person's selection for the study. In this letter, the researcher again provides information about the study, including its purpose, the time and place for sessions, and arrangements for compensation. This follow-up letter should be timed to arrive about a week before the session so that it serves as both confirmation and a reminder. Some researchers like to telephone the day before a session to confirm participation and answer any questions.

Ensuring Participation. Typically more people agree to participate in a focus group than actually show up for the session. Therefore, researchers usually overrecruit by as much as 40 or 50 percent. To help guarantee that enough people actually show up, they also offer certain inducements.

If more persons than are needed show up for a session, those who arrive last may be sent home. But like those who actually participate in the sessions, they will usually be served light refreshments. Both they and the participants will also be reimbursed for any parking fees and for transportation if they used mass transit or had to drive more than a few miles to reach the session. Both those who actually participate in the sessions and those who are sent home will also be paid.

Although paying people to complete a survey questionnaire is considered problematic, paying those who show up for a focus group is standard because of the greater inconvenience and time commitment. Payments may be in cash or in products or services. But as a general rule, people selected for common characteristics receive between $25 and $50 for a single session lasting between 90 minutes and 2 hours. People with rare characteristics and those recruited for longer sessions or to discuss sensitive topics will receive more. Hard-to-recruit groups such as medical doctors and lawyers may receive as much as $500.

Groups

The screeners used during the recruitment process help eliminate persons who do not possess the desired characteristics, but the information gathered during screening is also necessary for creating the groups of people who will meet for the focus group sessions. Here the most important things to take into account are the size, number, and composition of the groups.

Size. Although minigroups are common with focus groups involving children, those who conduct focus groups regularly recommend that each adult group should have between six and twelve members. With fewer than six, it may be difficult to get a discussion started or to sustain it. There is also increased risk that the group will not generate enough useful information.

The risk that the discussion will produce too little information in the form of a range of ideas or opinions decreases with group size. However, as the number of participants approaches or exceeds twelve, the groups may become unwieldy. The larger the group, the greater the risk that participants may not get a chance to say everything they would like to say because of time constraints or that some members may be intimidated by the size of the group or by the personality of some of its members.

Number and Composition. Each group must be relatively homogeneous so that members will be comfortable enough with each other to express their thoughts candidly. But if the overall composition is too homogeneous, sessions are less likely to produce a full range of information and ideas.

As a rule of thumb, researchers should assemble a minimum of two focus groups in each location where research is being conducted if the study is designed to elicit information on noncontroversial topics such as the evaluation of products and if the participants were recruited for characteristics that will most likely make them reasonably compatible. But more groups will be needed in each location if the topics for discussion are more controversial or if participants were recruited to represent a range of characteristics. In these situations, you must add additional pairs of groups to provide diversity to the study while also allowing for the kind of homogeneity within each group that will encourage discussion.

To minimize the possibility of miscommunication and/or the danger that some group members will find others in the group intimidating, it is usually best to separate people with obviously high socioeconomic status from those with much lower levels of education or income and senior citizens from teenagers. Staff and volunteers, workers and their superiors, and employees and clients should also have their own groups. For some purposes it may also be necessary to create separate groups on the basis of race, gender, lifestyle, or ideology.

Although group homogeneity generally encourages discussion, too much homogeneity can discourage candor. To diminish pressures to conform to group norms, you should separate friends, family members, colleagues, and co-workers instead of working with in-tact groups.

By working with pairs of groups, the design for the study takes on the characteristics of the quasi-experiment. Matched pairs of groups make comparisons across groups and between sets of groups possible. This helps protect against relying too much on what may be truly idiosyncratic viewpoints or falsely attributing them to group characteristics. If any two groups produce too little information or if one group in a pair produces information that cannot reasonably be reconciled with that from the other group, additional groups can easily be added.

The Moderator

Because the actions of the moderator can so easily influence the discussion, it is usually best to select someone who has a proven track record. In areas where professional moderators are unavailable or if hiring one is too expensive, hiring and training a "semipro" is a better option than letting someone from the organization sponsoring the research act as moderator. College professors and graduate students specializing in small-group communication, the social sciences, advertising, or marketing often have the requisite skills. High school teachers can also be a good choice.

Requisite Skills. Moderators should be comfortable with the groups they will be working with and acceptable to the group. Although they should be comfortable with the subject for discussion, they should not have strong opinions about it. They must be able to listen and probe for information, draw out those who are less fluent or more hesitant to speak up while simultaneously controlling those who would monopolize the conversation. They must be pleasant and flexible enough to follow the discussion wherever it leads, yet firm enough to keep it on track and bring it to closure within the allotted time.

In general, if the goal is to collect opinions on issues or program priorities, it usually works best if the moderator matches group members in gender and race and appears to be of the same or just slightly higher socioeconomic status. For discussions of products or services, matching by race, gender, or status may or may not be advantageous. Similarly, in some situations it may be best to have a moderator who is totally familiar with the product or service. In others, one who can plausibly "play dumb" may be able to elicit more information.

Training Moderators. Even the best professional moderators will need some training each time they undertake a new assignment. In order to do a good job, less-experienced moderators will also need a chance to practice.

Before they attend a training session, all moderators will need time to become familiar with the purpose of the focus groups, the points to be covered in the discussion, and any other activities that will occur during the sessions. The training sessions provide the opportunity for moderators to ask questions and to review the purpose of the research, the points to be covered during the sessions, and general principles for leading the discussion and for dealing with situations that may arise during sessions.

Number of Moderators. Moderating a group is a demanding, high-energy job. Few professional moderators will agree to work with more than two groups in a single day.

> **Box 9.1.** *Dealing with difficult participants*
>
> **The reticent participant.** Draw into the conversation by addressing this person directly. Praise any contributions, but don't actually support them.
> "Bob, how do you feel about that?"
> "Thank you for sharing your feelings."
> **The dominator.** Firmly insist others get a chance to speak.
> "Sue, let's hear from Mary first."
> **The interrupter.** Insist on turn taking and courtesy to others.
> "Sam, Fred is speaking now. We need to hear what he has to say. We'll hear from you next."
> **The rambler.** Insist on a concise answer. If these participants still talk too long, you may be able to cut them off by calling on an active, somewhat assertive participant you can count on to start talking.
> "Dick, please wrap up your thoughts in a sentence or two."
> As a last resort, "Dick, we're running a bit behind schedule, and we really must hear from others. Linda, how do you feel about the possibility of publishing the newsletter monthly?"
> **The really obnoxious one.** Call attention to the problem. If that doesn't work, invite the person to leave.
> "Jane, we're here to get a variety of opinions about the newsletter, so you must confine your comments to the newsletter. I'm not interested in your opinions about other members of the group."
> "Jane, you don't seem to want to talk about the newsletter, so I'll let you leave now."
> If someone still undermines the group by constantly interrupting, making sarcastic or demeaning comments, or otherwise behaving in unacceptable ways, leave the room to find a research assistant or company representative. Tell the assistant or representative to summon the obnoxious participant to come get a message or take a "phone call." Once the participant is outside the focus group room, the assistant or representative can dismiss the person: "Jane, I'm sure you have many valuable things to say, but your comments are affecting others. Thank you for coming, but we must insist you leave now." Be sure to pay Jane and give her any other inducements you promised. Giving her a bit extra may also make up for any hurt feelings.

Therefore, you will have to hire extra moderators if the schedule for completing all sessions requires holding more than two sessions on a single day or holding sessions in scattered locations.

If you must use several moderators, it is usually best to have each one conduct just one session from each pair of sessions. That way, if the information from both groups is very similar, moderator influence can be eliminated as the cause. If the results are very different, more groups can be added to the study to help determine whether differences were due to the groups themselves or to moderator influence.

The Materials

Although focus groups seem disarmingly simple, making sure that they provide the necessary information requires creating materials to assist with data collection. These include the moderator's guide and any supplementary materials that may be needed for data collection.

The Moderator's Guide. This guide is standard for all focus groups. Its main part consists of the questions or topics to be covered during the discussion and information on any other activities that subjects will engage in during the session. Most guides also contain a timeline and checklist of tasks the moderator must complete before and after a session and ground rules for handling problems such as broken recording equipment or dealing with difficult participants. Especially in cases where nonprofessionals are used as moderators, the guide may also contain tips for asking questions and eliciting responses without influencing the outcome.

Although the questions or topics are usually arranged in an order that seems logical, the order is not fixed. Following each question or topic there may be suggested branching routes to help the moderator keep the discussion focused regardless of the kinds of comments participants provide. The key here is that the list of questions and topics should be complete and ordered in some logical way so that the moderator can easily refer to it and check off items as they are covered.

In some cases, the moderator's guide will also need to have a section devoted to directions for completing other activities that subjects may engage in as part of a focus group session. But if these activities are fairly simple or self-explanatory, they may simply be listed as part of the timeline.

The main purpose of a timeline is to provide moderators with a rough guide as to how much time they should allow for each activity or each question or topic for discussion. This timeline tells whether each activity is mandatory or optional, about how much time should be allocated to it, and whether it should be completed before, during, or after the discussion. If no activities other than the discussion are planned, the timeline may simply provide the moderator with an estimate of how much time may be allotted to each question or topic.

Supplementary Materials. These are of two types: materials necessary to spark the conversation and ones used to collect additional data.

Materials necessary as conversation starters may include things such as samples of products, clips, or mock-ups of media content or equipment or computer software that participants can see, perhaps use, and then react to.

Although the discussion is the main source of raw data, both group and individual activities can also be used to collect additional information.

Activities such as role playing can be incorporated into a session. Others such as usability tests may be administered immediately before it, while still others such as ranking or sorting exercises can be administered before, during, or after the session itself. Many times it will also be desirable to prepare a self-administered questionnaire for participants to complete before or after the session.

Having people complete a questionnaire on the topic before discussing it can help them focus their thoughts. It may also help subjects commit to a position and thus resist group pressure. In any case, it gives the researcher a way to separate subjects' initial reactions from those that develop later as a result of group interaction.

Having subjects complete the questionnaire at the end of a session is an easy, effective way to gather information on participants' impressions of the sessions and what transpired in them. Used this way, questionnaires can also serve as a check on group opinion in situations where some subjects may have been uncomfortable voicing a minority viewpoint or standing up to the group know-it-all.

Logistics

The value of the information collected by means of a focus group depends primarily on the nature of the participants in each group, the skill of the moderator, and the adequacy of the list of questions or topics to be addressed and of the supplementary materials for use in data collection. However, the quality of the work can also be affected by logistical considerations such as when and where the sessions are held.

One of the major problems with a focus group is getting the desired participants to show up for sessions. Therefore, sessions must be scheduled at a time and place that will be convenient for them. Although sessions may be held almost anywhere, the facilities at the location where the group will meet should be adequate for the purpose. Safety and transportation options also deserve consideration.

Scheduling Sessions. Asking people to take time off from work, interrupt family life, or cancel plans they have already made is unreasonable. Therefore, sessions should be scheduled and participants recruited two to three weeks in advance so they can plan accordingly.

In scheduling sessions for several comparable groups, it is usually best to arrange for them to meet at slightly different times such as morning and afternoon or afternoon and evening. Holding sessions at different times or on different days such as a Monday and Tuesday will also make it possible to include some people who otherwise would not have been able to participate.

Ordinarily groups should not be scheduled for Saturday or Sunday because most people are reluctant to give up their weekend to participate in someone's research project. But weekend sessions can work well in some cases.

Groups composed of an organization's employees or clients may be scheduled during normal business hours, but sessions made up of people who cannot be expected to receive time off from work to participate may appreciate evening or weekend sessions. Ones for homemakers, retired persons, and young children are usually scheduled in the late morning or early afternoon.

Sessions for school-age children are best scheduled during summer vacation. If that is impossible, a Saturday or a minor vacation day such as Columbus Day can be a good choice. Short sessions in the late afternoon or early evening can also work.

If participants will use public transportation to reach the session, as may be the case with senior citizens and low-income groups, starting and ending times should be set to coincide with mass-transit schedules. Even when people can be expected to use their own transportation, factors such as drive time and safety should be taken into account. Women and senior citizens may be reluctant to participate in sessions that will force them to be out late at night. No one wants to fight rush-hour traffic. Although few people will willingly give up their dinner hour, many employed persons like a 5:30 p.m. start time so they can attend the session first and then have the rest of their evening free for other activities.

Location. Firms that specialize in focus group research often have meeting rooms designed just for the purpose. Using these rooms is an attractive option because they will have all of the required facilities for recording or observing the sessions. Many have viewing rooms equipped with one-way mirrors so that researchers or research sponsors can watch sessions and take notes without being observed by participants. Typically these facilities also have catering arrangements that make it easy to serve the refreshments or light meals that are an expected part of most focus group sessions.

However, these specialized facilities are not available everywhere. Even where available, they may not always be the best option. Some subjects may find them intimidating. Facilities located in suburban business parks may be beyond the reach of mass transit. Those located downtown may be in or near an area that is unsafe after dark. In any case, driving and parking can be difficult in the central city.

Therefore, it is often better to forgo the amenities of a specially designed facility and instead choose a setting that is safer, easier to reach, or more familiar and hence less intimidating to the desired participants.

Meeting rooms in a research client's place of business are a good choice for sessions with the organization's employees, clients, and customers. For many purposes, meeting rooms at a hotel will work well. They almost always have catering capabilities and may also have facilities for recording sessions. However, hotels may be inconveniently located or seem intimidating to some subjects. Therefore, other attractive options include meeting rooms in community centers, senior citizens' centers, schools, or, for some purposes, churches.

The only real requirements for a meeting place are that it be acceptable to participants and that it have the amenities you will need to record the sessions and to use any equipment necessary for the supplementary activities you have planned.

Recording Provisions. A skilled moderator will be able to jot down some brief notes during the session, but it is almost impossible for even the best moderators to record the kind of detailed information necessary for data analysis. Therefore, you will need some way to record the session.

Audiotaping is standard. Although it is more expensive, videotaping is preferable because participants' facial expressions and body language provide valuable information.

Because tapes may fail or fail to capture everything that occurred during a session, it is also standard for the moderator to write up a detailed report at the end of each session. This report typically includes a summary of the discussion, the moderator's impressions of the session, and information on any problems that occurred.

As a backup to both the tape and the moderator's summary, many researchers like to hire an assistant to take notes during each session. It is also fairly standard to have the moderator or a research assistant debrief some or all participants. These debriefing sessions may be taped, the researcher may simply take detailed notes on them, or, as an alternative, participants may be asked to fill out a questionnaire or write up a brief summary of the session and their impressions of it.

Alternatives to the True Focus Group

Typically focus group participants meet at a central location for what amounts to a brainstorming session. However, at times the purpose for the session requires modifying the way it will be conducted or pragmatic concerns such as cost and convenience require a change in the delivery mode.

The primary purpose for focus group research is to bring to light ideas, issues, concerns, and possibilities that might otherwise remain unknown to the researcher or sponsoring organization. For that reason, consensus among group members or across groups is unimportant and may even be undesirable.

But organizations may need to find out which issues or services its publics think are most important. They may also find it necessary to have employees or volunteers meet to set priorities or solve a problem. In these situations, the nominal group, Delphi technique, and Q-methodology are attractive alternatives to the true focus group.

In cases where time or budgetary constraints make meeting impractical or when face-to-face interaction may be intimidating, the Delphi technique and various technology-based alternatives to the true focus group work well.

The Nominal Group Technique

As with the traditional focus group, the nominal group technique involves six to twelve people who meet with a leader or moderator. However, with this technique discussion is orderly. It is also guided so that it culminates in a group decision.

Sessions begin with the group leader or moderator explaining the purpose for the group meeting and the ground rules for the discussion and for reaching a decision. The session then proceeds through five steps:

1. Idea generation. Without consulting other group members, each person in the group writes a list of ideas, possibilities, or solutions on a sheet of paper.
2. Idea sharing. Each group member in turn offers one idea from his or her list, and the leader or moderator writes that idea on a chalk board or flip chart. This process continues until all of the ideas generated in Step 1 are posted on the master list. At this stage, group members may, with the consent of the proposer of each item, combine ones that seem repetitive.
3. Comment. Each member, in turn, comments on the first item on the master list, then on the second item, and so on until everyone has had a chance to comment on each item.
4. Evaluation. Each member, in turn, makes a case for which items on the list should receive the top priority and which ones should rank lower or even be eliminated. After each member has had a say, turn taking can be extended so that those who want to add information or revise their initial judgments may do so. During the evaluation stage, group members are not allowed to criticize or argue against the speaker's judgments. However, they may ask for clarification or request more information.
5. Conclusion. After the evaluation phase is complete, each group member votes by secret ballot. Consensus is derived mathematically by tallying the results according to the rules explained and agreed upon at the start of the session.

The nominal group technique works equally well with highly educated or experienced "experts" and with people with low verbal or cognitive skills. Groups do not need to be as homogeneous as with the typical focus group. Experts and average people, service providers and clients, and people with different demographic, lifestyle, or attitudinal characteristics can usually be included in the same group.

Group members may sometimes defer to the opinion of "experts" when they cast their final ballot. But because the final vote is taken by secret ballot, they are less likely to be influenced by the moderator or by the group member who is more outspoken or glib than they might be in a true focus group.

Because everyone gets an equal chance to offer ideas, comment on them, and then vote, participants usually voice satisfaction with the process and accept the outcome. Problems occur primarily in those situations where participants do not receive adequate assurance that voicing their opinions cannot lead to any form of retaliation or where they suspect or subsequently discover that those with the ultimate authority to put their recommendations into practice did not accept the group's decision or at least take it seriously.

The Delphi Technique

Like the nominal group technique, the Delphi technique also aggregates individual opinions to reach a final consensus as to priorities or solutions to a problem. However, those who do the prioritizing and ranking never meet and, in most cases, are unknown to each other.

With this procedure, a staff group generates an initial questionnaire and then circulates it to a respondent group. Respondents answer the initial questions, add comments, and make other suggestions on the questionnaire before returning it to the staff group. The staff then prepares a summary of the respondents' answers and comments, generates a new set of questions, and sends the summary and new questionnaire back to the respondent group. The respondents again answer the questions and provide additional feedback. This back-and-forth process continues until some kind of closure is reached. Closure usually comes from mathematically aggregating respondents' final prioritized lists or their votes for the top priority.

The Delphi technique can be used to gather information from diverse groups of subjects, but all respondents must have higher cognitive, reading, and writing skills than is true for participants in a focus group or nominal group. Even though it is less intrusive on respondents' time, the Delphi technique takes much more staff time than the nominal group technique. It also takes longer to produce results.

Respondents may also find the Delphi process less satisfactory than the nominal group technique. Because they never meet with each other or with the staff that generates the questionnaires and reports, respondents have no way to clarify ambiguities or judge the level of expertise of those offering suggestions or comments. They may also be much more suspicious of the completeness or fairness of the interim reports and thus of the ultimate outcome than is usually true for participants in nominal groups who can see and judge the entire process for themselves.

For these reasons, the technique is best used in those situations where meeting as a group would be impractical or impossible, as might be true if a company needed to reach consensus among staff located in branch offices around the world. However, the technique is also appropriate when subjects need more time to reflect on information and ideas than they would have during a single group meeting, but a series of meetings might lead to undue lobbying or other pressures on participants.

Q-methodology

This method is a variant of and a cross between the Delphi technique and questionnaire research. In the Delphi technique, people work alone to provide open-ended responses to statements or answers to questions. With Q-methodology, people also work alone, but they respond to statements by ranking them much as they might rank lists of items in response to a survey question.

Q-methodology cannot be used as a tool for consensus building, but the individual rankings obtained using Q-methodology make it better than the nominal group and Delphi techniques for determining individual people's priorities. Researchers can also get some sense of group priorities by analyzing the set of all individual prioritizations qualitatively or quantitatively.

This possibility of quantitative analysis sets Q-methodology apart from other techniques discussed in this chapter. We include it here because one of its most common uses is as a follow-up to a traditional focus group.

Advertising researchers frequently use Q-methodology that way because it can be a powerful technique for understanding the structure and logic behind people's attitudes, beliefs, and opinions about themselves, brands and products, advertisers, and advertising practices. However, it can be used whenever there is a need to gain similar kinds of insight into people's thought processes or their feelings about a referent such as a group, an organization, a particular medium, or a kind of media content. Q-methodology is also useful in the early stages of research to develop measures for use with other research methods.

The first step in Q-methodology is to generate a large number of statements that, in one way or another, are at least tangentially related to the problem at hand. Transcripts of focus groups or brainstorming sessions are a common source for these statements, but they may also come from previous research, from the researcher's personal experience, or documents such as media accounts or minutes of meetings.

Once this generation phase is complete, all statements are merged into a single set representing the total population of all possible statements. The researcher then edits this list of statements by revising unclear statements, combining redundant ones, and eliminating those that seem irrelevant to the purpose for the study. After that, the researcher checks the list to make sure the set of statements covers the full range of possibilities and that each item in the set possesses face validity.

Typically in Q-methodology this editing procedure will result in a set of between 50 and 100 items. Each item will then be printed on a card. The researcher then gives a set of these cards to a small group of chosen subjects and instructs each person to use either an unforced or forced sort procedure to arrange the statements according to some criterion.

Subjects might, for example, be told to arrange the statements from most important to least important, from best to worst, or from "most like" to "least like" the subject, a product, or an organization.

For an **unforced sort,** the researcher specifies the number of points on the continuum but leaves the subject free to decide the number of items to place at each point. With a **forced sort,** the researcher determines both the number of points on the continuum and the number of statements that may be placed at each point.

If the goal is to get a true ranking of all statements, researchers use a forced sort in which subjects are instructed to place just one item at each point on a continuum that contains the same number of points as the number of statements to be ranked. But because making the fine distinctions that procedure requires can be very difficult, they may sometimes allow subjects to create ties by placing as many items as they wish at points on the continuum.

Researchers may also use that kind of unforced sort when ranking is not the goal; they simply want to see how subjects arrange the statements. But because of the difficulty of making the kind of judgments necessary to cope with all the possible places to put each of a large number of statements, researchers frequently create a continuum with between seven and eleven points on it. For an unforced sort, subjects would still be free to place as many statements as they wish at each point on the continuum. For a forced sort, subjects typically would be required to place a few items at each end point and the most items at the midpoint so that the final sort approximates a bell-shaped curve. For example, with 70 statements and nine points on the continuum, the final arrangement of items for a forced sort might be:

4   6   9   10   12   10   9   6   4

Technology-based Techniques

To overcome problems of time and distance, researchers and their clients are increasingly taking advantage of modern communication technologies that make it possible for people to communicate and interact with each other even though they never actually meet. These technological fixes include the **conference telephone call, videoconferencing, e-mail,** and the **Internet chat room.**

With conference telephone calls, a moderator can lead small groups of people through a discussion much as if they were assembled in the same room. Videoconferencing can do the same thing but with the added advantage that participants can see the moderator and other participants as well as talk to them.

Use of e-mail can greatly speed up the Delphi process. E-mail can also be used to gather information and insight from small groups of people and let them comment on each other's contributions.

Like the conference call and videoconferencing, the chat room allows the discussion to take place in real time. But like e-mail, it is more convenient for participants. It also opens up the possibility of getting information that might be missed with other methods.

Both e-mail and chat rooms have the advantage that sessions may more easily occur over an extended period of time. Subjects can break away from the discussion to attend to other matters, to locate information, or to consult with others who are not a part of the group but whose contributions to it may be valuable. Upon returning to the group, they can easily read contributions that were made during their absence before rejoining the conversation. All of these technology-based delivery techniques have the advantage that recording the interactions among group members is both possible and relatively easy. Their disadvantage is that all require access to relatively sophisticated and expensive equipment.

Quality Control

In some ways the focus group is the antithesis of the experiment with its tight control over all factors that could affect the findings and also of the survey where findings are based on data gathered from a large probability sample of subjects. Nevertheless, findings from a well-done focus group can be just as "real" and, for some purposes, even more useful.

Reliability

Because of their very nature, it is almost impossible to set up focus groups so that each session will be conducted exactly like every other session that is part of the same study. The open-ended nature of the discussions virtually guarantees that no two sets of conversations, even two conducted by the same moderator on the same subject, will proceed in the same manner. Therefore, it would appear that focus group findings could never be reproducible. But that should not be the case.

Any reasonably well-trained moderator should be able to elicit essentially the same information from a group of people so long as the moderator makes sure that the sessions cover all of the required topics and does not behave or allow others to behave in ways that change the cues participants receive from one replication to another.

Giving the moderator a chance to become familiar with the moderator's guide and some time to practice using it will go a long way toward making sure that all sessions are conducted using the same ground rules and that they all cover the same material. The key to reliability, then, is finding or training good moderators and then providing them with a thorough, easy-to-use moderator's guide.

Internal Validity

Here the major problems will most likely stem from subject selection and various interactive effects between the moderator and participants, between participants and the setting, and among the participants. The same factors that help ensure reliability also contribute to internal validity.

Although there is no way to eliminate "guinea pig" effects completely, it is hard for people to hide their true feelings while simultaneously remaining consistent in their responses over the course of a lengthy discussion. Therefore, the primary data—subjects' own words taken in context—provide information that has good face validity. But beyond that, words can be checked and cross-validated using the verbal and visual cues recorded on tape and then cross-checked again with information from the moderator's report and debriefing reports and with the supplementary information from questionnaires or tests administered in conjunction with the sessions. Doing this checking and cross-checking goes a long way toward ensuring that you are measuring what you think you are.

External Validity

This is the one major weakness with focus group research. Although compelling evidence exists that findings from focus group studies can be very similar to those from large-scale surveys using probability sampling techniques, there is no guarantee this will be true for any particular study.

External validity may be enhanced by using some approximation of quota sampling to choose subjects with different characteristics and by working with paired groups. It can be built up over time by combining focus group findings with data gathered in other places and at other times and using other methods.

But because of the small number of subjects and the way they are chosen, results from any single focus group study simply are not generalizable. Relying solely on focus group findings to make decisions, particularly when the stakes are high, can be risky business.

Data Analysis

Focus group research is a qualitative method.
Because of the small number of subjects involved and the way they are selected and

then assigned to groups, any use of statistical techniques to draw conclusions is not appropriate. Focus groups simply are not designed for testing hypotheses or making inferences about a larger population. The one exception is the analysis of data collected by using Q-methodology.

Quantitative Analysis

True quantitative analysis should be done only with data derived from Q-methodology. It should not be used with other methods.

In Q-methodology, the variable is the person performing the sort, not the item the person is sorting. With this in mind, the quantitative analysis of data from a Q-sort usually begins with factor analysis to identify similarities and differences in the way subjects arranged the statements. The factor loadings provide an indication of their underlying opinion structures or their priorities. Loadings from the factor analysis can also be used as a first step in identifying statements that may be combined to create a composite measure. Directions for doing a factor analysis and creating composite measures are in Chapter 5.

Although quantitative analysis of data from a Q-sort usually begins with a factor analysis, the statistical procedures discussed in Chapters 6 and 7 can also be used.

Frequencies will provide information on the distribution of rankings for individual items as well as the distribution of characteristics of those doing the rankings. Because of the small number of subjects typically involved in a study using Q-methodology, analysis of variance is a good way to determine whether there are statistically significant differences in the way people with those characteristics sorted individual items.

Cross-tabulations can also be useful for showing how a characteristic such as gender may be related to rankings. As a measure of association, Spearman's rho is particularly appropriate for comparing rankings by two groups of subjects such as men and women or experts and novices. However, because of the small number of subjects and the potentially large number of rankings each statement might be given, you will probably need to do some recoding in order to get meaningful results. Even then, tests of statistical significance should be used with extreme caution. Because of the small number of subjects and the way they are chosen, drawing inferences about some larger population is unwarranted.

With the basic focus group, with nominal groups, and with the Delphi technique, it can be useful to count the number of subjects who provided a particular kind of information or held a particular viewpoint, but most researchers prefer to use general terms such as "a few," "some," or "most" in their reports. Any reporting of actual numbers should be minimal. Using percentages is even riskier. Both numbers and percentages invite inappropriate generalizations.

Qualitative Analysis

As qualitative research, analysis of findings from focus groups follows the inductive model. Therefore, the first rule is: Avoid beginning the analysis with preconceived notions of what you will find. Giving in to pressures from a client who hopes the data will provide certain findings or succumbing to expectations of what people will or should know or think is a sure way to miss important information.

The goal is not to find supporting information for a hypothesis. Rather, it is to extract all potentially useful information from the available record. To do this, initially at least:

- Give equal credence to input from all participants. Do not be unduly swayed by comments from the "expert" or from the person who seems nicest or speaks most fluently or persuasively and do not discount input from the obnoxious participant or the one who may have trouble expressing thoughts or who uses disclaimers such as "Maybe this isn't a good idea, but . . .".
- Give equal credence to input from all groups. Never make judgments about the value of information based on group characteristics. Remember that some of the most useful findings will be the similarities and differences in ideas that groups made up of people with different demographic, lifestyle, attitudinal, and/or behavioral characteristics will provide.
- Use all available information. One of the major strengths of the focus group is the availability of a variety of kinds of evidence. Take advantage of it.

With these ground rules in mind, the analysis may be done by the moderator, the person who planned the research project, or another researcher. The advantage of using a moderator is that this person will almost certainly know more than anyone else about the personalities of participants, the nuances of each discussion, and the problems that occurred during each session. This insight is useful and should not be ignored, but at the same time it may make it more difficult for the moderator/analyst to avoid preconceptions and treat all participants and groups equally. Although another person may not have all the firsthand experience with the groups and group members as a moderator will have, the advantage of using the research planner or another researcher to do the analysis is that this person should be better able to detect moderator behaviors that may have affected the findings.

But regardless of who does the work, the analyst must first assemble all relevant materials. These materials consist of the record of the discussion, any notes taken by the moderator during sessions, summaries and comments prepared by the moderator after each session, summaries and/or debriefing notes from participants, information from questionnaires or usability tests gathered in conjunction with the sessions, and notes or comments from any session observers.

The record of the discussion will ordinarily be on audio or videotape. These tapes may be examined, but it is very difficult to compare a taped record of one session to that from another. Therefore, it is usually advisable to prepare a written transcript of the discussion even though this will take time and increase costs.

Once all of the materials have been assembled and transcripts of sessions have been prepared, actual data analysis progresses through these steps:

1. Read/listen to/watch and take notes on all materials from all sessions. This provides a general overview that will help identify problem areas, strong opinions, trends, and other potentially useful information.
2. Reread/listen to/watch all material from one session, preferably at one sitting. Again note problem areas, strong opinions, and trends in the discussion. Also look for less common or weaker opinions.
3. Reread the transcript of the discussion. Mark comments relevant to the purpose of the research.

4. Consider the context. Identify comments that were generated spontaneously and those that were offered in response to a general question or to one directed to the speaker or to another group member.
5. Listen to/watch the tapes while reexamining the text. Make notes on the transcript to indicate any comments that may not be supported by tonal inflections, facial expressions, or body language.
6. Reexamine supporting materials for additional evidence of internal consistency or inconsistency within the information provided by individual participants and the group as a whole.
7. Categorize relevant comments by topic. Subcategorize them by strength and direction.
8. Extract the most relevant comments of each type, being sure to select both common comments and minority opinions as well as any contraindications for them.
9. Note the demographic, lifestyle, or other characteristics of the participants making each type of comment.
10. Prepare a summary of findings.
11. Repeat Steps 2 through 10 for all groups.
12. Compare the evidence across groups. Look for similarities and differences in the conduct of the groups; questioning patterns; content, strength, and direction of comments from group members; and the characteristics of persons offering each kind of comment.
13. Prepare a report based on the summaries of each individual group and the comparison across groups.

The amount of information available and the number of steps involved can make the task of data analysis seem overwhelming. Because so much of it is really an exercise in content analysis, the process of coding, retrieving, and comparing information can be made more manageable by utilizing some of the tips provided in the Qualitative Analysis section of Chapter 8.

Main Points

- In a focus group, between six and twelve people who possess characteristics of interest to the researcher meet to provide qualitative data as a moderator leads them, in a nondirective way, through a discussion of a specific topic.
- Focus groups are useful for probing beneath surface opinions to gain insight into people's real feelings and their reasons for opinions and behaviors. Other common uses are as a planning tool and as a way of cross-checking interpretations of findings from research using other methods.
- Ethical problems are minimal so long as participation is truly voluntary. Most stem from the difficulty of ensuring anonymity or even true confidentiality.
- Planning focus group research requires identifying and recruiting appropriate subjects, setting up the groups, finding and training good moderators, developing materials to be used during group sessions, and working out the logistics of where and when to hold sessions.
- Focus group participants are usually chosen because they possess certain demographic, attitudinal, or behavioral characteristics that indicate that they can provide information or insight related to the problem of interest.

- When and where focus group sessions will be held can affect people's willingness to participate in focus group research.
- Typically, more people agree to participate in focus groups than actually show up for the session. Therefore, researchers usually overrecruit by as much as 40 to 50 percent and offer inducements to help guarantee that people will show up.
- Each focus group must be large enough to allow for sustained discussion, but small enough to ensure that everyone has a chance to talk.
- Groups must also be homogeneous enough to ensure that members will be comfortable with each other, but not so homogeneous that a full range of ideas and information is not represented.
- Characteristics of good moderators include experience, the ability to listen and probe for details, and the ability to be pleasant and flexible yet firm. The moderator should also be free of strong feelings about the research topic.
- Because moderators must feel comfortable with and be accepted by the members of the group, it is often best to match the moderator to group members on characteristics such as gender, race, and socioeconomic status.
- During sessions, the moderator follows the plan outlined in a moderator's guide. This guide usually includes the questions or topics to be covered during the discussion, information on any other activities the subjects will be engaging in, a checklist of tasks the moderator should complete before the end of the session, ground rules for handling potential problems, tips for answering and eliciting responses without influencing the outcome, and a timeline.
- In order to get the detailed information needed for data analysis, sessions should be recorded if possible. Audiotaping is standard. Videotaping is preferable because participants' facial expressions and body language can provide valuable information to the researcher.
- Additional data may come from the moderator's notes, from questionnaires completed by group participants before or after sessions, and from debriefings with them.
- To overcome problems of time and distance, modified focus groups may be conducted by means of conference telephone calls, videoconferencing, e-mail, and Internet chat rooms.
- Other alternatives to a true focus group are the nominal group, Delphi technique, and Q-methodology.
- The nominal group and Delphi techniques are useful for developing consensus and setting group priorities.
- Q-methodology is good for learning about individuals' priorities or investigating the structure and logic behind their opinions and behaviors.
- In a focus group, the key to reliability is to find and train good moderators, provide them with a thorough, easy-to-use moderator's guide, and then give them a chance to practice using it to lead group discussions.
- Most internal validity problems are related to subject selection and various interactive effects between the moderator and participants, between participants and the setting, and among the participants.
- Internal validity can be enhanced by working with matched pairs of groups so that the research design mimics that of the quasi-experiment.
- Because data come from a small number of subjects selected through nonprobability methods, focus group findings are not truly generalizable.
- Representativeness may be improved by using quota sampling to select subjects and working with matched pairs of nonequivalent groups. However, external

validity can be established over time only by combining the findings from focus groups with data gathered in other places and at other times and using other methods.
- Data collected from focus groups and nominal groups and by means of the Delphi technique must be analyzed qualitatively. Data collected using Q-methodology is appropriately analyzed using quantitative techniques.
- Although other statistical procedures may be used, the quantitative analysis of data from a Q-sort usually begins with factor analysis to identify similarities and differences in the way subjects arranged the statements they sorted.
- The main goal in the qualitative analysis of focus group data is to extract all potentially useful information from the available record. To do this, researchers must use all the available information, giving equal credence to all participants and the input from all groups.

Terms to Know

matched group pair
moderator's guide
debriefing
nominal group technique
Delphi technique
Q-methodology

Q-sort
unforced sort
forced sort
technology-based technique
factor analysis

Questions for Discussion and Review

1. What are the relative strengths and weaknesses of focus groups? How do they compare to those for experiments, surveys, and content analysis?
2. What is the ideal size for a focus group? Why might it be a problem to have a larger group? A smaller group?
3. Explain the factors that influence the number of groups chosen for a particular study.
4. Explain the factors that should be taken into account when selecting a moderator for a focus group.
5. Explain how you might use the techniques for designing and conducting an experiment, a survey, and content analysis in designing and conducting focus group research.
6. Find an article in a communication journal that uses the focus group as the main method of data collection. For that study, (a) identify the research question addressed by the study, (b) describe the moderator(s) and their experience and/or training, (c) tell what questions were asked and identify others that you think were not asked but should have been included, (d) tell how the discussion was recorded (e.g., notes, audiotape, video), (e) describe any supplementary materials, and (f) explain how the data were analyzed. Finally, (g) tell whether there appear to be any ethical problems with the way the study was done and (h) assess its reliability, internal validity, and external validity.
7. Conduct an Internet search on Q-methodology and compile a bibliography of at least five works on the subject.

8. Describe and distinguish between the alternatives to the true focus group: the nominal group, the Delphi technique, and Q-methodology. What are the relative strengths and weaknesses of each of these methods?
9. Discuss the advantages and disadvantages of technological variants: the conference telephone call, videoconference, e-mail, and chat rooms. Explain when it would be appropriate and inappropriate to use each one.
10. Find an example of a study in a communication journal that uses the nominal group technique, the Delphi technique, or Q-methodology. For that study, (a) identify the research question addressed by the study, (b) tell how many subjects were used, how and why they were selected, and describe any strategies that were used to encourage their participation, (c) describe the procedures used to collect and analyze the data, (d) tell whether there appear to be any ethical problems with this study as it was conducted. Finally, (e) assess the reliability and internal and external validity of this study, and (f) tell whether it would have been possible to use a focus group or one of the other alternatives to the focus group instead of the method that was used.
11. For each of the following situations, tell whether you think it would be best to conduct a focus group or use one of the alternatives discussed in this chapter. Explain and justify your choice.
    a. The Guzzler Beer Company wants beer drinkers' reactions to its current advertising campaign for "lo-cal" beer and to potential themes for a new campaign.
    b. Westside Hospital needs to learn what board members, senior management, doctors, and nurses believe the hospital should do to reduce expenses in order to survive now that health-care costs are rising faster than allowable reimbursements from insurance companies and Medicare.
    c. Station managers at WKBN want to find out why their morning team isn't succeeding with its target listeners, 19- to 25-year-olds and what can be done to get people in this age group to listen.
    d. WKBN managers also want to know what songs should stay on its play list and which ones can be eliminated to make room for new music.
    e. The board of directors of Multinational Megabucks Corporation is interested in finding out what factors contribute to miscommunications between management and labor so it can begin to develop strategies to reduce the number of miscommunications.
    f. The Gooddeeds Community Center wants to know what projects members of its board of directors, its major contributors, and its clients think should be funded during the upcoming fiscal year.
13. Big City School District has just received bad news. The state tax board has just ruled that several major businesses in its district have overpaid their taxes. As a result, the school district received too much money and will have to refund almost $1 million—10 percent of its total budget—during the current fiscal year. As research director for the school district, it is now your job to conduct some research to help the district decide where it should cut expenses in order to pay back the money. Because of budget and time constraints, you will have to do your research as quickly and cheaply as possible.
    a. Design a plan for using focus group research to explore options for cutting back expenses. In your plan, include information on (a) the number and kind(s) of people you will want as subjects, where you will find them, and

how you will go about persuading them to participate, (b) the number of groups you will need and their size and composition, (c) where and when you will hold the focus group sessions, (d) who you will use as moderator(s) and what training you will provide, (e) topics and suggested questions for the focus group sessions, (f) any supplementary material you will need for the sessions, (g) how you will record the sessions, and (h) how you will analyze the data. Also, (i) assess the reliability, internal validity, and external validity of your research plan. Justify your answers.

    b. Now that you have gathered information on options for cutting expenses, plan a study using one of the alternative techniques (nominal group, Delphi technique, or Q-methodology) to help the school district set budget-cutting priorities. In your plan, (a) explain which technique you chose, (b) tell who your subjects will be and how many subjects you will need, (c) describe how you will collect and analyze the data, and (d) assess the reliability and internal and external validity of your research plan. Justify your answers.

14. Many marketing research companies that specialize in focus group methodology will have opportunities for people to volunteer to participate in one of their studies. Search the Internet to find one and then participate in the study. Report to your class on your experience and your evaluation of the study. **Caution:** Be sure to pick a company and a study with which you are comfortable. For screening purposes many studies will require you to sign up and fill out a fairly detailed questionnaire concerning demographic and other characteristics. Before you give out any information, check the privacy policy and other details of the study to be sure it is being conducted in accordance with accepted standards for human subjects research.

Readings for Discussion

Ayres, J., Keereetaweep, T., Chen, P. and Edwards, P. A. (1995). Communication apprehension and employment interviews. *Communication Educator,* 47(1):1–17. The authors use subjects selected from student volunteers. This report includes a copy of the research protocol.

Chapel, G., Peterson, K. M. and Joseph, R. (1999). Exploring anti-gang advertisements: Focus group discussions with gang members and at-risk youth. *Journal of Applied Communication Research,* 27(3):237–257. The authors collect data from three mixed groups.

Croft, C. A. and Sorrentino, M. C. (1991). Physician interaction with families on issues of AIDS: What parents and youth indicate they desire. *Health Values,* 15(6):13–22. The authors collect data from youth, their parents, and educators as part of a needs assessment.

Dewan, S. and Littrell, J. (1996). Impact of small group dynamics on focus group data: Implications for social work. *The Journal of Applied Social Science,* 20(2):195–206. This literature review points out strengths and weaknesses of focus group research.

Krause, N., Chatters, L. M., Meltzer, T. and Morgan, D. L. (2000). Negative interaction in the church: Insights from focus groups and older Americans. *Review of Religious Research,* 41(4):510–533. The authors' interests are primarily academic, but their work points toward practical applications of focus group findings while also calling attention to the method's limitations.

Liebnow, E., Branch, K. and Orians, C. (1993). Perceptions of hazardous waste incineration risks: Focus group findings. *Sociological Spectrum, 13*:153–173. The authors gather data from community members as part of a risk management process.

Livingstone, S. M. (1994). Watching talk: Gender and engagement in the viewing of audience discussion programmes. *Media, Culture and Society, 16*:429–447. This is an example of qualitative audience analysis.

Lunt, P. and Livingstone, S. (1996). Rethinking the focus group in media and communication studies. *Journal of Communication, 46*(2):79–89. The authors address issues and review uses within the critical studies tradition.

Mitra, A. (1994). Use of focus groups in the design of recreation needs assessment questionnaires. *Evaluation and Program Planning, 17*:133–140. This is an example of using focus groups to help develop measures for use with other research methods.

Morris, T. A. (1995/1996). Delphi study of marketing teacher competencies. *Marketing Educators' Journal, 21:*3–16. The authors use fifty-three members of various professional organizations to reach consensus on qualifications teachers should have.

Simmons, H. (1999). Media, police and public information: From confrontation to conciliation. *Communications and the Law, 21*(2):69–84. The author uses Q-methodology to compare the orientations of reporters and police.

Ward, V. M., Bertrand, J. T. and Brown, L. F. (1991). The comparability of focus group and survey results: Three case studies. *Evaluation Review, 15:*266–283. The case study approach helps clarify situations in which focus group and survey findings may and may not coincide.

References and Resources

Davis, J. J. (1997). *Advertising research: Theory and practice.* Upper Saddle River, NJ: Prentice Hall, pp. 487–504. These pages cover Q-methodology as it is used in advertising research.

Delbecq, A. L., Van de Ven, A. H. and Gustafson, D. H. (1975). *Group techniques for program planning: A guide to nominal group and delphi processes.* Glenview, IL: Scott, Foresman and Company. The authors of this thorough guide conducted much of the early research that went into the development of small-group techniques.

Edmunds, H. (1999). *The focus group research handbook.* Chicago: American Marketing Association. This thorough yet easy-to-read guide has good examples of screeners, discussion guides, and final reports.

Javidi, M., Long, L. S., Vasu, M. L., and Ivy, D. K. (1991). Enhancing focus group validity with computer assisted technology in social science research. *Social Science Computer Review, 9*:231–245. The authors provide a good overview of ways computers can be used to manage and analyze data from focus group research.

Krueger, R. A. and Casey, M. A. (2000). *Focus groups: A practical guide for applied researchers,* 3rd ed. Thousand Oaks, CA: Sage. This updated version of a classic in the field supplements the basics with sections on collaborative empowerment techniques, action research, and computerized data analysis.

Krueger, R. A, Morgan, D. and King, J. A. (1998) The focus group tool kit. Thousand Oaks, CA: Sage. Authors vary, but collectively this set of six little books covers all the basics. Titles in the series are *Focus group guidebook; Planning focus groups; Developing questions for focus groups; Moderating focus groups; Involving community members in focus groups;* and *Analyzing and reporting focus group results.*

Merton, R. K., Fiske, M. and Kendall, P. L. (1956). *The focused interview.* Glencoe, IL: Free Press. This is the pioneering work by the noted sociologist and his colleagues.

Moore, C. M. (1994). *Group techniques for idea building,* 2nd ed. Newbury Park, CA: Sage. Chapters cover the nominal group technique and interpretive structural modeling for analyzing data. The 1987 edition includes the Delphi technique.

Morgan, D. L. (1988). *Focus groups as qualitative research.* Newbury Park, CA: Sage. This is another good overview of issues, methodologies, and data-analysis procedures.

10

Observational Research

THERE IS NO perfect, universally agreed upon name for this kind of research. In anthropology, this research method is usually called "ethnography"—the examination through observation of patterns of interactions, symbols, and other cultural artifacts to identify the norms or rules that govern the behavior of specific cultural groups and the meanings members of those groups ascribe to their own behavior and that of others.

But with the exception of some sociologists, researchers from other fields rarely immerse themselves in a culture to the extent and for the length of time necessary to produce a true ethnographic study. Therefore, we have chosen to call this method "observational research" although some practitioners and texts prefer to call it "participant observation," "field research," "naturalistic inquiry," or just "qualitative research."

By whatever name, this method is an essentially qualitative, inductive approach in which the researcher collects data by observing people as they interact with each other and their environment. This procedure makes it the most natural of all data-collection methods. It is natural, first because the people are observed on their own "turf"—in their natural setting, doing what they naturally and normally do as they go about their daily routines. Second, it is natural because observing others is the way people typically learn about and make sense of their world. What sets the method apart from the kind of naive observation people do every day is purpose and intensity.

In contrast to the everyday act of merely noticing—looking or watching as the opportunity or need presents itself—observation, as the basic data-collection method in observational research, must meet certain criteria. It must be intentional, intensive, extensive, and at least semidetached. The observations must also be documented.

These requirements make the method labor-intensive. It may or may not be expensive depending on the location and length of observation. But even if the research is inexpensive because the location is nearby and the observation period relatively short, doing the study can be exhausting work. The researcher must go to the scene of the action and stay there, observing and listening long enough to understand the setting and the subjects, and then figure out what is typical and atypical, important and unimportant about the interactions in order to make sense of what is seen and heard.

The chief advantages of observational research are that the method is flexible and the findings are real.

As an essentially inductive technique, researchers need not plan every detail in advance. They can enter the field and then adjust their plan as necessary to overcome unforeseen problems and take advantage of serendipitous happenings. This makes the method very well suited for studying both sudden, sometimes fleeting, happenings such as employees' reactions immediately after a corporate downsizing and ongoing processes such as adjustments in work routines and employee interactions after two very different companies merge.

In contrast to the indirect evidence in the form of self-reports collected through experiments, surveys, and focus groups and the even more indirect evidence of a content analysis, observational research provides direct, virtually irrefutable evidence of how people actually behave and interact. The behaviors that are the raw data actually exist. Subjects cannot easily conceal their normal actions and reactions the way they can "forget," selectively withhold, or even lie in other kinds of research.

Nevertheless, some weaknesses exist, most notably a lack of external validity.

Some places and people will always be inaccessible. But even if all places within a geographic region or an organization were equally open to the researcher, it would be impossible for the researcher to be in more than one place at a time. Therefore, the researcher will almost always miss something that might be vitally important.

Moreover, the presence of the researcher, whether observing openly or covertly, may change what transpires in ways that cannot be accounted for. And, no matter how long the researcher observes a setting or how intimately connected to the group the researcher becomes, the mere act of being a researcher can create a certain distance that invites missing or misinterpreting what transpires. At the same time, the researcher runs the risk of "going native." That is, the researcher may become so intimately involved with the setting and the group under observation that it becomes difficult to detect subtle but important changes. Even worse, from a research perspective, a researcher who comes to identify too closely with a group may try to protect it by withholding potentially damaging information.

As these drawbacks suggest, ethical problems abound. Their severity depends on the purpose of the study and, more directly, on where and how the observation takes place.

Communication Applications

Observational research is most common in anthropology and sociology, but it also has applications in applied communication research, where it is a staple both for news reporting and evaluation research.

- In journalism, reporters use the techniques of observational research whenever they add descriptive details about a person or place to their stories. It is the primary method for news gathering whenever they write reviews of restaurants, movies, or stage performances or when they go undercover to find out firsthand whether discrimination exists, whether stores are selling tobacco or alcohol to minors, how a group such as the Ku Klux Klan operates, or what life is like in a high school, prison, or mental hospital.
- Businesses and nonprofits frequently hire consultants to observe work procedures and then make recommendations about how to improve things such as efficiency, customer service, or employee morale.
- With increasing pressures for accountability in education, some school districts now hire outside experts to observe teachers in the classroom and make recommendations for improving student performance or distributing merit pay.

- The Association for Education in Journalism and Mass Communication sends out teams of evaluators to observe and evaluate college journalism programs and then make recommendations about whether programs should be accredited or reaccredited. Other accrediting agencies conduct the same kind of on-site evaluations of schools and other institutions such as hospitals and nursing homes.
- Grocery stores, department stores, and shopping malls hire researchers to evaluate store layout and design by watching customers to see how they move through the store and which displays they linger over and which ones they ignore. They also hire "secret shoppers" to gather information on customer service.
- Businesses or their advertising and marketing agencies make it a practice to send observers into the community where they can interact with teenagers or members of other demographic groups to learn about their tastes in music, clothing, and other consumer products.
- In public relations, researchers may visit museums or special events to find out through observing and listening whether displays are effective or whether funding an event was money well spent.
- Researchers may also attend meetings or participate in a program to determine how they are functioning and whether they are meeting the needs of a particular public.

As the examples suggest, applied research using observation as the method can take place in many settings. Studies vary in duration and intensity. But as a method, observational research is appropriate whenever it is important to learn about naturally occurring behavior, examine the interaction of people with their particular environment or setting, identify communication networks or channels of influence, or preserve a record of whole events or situations as they unfold.

Ethical Concerns

Ethical problems vary depending on the location for the research, the way the work is conducted, and whether the researcher knows the identity of the persons being observed.

As a general rule, ethical concerns are greater when observation is conducted in places where people have at least some expectation of privacy and when subjects' identity is known to the researcher than when observation occurs in public places or when subjects are truly anonymous. Using deception or misrepresentation in order to gather information can be acceptable, but engaging in those practices must be done with care because they can create ethical, legal, and practical problems.

In subsequent sections of this chapter, we have much more to say about the ethics of conducting observational research. In this section we discuss issues of consent, confidentiality, and risk to the researcher.

Because observation is an intrusive kind of research, you should always obtain clearance from your employer and/or from the research sponsor. But requirements for obtaining informed consent from subjects varies with the setting.

For studies conducted in truly public places such as a park or at a public event such as a street festival, ethical problems are minimal. Consent is unnecessary. The settings are accessible to anyone for almost any purpose and everyone present is at least minimally aware that other people may be watching or even observing them. Because of the very publicness of the setting and because the researcher generally does not know or even care enough to find out who the people are, privacy is not an issue. Anonymity is pretty much ensured.

But in other settings, privacy is very much a concern. Although it is possible to conduct observations on private property or to study groups and individuals whose identity is known without first gaining consent, doing so is risky. Failure to gain appropriate consent can open a researcher up to criminal trespass charges as well as to civil suits for intrusion and public disclosure of private information.

To conduct observations on private property or to study people who have at least a minimal expectation that the people with whom they are interacting are a part of the group, you should obtain consent from the person with the authority to grant permission. In most cases that will be the owner or manager of public facilities or private businesses, the president or perhaps the board of directors of a private organization, or an adult who actually lives in a house, condominium, or apartment.

Although it is rarely possible to spell out all methodological details before observation begins, those whose permission for a study is required must be told the purpose of the study, its likely duration, whether observation will be overt or covert, and enough other details to understand the benefits and the risks of the observation.

In institutional settings, it usually is not necessary to get informed consent from those actually being watched such as employees, clients, and customers. In any case, getting it might not be possible or it might interfere with the ability to observe truly naturally occurring behaviors. However, when children are the targets of the observation or an ancillary part of the study, both organizational policy and law may require obtaining informed consent from parents or guardians.

Although obtaining informed consent from everyone being observed is rarely necessary or even possible, being observed in places other than those that are truly public can generate genuine concerns for privacy among those being observed. If subjects' identities are known to the researcher, rightly or wrongly they may fear embarrassment or retaliation if their identities are subsequently revealed. If they did not know but later find out that they were observed, they may feel both spied upon and betrayed by an employer or an organization they trusted.

These feelings can open up the organization, the researcher, and the researcher's employer or sponsor to lawsuits. To minimize this kind of risk, information must be collected and handled with care. In most cases, this means taking appropriate steps to ensure the privacy of those being observed.

In observational research it is standard practice to remove subjects' names from observational records. Instead of using a name, researchers may refer to subjects by only a code number, the key to which is kept separate from the records themselves. It is also standard practice to omit names and other details by which a person could easily be identified from any reports. Where a name is necessary, researchers create and use pseudonyms.

However, there may be limits to how much confidentiality a researcher can or should provide. Through observation, a researcher may become privy to evidence of unethical or illegal behavior. If that occurs, duty to the sponsoring agency, the subject's employer, or society may outweigh the privacy interests of subjects.

Because this duty to others may supersede duty to subjects, you must consider carefully the circumstances under which you will reveal any potentially damaging information you acquire while doing your research. You might, for example, decide you will voluntarily reveal the information to others. Or you might decide you will reveal it upon request, only under duress, or refuse to divulge it even at the risk of losing a research contract, causing harm to others, or, in an extreme case, being charged with contempt or convicted of it for refusing to divulge information even when subpoenaed by a court.

In addition to the rather remote possibility of facing contempt charges, you could face civil suits for invasion of privacy, for libel should your findings embarrass or displease the subjects or sponsors of your work, or for restitution if anything goes wrong while you are present at a site.

Some kinds of research may also place a researcher in physical danger and/or at risk of being arrested on criminal charges. Riding along with the police or studying gangs or other marginal groups are obvious examples of research that can put you in a dangerous situation. A disgruntled employee or client can turn a workplace into a battle zone; crowds at parades, rallies, demonstrations, or rock concerts can get too exuberant or turn ugly. In such situations, you could be injured or caught among those arrested simply as a result of being in the wrong place at the wrong time.

Therefore, it is important to think about what could happen and then take steps to minimize foreseeable risks. But risks cannot always be eliminated completely. Some cannot even be foreseen. So you should always consider carefully whether the value of the research project makes any possibility for harm a risk worth taking.

Planning for Observation

As a primarily inductive approach, the researcher conducting an observational study begins with a broad substantive, theoretical, or practical problem. Because neither the setting nor the activities being observed are under the researcher's control, research design is necessarily emergent. Initial plans may be revised or scrapped. Details will almost certainly have to be worked out once data collection begins. Still, you will need to make some initial decisions about the setting and sites for data collection, how to gain access to them, and what role and stance to assume.

Choosing Settings and Sites

Here we use **setting** to mean the geographic or physical location of interest and **site** to mean a vantage point or a particular place within the setting.

For some applied observational studies, the setting and site may be the same thing. Both may be so integral to the purpose of the study that no real decision making about them is necessary. If the purpose for the research were simply to examine customer service at a grocery store, no real decision making would be involved. You would obviously go to the store. Once there, you would pose as a customer so you could observe what transpires from a customer's perspective.

But for most studies, any number of settings or sites within them might be appropriate. Because a researcher or even a team of researchers cannot possibly be everywhere simultaneously, some decisions about where to collect data will be necessary. Even with an unlimited budget, choosing many settings and sites through probability sampling will usually be impractical. Therefore, researchers usually begin by choosing a single setting because it is reasonably accessible, it appears to be typical, or it is of theoretical or practical significance for the research problem at hand.

Either at the beginning of the research or at some later time, a researcher may add more settings. These locations are usually selected to facilitate comparisons or help

> **Box 10.1.** *Evaluating research settings and sites*
>
> Determining whether a setting or site is appropriate for your purpose depends on your answers to questions about:
>
> **Location:** Is the place big enough to provide some diversity in the people present and their activities? Is it small enough to be manageable but not so small that your presence will be unduly disruptive? Are there suitable vantage points?
>
> **Subjects:** Who will be present at the setting or site? How many will be present? What are the positions of and relationships among the people who will be present? How long and on what days and during what hours will the people you are most interested in observing be there?
>
> **Actions:** What is the extent of people's activities at the site? How involved will they be? Will their activities be typical, atypical, or a mixture? When will the activities you are most interested in observing take place? Will subjects expect you to join in or to be uninvolved? How likely is your presence to affect subjects' actions?
>
> **Access:** Can you gain entry? Will you need an informant or sponsor? If so, can you find one? Can you gain acceptance? Will becoming accepted or comfortable at the location require extensive socialization? Can you assume and maintain an appropriate role and stance?

establish typicality by creating a research design that mimics that of a quasi-experiment. By using matching principles, for example, you might select a second location because it is very similar to the first site, or you might choose settings that differ on one or two dimensions that you could use as blocking variables in data analysis.

Where setting and site are not synonymous, the same principles outlined for choosing settings can be used for selecting sites. However, except on the rare occasion where a single site or vantage point will allow observing an entire setting, it will be necessary to select several sites, each one chosen for what it will contribute.

To provide a complete study, the chosen sites should vary in terms of their location and in the kinds of people and behaviors that you may observe from each vantage point. If there are too many such sites within a setting or too many people at one of those sites to make observing them all feasible, you may purposely select only certain sites or subjects. Alternatively, you may use probability sampling even though the number of sites or people selected in this way will very likely be too small for meaningful statistical analysis.

But in choosing settings and sites, you also need to take time into account. The people present and the behaviors they engage in will vary by time of day, day of week, and perhaps also by week, month, or season. Therefore, it will be necessary to schedule the study itself and also schedule, and perhaps reschedule, observations at each site so that they capture useful information.

Gaining Access

Two kinds of access must be considered: access to settings and sites and access to the people who are the main subject for the observation.

If you are hired to study people's reactions to a public event, gaining access to the setting and site will be no problem because of their very publicness. In other situations access may be guaranteed because the organization sponsoring the research also controls the setting. When access to a setting isn't guaranteed, you may be able to create your own access by simply "hanging around" a setting or with group members until someone invites you to come in. Or sometimes you may be able to find someone among your friends or acquaintances who is a member of a group you want to study and who can arrange access for you.

Gaining access by finding someone to arrange access by acting as your sponsor is very common. So is working with an informant. Where the **sponsor** just paves the way for access by vouching for you as a person or a researcher, the **informant** also acts as a guide by providing you with background information, pointing you toward strategic sites, and helping you become comfortable with and accepted by group members. In many cases the informant will also perform an audit function by checking your information, assumptions, and interpretations of what transpires during the course of your work.

Although there are many advantages to using a sponsor and/or informant, relying solely or even primarily on just one person can be very limiting. It may also compromise the research. The sponsor or informant may steer you to certain sites or people just because they are easy or obvious choices or away from others because of what observation at those sites or talking to those people may reveal. Therefore, it is a good idea, once observation begins, to find several more people whose advice can supplement and also serve as a check on the information provided by the sponsor/informant who initially helped arrange access.

Selecting a Role and Stance

Gaining access to a setting may be a one-time task, but gaining access to sites and subjects is not. You will probably have to negotiate effective access many times with a variety of insiders as the appropriateness of and need for access to particular sites or sets of people becomes clear over the course of the research. The ease of those negotiations and how much access you may achieve often depends on your ability to develop and maintain a plausible and appropriate role and stance.

In observational research, **role** refers to whether the researcher will participate in the group under observation and its various activities and, if so, how fully and in what capacity. **Stance** refers to whether those who are observed know or do not know they are being observed as part of a research project.

Both role and stance may sometimes differ from site to site or change over the course of the study. Any decision about how to present yourself will have both ethical and pragmatic implications for data collection. Events may occur that necessitate involvement where none was planned, or you may find it impossible to maintain a particular role or to keep the observation secret. Choosing from among the range of options requires a balancing of interests. These include the advantages and disadvantages of each possibility as well as the purpose of the study, the length of time available for observation, the preference of your client or research sponsor, and your own skills and preferences.

Role. You may choose to observe what is going on without getting personally involved with the group under observation and its activities, or you may choose to join the group at some level and in some capacity. Any choice you make will open up some opportunities and foreclose others.

Each role and each position within it will provide a perspective different from the one afforded by another role or position. The participant's perspective will more closely match that of true members of the group being observed than will that of the nonparticipant, who will always be more of an outsider. A researcher posing as a high-level employee might have access to the entire workplace and the freedom to roam around it but could find it difficult to establish the kind of rapport with low-level employees that might be useful. A secretary would be confined to a desk for much of the day but could probably interact successfully with both high-level and lower-level employees.

Although researchers who participate will be better able than nonparticipants to understand behaviors and interactions from the perspective of group members, researchers who choose a more involved role, and particularly one at a high level, will usually find it harder to behave in character than those who choose less active roles and/or lower-level positions.

The participant is also more likely than the nonparticipant to affect others in ways that render their behaviors unreal. For that reason, the nonparticipant is less likely than the participant to be held liable for anything that may go wrong during the research. The nonparticipant may also find it easier to remain objective than the participant for whom the risk of "going native" is always a danger.

Stance. Regardless of role, a researcher may choose to observe overtly or covertly. And as with role, there are trade-offs. The decision as to which stance to assume must be made by balancing the risks and rewards of observing openly against those of observing surreptitiously.

In general, overt observation presents fewer ethical problems than covert observation. Openly observing is also more likely to set up reactive effects. But there will always be trade-offs and problems regardless of which stance you choose.

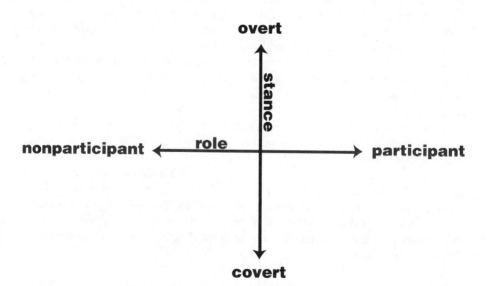

Figure 10.1. Typology of researcher presentation modes

In public places such as parks or museums, covert observation is no real threat to anyone because of the publicness of the site and because you are unlikely to know or care enough to ask about the identity of any individual person whom you are observing. In public places too openly observing may affect the naturalness of behaviors. Many people would also find it intrusive and at least mildly threatening were a researcher to approach them and explain how and why they are being observed.

Some groups will accept almost anyone as an overt observer because they welcome the opportunity to show outsiders what the group is really like. That kind of group will probably not adjust its behavior simply because members know they are being observed. But in other settings such as a workplace, group members may not accept an overt observer. They may, for example, refuse requests for information, purposely give misinformation, or try to adjust their actions to present themselves in a more favorable light.

By observing covertly, you will often be able to get a clearer, truer picture of customary behaviors. But covert observation also creates practical and ethical problems. Truly covert observers may not be able to roam a setting, ask certain kinds of questions, or gather supplementary material. They cannot take notes while observing as easily as overt observers can. They may also have more difficultly maintaining their cover or playing their chosen role successfully than will overt observers who do not have to guard every word or action lest a slip reveal their true identity and purpose.

With any kind of observation, people may come to feel used, threatened, or fearful for their safety if they know they are being watched and that the person doing the watching knows or could learn their identity. Even if they never find out they were watched, suspicions may be aroused when the covert observer suddenly disappears from the scene after group members have become accustomed to the observer's presence and may even have learned to depend on it. Therefore, researchers who observe covertly must devise an exit strategy that will let them withdraw gracefully from the scene at the end of the observation period.

Collecting and Recording Information

In observational research there are so many types of information and potentially so much of each type that collecting and managing it presents unique problems.

Types of Information

In observational research, data may be divided into three categories: data that result from observation, supplementary data generated as a result of researcher initiative, and supplementary data that comes from sources that exist independent of the research project.

Of the three data sources, data resulting from observation is fundamental to this kind of research. Most of the observational data come from what the researcher can see and hear. The observational data consist of physical descriptions of the research setting and site, the people who are naturally present, and accounts of their routines, behaviors, and interactions. It also includes records of overheard conversations between people naturally present at a site and interchanges between them and the researcher.

To supplement this kind of information, a researcher may gather additional information on activities that occur at a site by arranging for some subjects to keep diaries or use a video camera to record their activities during predetermined time periods. As an alternative, the researcher may contact people by telephone or pager and record or have the subjects record what they are doing at that time.

To find out about subjects' opinions, motives, and understanding, researchers may also distribute questionnaires, engage some subjects in lengthy focused conversations, or arrange for in-depth interviews or small-group discussions. Procedures for these techniques are covered in Chapters 7 and 9 on survey research and focus groups.

Additional data may come from artifacts and archival records. Typical artifacts include things such as samples, small gifts, or other handouts available to anyone who might visit a site or inquire about an organization or group and its activities. Archival records typically are documents that can be collected and analyzed using the content analysis techniques described in Chapter 8. These include newsletters and brochures, minutes of meetings, budgets and records of receipts and expenditures, policy statements, employee manuals, letters, memos, e-mail, and records of telephone conversations.

Determining What Information to Collect

With so much information available, deciding what to collect is a major problem in and of itself. Therefore, researchers usually begin by collecting as much information as possible. As they become comfortable with a site and its people, they begin to focus on just those typical or atypical occurrences that seem best suited to provide data relevant to the research problem. But at any time they may add items to their list of things to watch for or eliminate them as it becomes clear the information will be irrelevant. Work continues until the time or resources allotted to data collection run out or it becomes clear to the researcher that additional observation sessions are not producing much in the way of new or relevant information.

Techniques for Recording Data

Supplementary data provide no unusual problems. This kind of information can be collected using techniques outlined in previous chapters. Recording observational data, however, is a problem. Not only is there an overwhelming amount of it, but the information changes over time as people move in and out of a site and engage in or disengage in activities.

At least initially, recording as much of this observational information as possible is key to quality research. But even after observations have become focused, the amount of data that needs to be preserved for analysis can be huge. To preserve this information, researchers must rely primarily on their note-taking skills, although they may sometimes use various technological assists such as cameras and tape recorders.

Note Taking. Recording information is especially difficult in situations where the researcher is a covert, active participant. However, even when acting as an overt nonparticipant, simultaneously watching and taking extensive notes on what transpires is extremely difficult even for the best observational researchers. Therefore, most researchers will take **observation notes** and then create **field notes.**

Some observation notes may take the form of maps, floor plans, or drawings showing the arrangement of equipment or furnishings. But initially at least, most observation notes consist of brief notes jotted down during an observation session or immediately afterward. These notes are not meant to be a complete record of what transpires; rather, their purpose is to help the researcher remember what happened.

For simpler projects and at later stages in the research when the need for certain kinds of information becomes clear, observation notes may be structured. With this kind of note taking, the researcher creates a data-collection instrument similar to those used in survey research or content analysis and then uses it to check off, count, or otherwise keep track of things such as changes in the site or the people present, unusual occurrences, and/or the number, length, and nature of conversations or other activities.

Some information from structured observation notes may be transferred directly into a database suitable for quantitative analysis. However, most information from observation notes plus whatever else the researcher can remember from an observation setting will be used to create field notes. These notes are the extended record of what transpired during an observation session.

Because field notes are the most important source of data for analysis, they must be created as soon after an observation session as possible. To prevent information from becoming jumbled or lost as memories fade, researchers usually write up their field notes every day. In them they record physical descriptions of the setting, site, and people, accounts of what both the researcher and those naturally present on-site actually did or said, and notes on where, when, how, and from whom they collected or acquired any supplementary data.

The emphasis in these field notes is on objective information. But the researcher is not just the data collector. The researcher is a part of the scene. Therefore, it is crucial for researchers to record and preserve more subjective information such as their feelings, ideas for needed observations or information, and initial impressions or interpretations and questions about them.

Some researchers record this more subjective information in a **research diary.** Others prefer to incorporate it into their field notes, using notations such as a "?" to note questions, an "I" for ideas, or an "A" for their initial analysis or interpretation of what they said or heard. But whether they record this information in a separate diary or their field notes isn't important. What is important is that they record this information and record it in a way that precludes mixing up the two kinds of information when they analyze their data and then write up their findings.

Technological Assists. With so much information to record and preserve, tape recorders and still and video cameras can be a big help.

Still pictures and videotape of the data-collection site, important equipment, and people who figure prominently in the project can add to the record. Tape recorders are good for preserving natural language and detail from extended conversations and supplementary in-depth interviews. They can also be useful as an intermediary between observation notes and field notes. We have found, for example, that when working at several observation sites on a single day, it helps to keep a tape recorder in the car and then use it to record information while driving from one site to another.

But using technological assists can also create problems. Therefore, they are best thought of as a supplement to your observation notes and field notes, not as a substitute for them.

Equipment must be used, but using it may interfere with the more important work of observing and listening. Equipment may break or fail to record important information.

Action may move out of the range of a video camera. The typical still camera has a visual range of about 60°; the human eye has a field of vision just under 180°. Tape recorders may not pick up important conversations. Moreover, tapes must be transcribed. It can take six or more hours to transcribe just one hour of audio tape. For videotape, it will take even longer because of the need to create a record of the visuals as well as a transcript of the audio.

Cameras and recorders also inject another intrusive element into an already rather intrusive form of research. Recording in a public place or in a workplace setting in accordance with organization policy and appropriate permission is usually legal even if the people being recorded are unaware that they are being caught on audio- or videotape. Taping in other settings may very well be illegal.

Under federal law and in most states, consent to tape is needed from just one party to a conversation. Therefore, in most states you can legally record a conversation in which you are participating even if other parties to the conversation do not know they are being recorded. But in some states the law requires you to get consent from everyone taking part in the conversation. And it is illegal everywhere to record, without consent, any conversation if you are not present and participating in the conversation.

Managing Data

Developing a workable system for keeping track of observation notes, field notes, and other material is at least as important as creating the notes themselves. Managing the constant creation of data is basically a matter of coding and filing information during the data-collection phase.

Observation Notes and Field Notes. Filing observation notes and field notes each day in a master file will prevent your misplacing or accidentally destroying crucial evidence. Filing the notes chronologically creates a complete, easy-to-trace record of the work. But retrieving information from a master file can be difficult, so most researchers will also create a series of subfiles, each containing a specific kind of information that has been extracted from the field notes stored in the master file. The number and nature of categories for these files may develop over time, but creating files by site, person or small group, type of activity, or theme is common.

Copies of the materials for these subfiles can be created manually by cutting apart hard copy of field notes and filing the portions of copy chronologically in the appropriate binder or file drawer. These paper records can be further coded by placing symbols on them that will facilitate information retrieval. An "A," for example, could indicate that an activity record includes information about an atypical occurrence; "L" might indicate that one party to an activity is a group leader.

However, using a word-processing program to create the field notes will simplify the process of creating subfiles. It will also eliminate the problem of being swamped by binders, boxes, or file cabinets full of slowly disintegrating pieces of paper. With word processing, computer files of observation and field notes can be labeled by date and saved to a "master file" directory. Relevant portions from the master file can be extracted and saved to other directories. Files from all directories can then be directly imported into software programs where they can be coded or annotated to facilitate information retrieval and data analysis.

Supplementary Materials. Appropriately labeled and dated notes from supplementary interviews and some archival records may be filed along with observation notes and field notes as hard copy or in computerized form. Information from questionnaires and some kinds of archival records may best be preserved by entering it directly into computer files for quantitative data analysis. But many artifacts and some archival records can be preserved only by labeling and dating them and placing them in boxes, on shelves, or in file drawers reserved for storing odd-shaped, bulky, or otherwise unwieldy items.

Quality Control

Many qualitative researchers use terms such as "objectivity," "trustworthiness," and "credibility" to assess the quality of their work and that of others. These terms overlap with and encompass the concerns that, for consistency, we shall address under the rubric of "reliability" and "validity."

Reliability

This can be a weak spot. One of the most common criticisms of observational research is that the work cannot be reliable because the emergent and reactive nature of the research makes it almost impossible for anyone to replicate the study. But even if the study cannot be replicated, the results should be reproducible.

The work should, in the language of the qualitative researcher, be trustworthy. Reproducibility—trustworthiness—stems from careful documentation. Keeping complete records of everything that happens and everything you did and then making those records available to others offers assurance that whatever you said transpired actually happened. By examining your records, others should be able to agree that, "Yes. If I had been there, I could have done that, and so I, too, would have observed that."

Internal Validity

The same documentation that creates some assurance of at least hypothetical reproducibility of results also goes a long way toward ensuring internal validity—trustworthiness and also objectivity and credibility in the language of qualitative researchers. Here, the two main threats are researcher bias and reactivity effects between the researcher and the research subjects.

A study can be judged objective and, therefore, internally valid if the researcher does not enter the field with personal biases such as prejudices, fears, or strong positive or negative feelings about a site or a group or with preconceived notions about what will happen or about the need or lack thereof for change. These kinds of biases and preconceived notions can easily lead you to overlook, overvalue, or misinterpret some information.

To avoid this kind of problem, some researchers refuse to conduct observational studies in settings or with groups to which they have personal ties or about which they have strong feelings. Others prefer to team up with an outsider who has no ties to the setting or group or strong feelings about them. If strong feelings should develop during the course of a research project, they make those feelings "transparent" by recording and reporting them so that others may judge the likely impact for themselves.

Although strong feelings and preconceived notions must be avoided, some subjectivity is inherent in the method. These include judgments about where and when to study, what to observe and what to ignore, when to act or refrain from intervening in activities, and how to interpret what transpires. Because the researcher is on the scene and can come to understand the subjects in a way researchers using less direct methods of data collection cannot, making these kinds of subjective judgments helps produce findings that possess internal validity.

Your results will be judged real, trustworthy, and credible if there is sufficient evidence in the field notes and supplementary data for your observations and interpretations so that others can judge for themselves how your decision making may have contributed to your findings.

Some reactivity effects are also inherent in this kind of research. But they, too, can be a plus so long as the researcher is aware of them and accounts for them in the findings. By being on the scene, interacting with people, reacting to them, and being reacted to, the researcher can get a feel for how things really are—how people react to various stimuli and how they think and feel. Therefore, closeness to the scene and being a part of it creates face validity—a feeling that, "Yes, these findings are real."

But as a further step in quality control, you can and should triangulate your evidence whenever possible to help ensure the internal validity—the trustworthiness and credibility of your findings. **Triangulation**—comparing evidence collected in one way with evidence collected in other ways—ensures concurrent validity. Observations and interpretations can, for example, be checked against information from in-depth interviews or surveys; information gathered in those ways can be checked against artifacts and archival records. Everything can be further checked by having an informant and/or a few subjects whom you have come to trust and value for their honesty and insight assess findings and interpretations for completeness and reasonableness. For a more formal audit, you can also ask other researchers to examine the paper trail of field notes and other material to make sure there are no evidentiary gaps and that conclusions flow naturally and logically from the evidence.

Checking with informants or other researchers does not mean that you will or should make changes to protect the sensibilities of subjects or to match the preferred interpretations, conclusions, or recommendations of informants, subjects, or other researchers. It does mean taking their comments seriously by going back and filling in any gaps in the evidence, reconsidering interpretations, and/or finding ways to strengthen the link between your evidence and your recommendations or conclusions.

External Validity

Like focus groups, observational research is not intended to produce findings that can be generalized to other settings or groups. At best, external validity can only be built up over time by putting findings from one observational study together with findings from other studies. But this lack of external validity is not really a problem.

The strength of this method is in its ability to provide a thorough, accurate picture of how things are in one setting or with one group at a particular time. This makes the method useful in applied research as a way of assessing situations, finding solutions, or making recommendations that are applicable to and appropriate for a particular setting or site.

With that in mind, external validity is important only insofar as there is a need to make sure the evidence gathered fairly represents an entire setting or all members of a group. To ensure this more limited kind of external validity, you must choose your sites carefully so they provide the best possible vantage points for studying an entire setting and then conduct observations at various times to capture the full range of activities.

Whenever possible, you should also use supplementary data-collection techniques such as survey questionnaires that can be administered to an entire group or a subset of it.

Data Analysis

Like focus groups, observational research is intended as an inductive, qualitative research method. However, some quantitative analysis may be appropriate.

Quantitative Analysis

Although those who conduct observational research rarely use statistics to analyze their data, you can use statistics for some purposes. You might, for example, decide to quantify some measures and then use the tests of difference discussed in Chapter 6 to see whether any difference you detected between sites or groups of people through qualitative analysis is real. You might also appropriately use the measures of association discussed in Chapter 7 if you conducted a survey as part of your field work, or you could use the techniques described in Chapter 8 if you decide to do a formal content analysis of some documents you collect as part of your field work.

But in most cases, quantitative analysis of data collected during observational research is minimal. Researchers simply search their observation notes and field notes for evidence of the number of people present or absent at a site, the number who participate in various activities or kinds of interactions, and the frequency or duration of those activities or interactions.

They also look for and then report contingencies such as the number of men and women who initiate conversations with persons of the opposite sex. Usually they locate evidence for these contingencies by manually searching through their notes or using the "select" function in a qualitative data-analysis program to retrieve records that match on two or more keywords such as "sex" and "conversation." But even if they use computer programs for quantitative analysis, they do not really use or report the statistics because, in most cases, they would be misleading because of the way they select their settings, sites, and subjects and because of the small numbers involved.

Qualitative Analysis

With observational research, qualitative analysis is a continuous project. From the first day in the field through the final writing of the report, the researcher must continually be asking observational questions such as:

- What am I seeing? hearing?
- Who is present/absent?
- What is each person or small group doing?

and interpretive questions such as:

- Why are people doing what they are doing? Is it because of the environment, because of who is/isn't present, or because of me? Could it be just a habit, or is there a purpose or motivation?
- What does it all mean?

After each data-collection session, they write answers to observational questions into their field notes or research diaries. They enter answers to their interpretive questions as an observation if those answers come from what can be seen or heard. Otherwise they enter them as an analysis or as a question for further investigation. As they write up this information, they also begin coding the data or adding keywords that suggest meaningful categories or themes.

These initial keywords or codings guide subsequent data collection. Therefore, they will vary from project to project. Even within a single project, they may change over time. But by the end of data collection, the codes should have evolved into meaningful conceptual definitions. Proposed explanations will have become the research questions or hypotheses that can be answered by using the explanations and answers to the questions gathered through observations or by collecting supplementary material.

Actual data analysis begins with the researcher becoming fully immersed in the data by reading and rereading all observation notes, field notes, research diaries, and supplementary material and reflecting on their content. If the hypotheses that have evolved and the categories or keywords that have been used to code the notes no longer seem appropriate, revision will be necessary. But if things go as they should, everything should be in place for doing the final data analysis.

Either manually or with the assistance of a computer program designed for qualitative data analysis, the process begins by locating the records reflecting each concept specified in a potential explanation, checking those sets of records to make sure that nothing was misfiled, and then looking for the expected relationship among the concepts. In looking for these relationships, counting instances or durations and/or doing some more recoding to count intensities helps immeasurably in determining whether some observed behavior is typical or atypical.

Once the records that provide the evidence necessary to begin answering the research questions or testing hypotheses have been located and retrieved, the next steps are to

1. Search the master file and subfiles for evidence of chains of events, changes in the environment, and naturally occurring verbal accounts that could more fully explain, support, or refute basic findings.
2. Corroborate the findings with any available supplementary evidence such as that from in-depth interviews, questionnaires, artifacts, and archival records.
3. Extract excerpts from the master file, subfiles, and other data sources for those observations, quotes, or other materials that
   - best illustrate the basic findings.
   - indicate the strength of the findings.
   - provide countervailing evidence or illustrate atypical occurrences.

Although qualitative analysis is an ongoing, iterative process that begins on the first day of research and continues throughout the project, we find it can be helpful to take a break between data collection and immersion and again between immersion and the final analysis. Because the researcher is so intimately involved in all phases of the work, taking these little breaks makes it easier to see new connections and also to spot problems with the data and/or the proposed explanations, conclusions, and recommendations.

Main Points

- Observational research requires researchers to gather data in a natural or real-world setting through the observation of individuals as they interact with one another and their environment.
- Observational research is usually inductive rather than deductive.
- Observational research is appropriate whenever it is important to learn about naturally occurring behavior, examine the interactions of people with their environment, identify communication networks or channels of influence, or preserve a record of whole events or situations as they unfold.
- Ethical problems in observational research vary depending on the location for the research, the way the work is conducted, and whether the researcher knows the identity of the persons being observed.
- For studies conducted in public places, ethical problems are minimal and consent is usually unnecessary. In other settings, privacy is a major concern. Failure to gain consent can lead to problems.
- To protect subjects' privacy, it is standard practice to remove subjects' names from observational records and to use pseudonyms for subjects and locations or some other technique to mask identities in reports.
- The setting for observational research refers to the geographic or physical location of interest, and the site refers to a vantage point or particular place within a setting.
- Settings are often chosen for their theoretical or practical significance or because they appear to be typical or atypical.
- Sites are usually selected so that they vary in their location and the people and behaviors that can be observed from each vantage point.
- Researchers must work to gain access to settings, sites, and the people who are the main subjects of the observation. Access is often obtained through the use of sponsors and/or informants.
- The role that a researcher plays can vary from complete participant to complete nonparticipant. Each possible role has advantages and disadvantages. Each role will affect the researcher's degree of access and provide the researcher with a different perspective.
- The data in observational research can be divided into three categories: data that result from observation, supplementary data generated as the result of researcher initiative, and supplementary data that come from sources that exist independent of the research project.
- Data resulting from observation include physical descriptions of the research setting, site, and people involved; accounts of the subjects' routines, behaviors, and interactions; and records of overheard conversations and interchanges between them and the researcher.
- Note taking forms the basis for observational data.
- Observation notes consist of brief notes jotted down during an observation session or immediately afterward. Cameras and tape recorders can be used as a backup to written notes.
- Observation notes are used as the basis for field notes that are longer, detailed descriptions of what transpired.
- Supplementary materials include researcher-generated in-depth interviews and surveys as well as pre-existing documents and artifacts that are available at a setting or site.

- A reliable observational study is one that is trustworthy.
- The two main threats to internal validity in observational research are bias and reactivity effects between the researcher and the research subjects. Both can be guarded against by triangulating evidence.
- Observational research is not usually intended to produce findings that can be generalized to other groups and settings. At best, external validity can only be built up over time as findings from other studies are examined.
- Observational research is a qualitative method. The analysis is an ongoing, iterative process.
- Thorough documentation and description help make reports of observational research credible and trustworthy.
- Although some numbers may be included in reports, quantitative analysis is usually used only with supplementary data such as surveys or documents.

Terms to Know

ethnography
"going native"
induction
setting
site
sponsor
informant
role
stance

overt observation
covert observation
observation note
field note
structured note
unstructured note
research diary
triangulation

Questions for Discussion and Review

1. Explain how observational research differs from the naïve observations people engage in on a day-to-day basis.
2. Describe the strengths and weaknesses of observational research, and tell how those strengths and weaknesses compare to the ones for experiments, surveys, and focus groups. In your answers, be sure to discuss time, cost, ethics, reliability, and validity.
3. Explain how the purpose of the study, the location of the study, and the stance and role of the researcher may affect the severity of ethical problems surrounding observational research, and tell what a researcher can do to safeguard the subjects of an observational research study and protect the integrity of the research project.
4. Look at your state's law relating to intrusion and public disclosure of private information and also federal and state law relating to recording conversations and monitoring employees and clients in a workplace setting. Report your findings.
5. Describe the advantages and disadvantages of each role (participant, nonparticipant) and stance (overt observer, covert observer) observational researchers can assume, and give an example of when each would be appropriate and inappropriate.

6. Explain what is meant by "going native" and tell how "going native" might affect the reliability and/or internal validity of an observational study.
7. Describe the role of informants and sponsors in observational research. Explain how the choice of a sponsor or informant may affect reliability and validity.
8. Define and differentiate among observation notes, field notes, and a research diary.
9. Find an example of an observational study in a communication journal. For that article, (a) explain the main purpose of the study and tell whether the purpose could have been met by using some other research method. Then (b) describe the setting and site for the research, and tell why the researchers chose them, (c) tell whether the researchers used a sponsor, informant, or some other method to gain access to the site and to individuals and why they chose the strategy they used, (d) explain the role and stance the researchers chose and why they chose them, (e) tell whether the researchers used structured or unstructured observations and whether they relied on observation notes and field notes or also used recording devices and/or supplementary data, and (e) explain how they analyzed their data and presented their findings. Finally, (f) evaluate their work by taking into account how well the evidence presented in the article supports the findings and conclusions drawn in this study as well as any ethical and pragmatic problems the research raised. Be sure to note the ways in which the choice of sites, settings, informants, and/or sponsors and the stance and role of the researcher may have affected reliability and validity and whether the researchers could and should have done their work differently.
10. Develop three communication research questions for which observational research would be an appropriate approach, and explain why you would use this method instead of other methods discussed in this text.
11. For each of the questions you developed in question 10, explain (a) what you will choose as your setting and site(s), (b) how you will gain access to them, (c) what role and stance you will assume, (d) what kinds of information you will be looking for and what supplementary data you will want to collect, and (e) what practical or ethical problems you foresee in conducting this research.
12. You have decided that you want to do a study of gender differences in job-related communication behavior but are afraid that if you simply ask questions about communication behavior, people either won't tell or don't know the truth. Therefore, you have decided that observational research is the best method for studying naturally occurring gender differences in communication behavior. For the purposes of this question, communication is defined broadly as "any activity involving at least one male and one female during which some gathering or exchange of information occurs."
    a. Spend at least 20 minutes on two separate occasions observing and taking notes on communication in a job-related situation. During your observation periods, make observation notes and then make field notes. Make sure your notes contain the location of the observations; the day, date, and time of each observation; a description of the subjects; a description of your role and/or stance; whether your observations were covert or overt; and information on what actually occurred. The latter might include a transcript of the actual communication exchange, body language, and anything else that could affect the communication.

b. Prepare a summary of your observations and a list of hypotheses and/or research questions that deserve further investigation.
c. Compare your tentative findings and your hypotheses and/or research questions to those of others in your class. If all results are very similar, explain the steps you might take to make sure your findings are trustworthy. If the results differ, explain how you would go about resolving the differences so that the final results of your observational research will be credible and trustworthy.
13. Use one of the Internet search engines and the keywords "qualitative data analysis software" to find information about computer programs for analyzing qualitative data. Report your findings.

Readings for Discussion

Allen, D. (2000). Doing occupational demarcation: The "boundary-work" of nurse managers in a district general hospital. *Journal of Contemporary Ethnography, 29*(3):326–356. A nurse overtly acting as a researcher examines rhetorical strategies used by doctors and nurses in the workplace.

Boeyink, D. E. (1994). How effective are codes of ethics? *Journalism and Mass Communication Quarterly, 71*(4):893–904. The author studies decision making in three newsrooms.

Brown, J., Dykers, C., Steele, J. and White, A. (1994). Teenage room culture. *Communication Research, 21*(6):813–827. The authors create a scheme for categorizing teens into types on the basis of artifacts on display in their bedrooms.

Eveslage, S. and Delaney, K. (1998). Talkin' trash at Hardwick High: A case study of insult talk on a boys basketball team. *International Review for the Sociology of Sport, 33*(3):239–254. An insider and an outsider team up to conduct this study.

Fuller, L. K. (1983). "The Recorder": An observational study of a community newspaper. *Newspaper Research Journal, 4*(3):25–32. The author combines insider knowledge with nonparticipant observation to explore causes of worker harmony and potential problems of charismatic leadership.

Gibson, M. K. and Papa, M. J. (2000). The mud, the blood, and the beer guys: Organizational osmosis in blue-collar work groups. *Journal of Applied Communication Research, 28*(1):68–88. The authors use assimilation theory as a framework for examining how newcomers become part of a work group in a manufacturing plant.

Harrell, W. A. (1991). Factors influencing pedestrian cautiousness in crossing streets. *The Journal of Social Psychology, 13*(3):367–372. Authors use nonparticipant observation to assess risk-taking behavior at different times and locations and under varied conditions.

May, R. A. B. (1999). Tavern culture and television viewing: The influence of viewing culture on patrons' reception of television programs. *Journal of Contemporary Ethnography, 28*(1):69–99. The author uses the bartender to gain entry into a middle-class African American community.

Pardun, C. and Krugman, D. (1994). How the architectural style of the home relates to television viewing. *Journal of Broadcasting and Electronic Media, 38*(2): 145–162. The authors use floor plans and furniture arrangements to help explain viewing behavior.

Souza, T. J. (1999). Communication and alternative school student socialization. *Communication Education, 48*(3):91–108. The author observes openly as a nonparticipant.

References and Resources

Abrams, B. (2000). *The observational research handbook: Understanding how consumers live with your product.* Chicago: American Marketing Association. This is a practical introduction to the use of observational research in the field of marketing.

Anderson, J. A. (1987). *Communication research: Issues and methods.* New York: McGraw-Hill, Ch. 9–12. Illustrated with many examples from the author's own research, these chapters cover both the theoretical underpinnings and practical tips for conducting observational research.

Emerson, R. M., Fretz, R. I. and Shaw, L. L. (1995). *Writing ethnographic fieldnotes.* Using actual unfinished working notes as examples, the authors provide guidelines and suggestions for taking field notes.

Emmison, M. and Smith, P. (2000). *Researching the visual: Images, objects, cultures and interactions in social and cultural inquiry.* Thousand Oaks, CA: Sage. This book provides a thorough introduction to the use of images and artifacts in ethnographic and cultural research.

Guba, E. (1978). *Toward a methodology of naturalistic inquiry in educational evaluation.* Los Angeles: University of California Center for the Study of Evaluation. This helpful little monograph is a precursor to the more thorough Lincoln and Guba text.

Lincoln, Y. C. and Guba, E. (1985). *Naturalistic inquiry.* Beverly Hills, CA: Sage. This text provides a thorough introduction to the grounded, inductive approach to observational research.

Lindlof, T. R., ed. (1987). *Natural audiences: Qualitative studies of media uses and effects.* This volume is a handy source of examples of qualitative studies, some of which employ observational techniques.

Schensul, J. J., LeCompte, M. D., Hess, G. A. Jr., Nastasi, B. K., Berg, M. J., Williamson, L., Brecher, J. and Glasser, R. (1999). *The ethnographer's tool kit.* Thousand Oaks, CA: Sage. This set of seven books with varying authors covers all the basics plus some more advanced techniques. Titles include *Designing and conducting ethnographic research; Essential ethnographic methods; Enhanced ethnographic methods; Mapping social networks, spatial data, and hidden populations; Analyzing and interpreting ethnographic data;* and *Research roles and research partnerships.*

Van Massen, J. (1988). *Tales of the field: On writing ethnography.* Chicago: University of Chicago Press. The author describes and evaluates narrative conventions associated with writing about culture.

American Sociological Association (ASA) Code Of Ethics

THE FOLLOWING EXCERPTS, which deal with the conduct of research, are similar to those found in other professional codes of ethics. They are reproduced with permission of the American Sociological Association. Other parts of the ASA code cover standards for competence, social responsibility, data sharing, publication process, manuscript reviewing, teaching and training others, contractual and consulting services, and adherence to the code.

Ethical Standards

1. Professional and Scientific Standards

Sociologists adhere to the highest possible technical standards that are reasonable and responsible in their research, teaching, practice, and service activities. They rely on scientifically and professionally derived knowledge; act with honesty and integrity; and avoid untrue, deceptive, or undocumented statements in undertaking work-related functions or activities.

11. Confidentiality

Sociologists have an obligation to ensure that confidential information is protected. They do so to ensure the integrity of research and the open communication with research participants and to protect sensitive information obtained in research, teaching, practice, and service. When gathering confidential information, sociologists should take into account the long-term uses of the information, including its potential placement in public archives or the examination of the information by other researchers or practitioners.

11.01 Maintaining Confidentiality

(a) Sociologists take reasonable precautions to protect the confidentiality rights of research participants, students, employees, clients, or others.

(b) Confidential information provided by research participants, students, employees, clients, or others is treated as such by sociologists even if there is no legal protection or privilege to do so. Sociologists have an obligation to protect confidential information, and not allow information gained in confidence from being used in ways that would unfairly compromise research participants, students, employees, clients, or others.

(c) Information provided under an understanding of confidentiality is treated as such even after the death of those providing that information.

(d) Sociologists maintain the integrity of confidential deliberations, activities, or roles, including, where applicable, that of professional committees, review panels, or advisory groups (e.g., the ASA Committee on Professional Ethics).

(e) Sociologists, to the extent possible, protect the confidentiality of student records, performance data, and personal information, whether verbal or written, given in the context of academic consultation, supervision, or advising.

(f) The obligation to maintain confidentiality extends to members of research or training teams and collaborating organizations who have access to the information. To ensure that access to confidential information is restricted, it is the responsibility of researchers, administrators, and principal investigators to instruct staff to take the steps necessary to protect confidentiality.

(g) When using private information about individuals collected by other persons or institutions, sociologists protect the confidentiality of individually identifiable information. Information is private when an individual can reasonably expect that the information will not be made public with personal identifiers (e.g., medical or employment records).

11.02 Limits of Confidentiality

(a) Sociologists inform themselves fully about all laws and rules which may limit or alter guarantees of confidentiality. They determine their ability to guarantee absolute confidentiality and, as appropriate, inform research participants, students, employees, clients, or others of any limitations to this guarantee at the outset consistent with ethical standards set forth in 11.02(b).

(b) Sociologists may confront unanticipated circumstances where they become aware of information that is clearly health- or life-threatening to research participants, students, employees, clients, or others. In these cases, sociologists balance the importance of guarantees of confidentiality with other principles in this Code of Ethics, standards of conduct, and applicable law.

(c) Confidentiality is not required with respect to observations in public places, activities conducted in public, or other settings where no rules of privacy are provided by law or custom. Similarly, confidentiality is not required in case of information available from public record.

11.03 Discussing Confidentiality and Its Limits

(a) When sociologists establish a scientific or professional relationship with persons, they discuss (1) the relevant limitations on confidentiality, and (2) the foreseeable uses of the information generated through their professional work.

(b) Unless it is not feasible or is counterproductive, the discussion of confidentiality occurs at the outset of the relationship and thereafter as new circumstances may warrant.

11.04 Anticipation of Possible Uses of Information

(a) When research requires maintaining personal identifiers in data bases or systems of records, sociologists delete such identifiers before the information is made publicly available.

(b) When confidential information concerning research participants, clients, or other recipients of service is entered into databases or systems of records available to persons without the prior consent of the relevant parties, sociologists protect anonymity by not including personal identifiers or by employing other techniques that mask or control disclosure of individual identities.

(c) When deletion of personal identifiers is not feasible, sociologists take reasonable steps to determine that appropriate consent of personally identifiable individuals has been obtained before they transfer such data to others or review such data collected by others.

11.05 Electronic Transmission of Confidential Information

Sociologists use extreme care in delivering or transferring any confidential data, information, or communication over public computer networks. Sociologists are attentive to the problems of maintaining confidentiality and control over sensitive material and data when use of technological innovations, such as public computer networks, may open their professional and scientific communication to unauthorized persons.

11.06 Anonymity of Sources

(a) Sociologists do not disclose in their writings, lectures, or other public media confidential, personally identifiable information concerning their research participants, students, individual or organizational clients, or other recipients of their service which is obtained during the course of their work, unless consent from individuals or their legal representatives has been obtained.

(b) When confidential information is used in scientific and professional presentations, sociologists disguise the identity of research participants, students, individual or organizational clients, or other recipients of their service.

11.07 Minimizing Intrusions on Privacy

(a) To minimize intrusions on privacy, sociologists include in written and oral reports, consultations, and public communications only information germane to the purpose for which the communication is made.

(b) Sociologists discuss confidential information or evaluative data concerning research participants, students, supervisees, employees, and individual or organizational clients only for appropriate scientific or professional purposes and only with persons clearly concerned with such matters.

11.08 Preservation of Confidential Information

(a) Sociologists take reasonable steps to ensure that records, data, or information are preserved in a confidential manner consistent with the requirements of this Code of Ethics, recognizing that ownership of records, data, or information may also be governed by law or institutional principles.

(b) Sociologists plan so that confidentiality of records, data, or information is protected in the event of the sociologist's death, incapacity, or withdrawal from the position or practice.

(c) When sociologists transfer confidential records, data, or information to other persons or organizations, they obtain assurances that the recipients of the records, data, or information will employ measures to protect confidentiality at least equal to those originally pledged.

12. Informed Consent

Informed consent is a basic ethical tenet of scientific research on human populations. Sociologists do not involve a human being as a subject in research without the informed consent of the subject or the subject's legally authorized representative, except as otherwise specified in this Code. Sociologists recognize the possibility of undue influence or subtle pressures on subjects that may derive from researchers' expertise or authority, and they take this into account in designing informed consent procedures.

12.01 Scope of Informed Consent

(a) Sociologists conducting research obtain consent from research participants or their legally authorized representatives (1) when data are collected from research participants through any form of communication, interaction, or intervention; or (2) when behavior of research participants occurs in a private context where an individual can reasonably expect that no observation or reporting is taking place.

(b) Despite the paramount importance of consent, sociologists may seek waivers of this standard when (1) the research involves no more than minimal risk for research participants, and (2) the research cannot practicably be carried out were informed consent to be required. Sociologists recognize that waivers of consent require approval from institutional review boards or, in the absence of such boards, from another authoritative body with expertise on the ethics of research. Under such circumstances, the confidentiality of any personally identifiable information must be maintained unless otherwise set forth in 11.02(b).

(c) Sociologists may conduct research in public places or use publicly available information about individuals (e.g., naturalistic observations in public places, analysis of public records, or archival research) without obtaining consent. If, under such circumstances, sociologists have any doubt whatsoever about the need for informed consent, they consult with institutional review boards or, in the absence of such boards, with another authoritative body with expertise on the ethics of research before proceeding with such research.

(d) In undertaking research with vulnerable populations (e.g., youth, recent immigrant populations, the mentally ill), sociologists take special care to ensure that the voluntary nature of the research is understood and that consent is not coerced. In all other respects, sociologists adhere to the principles set forth in 12.01(a) - (c).

(e) Sociologists are familiar with and conform to applicable state and federal regulations and, where applicable, institutional review board requirements for obtaining informed consent for research.

12.02 Informed Consent Process

(a) When informed consent is required, sociologists enter into an agreement with research participants or their legal representatives that clarifies the nature of the research and the responsibilities of the investigator prior to conducting the research.

(b) When informed consent is required, sociologists use language that is understandable to and respectful of research participants or their legal representatives.

(c) When informed consent is required, sociologists provide research participants or their legal representatives with the opportunity to ask questions about any aspect of the research, at any time during or after their participation in the research.

(d) When informed consent is required, sociologists inform research participants or their legal representatives that their participation or continued participation is voluntary; they inform participants of significant factors that may be expected to influence their willingness to participate (e.g., possible risks and benefits of their participation); and they explain other aspects of the research and respond to questions from prospective participants. Also, if relevant, sociologists explain that refusal to participate or withdrawal from participation in the research involves no penalty, and they explain any foreseeable consequences of declining or withdrawing. Sociologists explicitly discuss confidentiality and, if applicable, the extent to which confidentiality may be limited as set forth in 11.02(b).

(e) When informed consent is required, sociologists keep records regarding said consent. They recognize that consent is a process that involves oral and/or written consent.

(f) Sociologists honor all commitments they have made to research participants as part of the informed consent process except where unanticipated circumstances demand otherwise as set forth in 11.02(b).

12.03 Informed Consent of Students and Subordinates

When undertaking research at their own institutions or organizations with research participants who are students or subordinates, sociologists take special care to protect the prospective subjects from adverse consequences of declining or withdrawing from participation.

12.04 Informed Consent with Children

(a) In undertaking research with children, sociologists obtain the consent of children to participate, to the extent that they are capable of providing such consent, except under circumstances where consent may not be required as set forth in 12.01(b).

(b) In undertaking research with children, sociologists obtain the consent of a parent or a legally authorized guardian. Sociologists may seek waivers of parental or guardian consent when (1) the research involves no more than minimal risk for the research participants, and (2) the research could not practicably be carried out were consent to be required, or (3) the consent of a parent or guardian is not a reasonable requirement to protect the child (e.g., neglected or abused children).

(c) Sociologists recognize that waivers of consent from a child and a parent or guardian require approval from institutional review boards or, in the absence of such boards, from another authoritative body with expertise on the ethics of research. Under such circumstances, the confidentiality of any personally identifiable information must be maintained unless otherwise set forth in 11.02(b).

12.05 Use of Deception in Research

(a) Sociologists do not use deceptive techniques (1) unless they have determined that their use will not be harmful to research participants; is justified by the study's prospective scientific, educational, or applied value; and that equally effective alternative procedures that do not use deception are not feasible, and (2) unless they have obtained the approval of institutional review boards or, in the absence of such boards, with another authoritative body with expertise on the ethics of research.

(b) Sociologists never deceive research participants about significant aspects of the research that would affect their willingness to participate, such as physical risks, discomfort, or unpleasant emotional experiences.

(c) When deception is an integral feature of the design and conduct of research, sociologists attempt to correct any misconception that research participants may have no later than at the conclusion of the research.

(d) On rare occasions, sociologists may need to conceal their identity in order to undertake research that could not practically be carried out were they to be known as researchers. Under such circumstances, sociologists undertake the research if it involves no more than minimal risk for the research participants and if they have obtained approval to proceed in this manner from an institutional review board or, in the absence of such boards, from another authoritative body with expertise on the ethics of research. Under such circumstances, confidentiality must be maintained unless otherwise set forth in 11.02(b).

12.06 Use of Recording Technology

Sociologists obtain informed consent from research participants, students, employees, clients, or others prior to videotaping, filming, or recording them in any form, unless these activities involve simply naturalistic observations in public places and it is not anticipated that the recording will be used in a manner that could cause personal identification or harm.

13. Research Planning, Implementation, and Dissemination

Sociologists have an obligation to promote the integrity of research and to ensure that they comply with the ethical tenets of science in the planning, implementation, and dissemination of research. They do so in order to advance knowledge, to minimize the possibility that results will be misleading, and to protect the rights of research participants.

13.01 Planning and Implementation

(a) In planning and implementing research, sociologists minimize the possibility that results will be misleading.

(b) Sociologists take steps to implement protections for the rights and welfare of research participants and other persons affected by the research.

(c) In their research, sociologists do not encourage activities or themselves behave in ways that are health- or life-threatening to research participants or others.

(d) In planning and implementing research, sociologists consult those with expertise concerning any special population under investigation or likely to be affected.

(e) In planning and implementing research, sociologists consider its ethical acceptability as set forth in the Code of Ethics. If the best ethical practice is unclear, sociologists consult with institutional review boards or, in the absence of such review processes, with another authoritative body with expertise on the ethics of research.

(f) Sociologists are responsible for the ethical conduct of research conducted by them or by others under their supervision or authority.

13.02 Offering Inducements for Research Participants

Sociologists do not offer excessive or inappropriate financial or other inducements to obtain the participant of research participants, particularly when it might coerce participation. Sociologists may provide incentives to the extent that resources are available and appropriate.

13.03 Reporting on Research

(a) Sociologists disseminate their research findings except where unanticipated circumstances (e.g., the health of the researcher) or proprietary agreements with employers, contractors, or clients preclude such dissemination.

(b) Sociologists do not fabricate data or falsify results in their publications or presentations.

(c) In presenting their work, sociologists report their findings fully and do not omit relevant data. They report results whether they support or contradict the expected outcomes.

(d) Sociologists take particular care to state all relevant qualifications on the findings and interpretation of their research. Sociologists also disclose any underlying assumptions, theories, methods, measures, and research designs that might bear upon findings and interpretations of their work.

(e) Consistent with the spirit of full disclosure of methods and analyses, once findings are publicly disseminated, sociologists permit their open assessment and verification by other responsible researchers with appropriate safeguards, where applicable, to protect the anonymity of research participants.

(f) If sociologists discover significant errors in their publication or presentation of data, they take reasonable steps to correct such errors in a correction, a retraction, published errata, or other public fora as appropriate.

(g) Sociologists report sources of financial support in their written papers and note any special relations to any sponsor. In special circumstances, sociologists may withhold the names of specific sponsors if they provide an adequate and full description of the nature and interest of the sponsor.

(h) Sociologists take special care to report accurately the results of others' scholarship by using correct information and citations when presenting the work of others in publication, teaching, practice, and service settings.

14. Plagiarism

(a) In publications, presentations, teaching, practice, and service, sociologists explicitly identify, credit, and reference the author when they take data or material verbatim from another person's written work, whether it is published, unpublished, or electronically available.

(b) In their publications, presentations, teaching, practice, and service, sociologists provide acknowledgment of and reference to the use of others' work, even if the work is not quoted verbatim or paraphrased, and they do not present others' work as their own whether it is published, unpublished, or electronically available.

15. Authorship Credit

(a) Sociologists take responsibility and credit, including authorship credit, only for work they have actually performed or to which they have contributed.

(b) Sociologists ensure that principal authorship and other publication credits are based on the relative scientific or professional contributions of the individuals involved, regardless of their status. In claiming or determining the ordering of authorship, sociologists seek to reflect accurately the contributions of main participants in the research and writing process.

(c) A student is usually listed as principal author on any multiple authored publication that substantially derives from the student's dissertation or thesis.

B

Evaluating Research Reports

THE FACT THAT a report of research findings has been made available or that an article based on the research has been published does not guarantee that the work was done well or that it will be useful for your purposes. Regardless of how the findings were made available to the public or where a study was published, both the quality of the report and the quality of the work on which it was based may vary widely. Even the best research may have some flaw that renders it useless for a particular purpose; the worst may have a kernel of useful information. Therefore, it is always a good idea to evaluate each report critically before deciding whether to adopt or adapt the methodology or trust the findings.

The information on reliability and validity in the chapters on sampling and measurement and the Ethical Concerns and Quality Control sections of the chapters on basic research methods provide the basic information you will need to evaluate published reports of research findings. This appendix condenses that information into a list of points to consider as you work your way through the evaluation process.

That process begins with a background check. This check can alert you to potential problems that may affect the quality of the research and/or the presentation of findings or recommendations. For a background check, you should:

1. Determine why and by whom the research was done.
   - Determine the source of funding for the research project: whether it was conducted in-house, commissioned, done on a grant from some funding agency, or "self-funded" as is true of much research done by faculty and graduate students.
   - Think about how the source of funding, or lack thereof, may have affected the purpose or conduct of the work and/or the presentation of findings.
   - Assess the experience and reputation of the researcher and/or the firm responsible for conducting the research.
2. Evaluate the manner and place of publication.
   - Determine whether the report qualifies as a primary or a secondary source.
   - Determine whether the report appears in a book, technical report, or other publication released by an established publishing house or whether it was self-published by the researcher, sponsor, or client organization.

- Consider whether the publisher or publication may have theoretical or methodological preferences and/or ideological biases.
- Assess the reputation of the publisher and/or the publication for quality.
- Determine whether reports are peer-reviewed prior to publication or, for trade publications and mass media, whether there are procedures for checking submissions for factual accuracy or quality.

Next, examine the report itself. The items listed here follow the outline for a journal article. However, the amount of information you can reasonably expect under each point will vary depending on whether the study was conducted as proprietary research, where and by whom the report was published, and whether the report qualifies as a primary or secondary source. In general, you should expect the most thorough information on all points in dissertations, theses, and books that qualify as primary sources. You should expect somewhat less but still very complete information in peer-reviewed journal articles and much less in most secondary sources and in articles from trade publications and the mass media.

1. Examine the literature review.
   - Determine whether the literature cited is relevant for the purpose of the study.
   - Look for obvious omissions of foundational, key, or recent work.
   - Assess presentations of information from the literature for accuracy.

Note: For secondary sources and articles in the trade press and the mass media, literature reviews are rare. Any information showing how the work draws on or fits with findings from other studies is a plus.

2. Consider the hypotheses and/or research questions.
   - Assess whether these address the stated purpose of the study and whether they flow naturally and logically from the literature review.
   - Determine whether hypotheses are stated clearly and properly so they can be tested and potentially proven false. Determine whether research questions are clear and answerable through social science research.
   - Check the methodology section to see whether there are clear conceptual and operational definitions for any evaluative terms (e.g., "prosocial" or "improve") used in the hypotheses or research questions.

Note: Hypotheses and/or research questions may not be stated explicitly in secondary sources, trade publications, or the mass media, but you should be able to infer them from other information in the report. You should also expect at least an implicit definition of evaluative terms.

3. Scrutinize the methodology very carefully.
   - Determine whether you are being given all the information you would need to conduct a reasonably faithful replication of the study without contacting the researcher. For a replication, you would definitely need to know:
       the basic method
       the population
       the number of subjects
       the procedures used to select subjects
       the kind of data that were collected
       how the data were analyzed.

Note: Except for content analysis, you will also need to know where and when the data were collected. For all quantitative studies, you will need clear conceptual and operational definitions for all but the most standard terms. For more qualitative studies, you will need a clear understanding of the kinds of things the researcher was looking for and what was counted as evidence.
- Assess the methodology for general conformity to the basic requirements for scientific research.
- Assess the methodology for general conformity to ethical standards.
- Decide whether the basic method used in the study is the most appropriate one or whether another method or methods would have been feasible and more appropriate.
- Decide whether the population, number of subjects, and subject-selection procedures are appropriate for the purpose of the research or whether another population, a different number of subjects, or a different procedure for selecting them would have been feasible and more appropriate.
- Decide whether the measures and/or kinds of observations make sense and whether they are appropriate for the purpose of the research or whether different methods or observations would have been feasible and more appropriate.
- For quantitative research, determine whether the appropriate statistics are being used. For qualitative studies, check to see whether procedures for analyzing information make sense and are consistent with those used in similar studies.
- Check to see whether the researcher took precautions to ensure the reliability and validity of the study.
- Decide whether the research is reliable and valid.

Note: Again, you can expect much less information in most secondary sources and in articles from the trade press and the mass media. However, at the very least, you should be told the basic method, population, number of subjects, how they were selected, and where and when the work was done if place and time could affect the findings. From the information presented, you should be able to infer most measures or procedures for collecting and analyzing data. Reports intended for a general audience should also tell you the sampling error associated with any probability sample and what sampling error means; you should be cautioned that results from studies using other kinds of sampling are not generalizable.

4. Check the findings.
- Determine whether the data that are presented provide a complete, meaningful test for the hypotheses and/or a complete, meaningful answer to any research questions.
- Assess the strength of the findings.

Note: For a quantitative study, expect to find information on the strength and statistical significance of any relationships between variables in the text and additional supporting data for key hypotheses or questions in tables or graphs. For a qualitative study, look for clear descriptions of the research setting and subjects and descriptions of behaviors or excerpts from documents or transcripts that let you see for yourself.
- Check to see how the findings fit with findings from studies cited in the literature review or with findings from other work you may know about.
- Ask yourself what else, besides the factors considered in the study, might account for the findings.

- Double-check to see whether information that might refute a hypothesis, create ambiguous findings, or embarrass a client or sponsor might have been ignored or glossed over.
5. Consider the conclusions.
    - Check to see whether the conclusions flow logically from the findings.
    - Make sure there are no generalizations beyond what the methodology and findings will support.
    - Check to see whether the author reveals or seems sensitive to any limitations in the research.
    - Check to make sure any recommendations flow logically from the data.
    - Determine whether the recommendations seem reasonable and feasible.

Finally, consider the extent to which you can or will use the study for your own purposes. But as you read, remember that the usefulness of a particular report is not always directly related to the quality of the research on which it is based. Even the best study may not provide much information you can use for a particular purpose. Even some very weak studies may contain a useful idea or piece of information.

Usefulness is a judgment call. If you are reading for specific facts or pieces of information, you will want to look for evidence of quality and care in the work. But beyond that, the factors you should consider carefully depend on whether you are reading an article primarily to get methodological guidance or to uncover findings that may eliminate the need to do your own research or that could help you put your findings in a broader perspective.

If you are reading for methodological guidance:

- Decide whether the basic method is well suited to your research problem.
- Look carefully at the number of subjects and how they were selected to see whether the number of subjects and the method used to find and choose them will give you the kind of generalizability you need.
- Determine whether the sampling frame and selection procedure will work with the population you will be studying.
- Evaluate the measures to see whether they have sufficient reliability and internal validity for your purpose.
- Determine whether the measures will work with the population you will be studying.
- Decide whether the statistics or other data analysis techniques are appropriate for your study.
- Evaluate the entire methodology or the parts of it you are considering using to make sure you have the expertise, resources, and time you will need for the work.

If you are reading for findings that apply to your problem:

- Check the external validity of the study.
- Decide whether the population, setting, and time when data were collected are similar enough to the population, setting, and time period of interest to you that you can reasonably expect the findings to apply to your situation.
- Consider whether other factors that may not have been taken into account could affect the applicability of findings to your situation.
- Decide whether recommendations make sense and are likely to work in your situation.

C

Politics, Religion, and Media Use Survey 2000

This first series of questions asks you about your interests in the upcoming presidential election and your political activities and opinions.

1. How interested are you in the presidential election?

    Very interested 4
    Somewhat interested 3
    Not very interested.. 2
    Not at all interested... 1
    NO RESPONSE/DON'T KNOW 9

2. And how important would you say the presidential election is to you personally? Is it . . .

    Very important 4
    Somewhat important 3
    Not very important.. 2
    Not at all important... 1
    DON'T KNOW/NO RESPONSE 9

3. Now please tell me what you think are the most important issues in the presidential election campaign. (RECORD FIRST TWO ISSUES MENTIONED. IF NO RESPONSE/DON'T KNOW, ENTER 999).

    ISSUE: _____

    ISSUE: _____

Now I am going to read you a list of political issues some people think are important. Please tell me how important you think each of these issues is.

	very important	somewhat important	neither important nor unimportant	not very important	not at all important	NR
4. First, the issue of gun control, is it......	5	4	3	2	1	9
5. abortion, is it	5	4	3	2	1	9
6. crime....................	5	4	3	2	1	9
7. the environment...	5	4	3	2	1	9
8. Social Security	5	4	3	2	1	9
9. taxes.....................	5	4	3	2	1	9
10. health care	5	4	3	2	1	9
11. education	5	4	3	2	1	9

Next is a list of political opinions some people hold. Please tell me whether you <u>strongly agree, agree, neither agree nor disagree, disagree, or strongly disagree</u> with each opinion.

	strongly agree	agree	neither agree nor disagree	disagree	strongly disagree	NR
12. The government should provide tuition assistance to parents whose children attend church schools. Do you	5	4	3	2	1	9
13. The government in Washington ought to reduce the income difference between the rich and poor..................	5	4	3	2	1	9
14. A pregnant woman should be able to obtain a legal abortion for any reason if she chooses.................	5	4	3	2	1	9
15. Government regulations protecting the environment should be stricter and better enforced even if it results in higher energy costs for people..............	5	4	3	2	1	9
16. Public officials have an obligation to be directed by the moral teachings of the church.................	5	4	3	2	1	9

17. Members of the clergy
    should come out in
    support of a presidential
    candidate   5   4   3   2   1   9

18. Have you ever taken a position on a political issue because of your religious beliefs?

    YES 1
    NO 2

    IF YES: What issue was that? (IF NO RESPONSE, ENTER 999)

    ISSUE: _____

    POSITION: _____

19. Do you plan to vote in the presidential election on November 7? Would you say

    Yes, definitely 4
    Yes, probably 3
    No, probably not 2
    No, definitely won't vote 1
    NO RESPONSE 9

20. IF RESPONDENT MAY VOTE (#2, 3, OR 4 IN Q#19: If the presidential election were held today, for whom would you vote?

    GEORGE W. BUSH 1
    AL GORE 2
    OTHER (WHO? _____) 3
    UNDECIDED: WHOM ARE YOU LEANING TOWARD?
        GEORGE W. BUSH 4
        AL GORE 5
        OTHER (WHO? _____) 6
    DON'T KNOW/NO RESPONSE 9

21. In general, how would describe your overall political views? Would you say you are:

    Strongly conservative 1
    Conservative 2
    Middle of the road 3
    Liberal 4
    Strongly liberal 5
    NO RESPONSE 9

22. And do you usually think of yourself as Republican, Democrat, an Independent, or a member of another party?

    REPUBLICAN 1
    DEMOCRAT 2
    INDEPENDENT/NO PARTY PREFERENCE 3
    OTHER PARTY: (WHICH ONE? _____) 4
    NO RESPONSE 9

OK. The next series of questions is about sources some people use to get political information.

23. First, please tell me how much attention you pay to stories about the presidential election that you come across in the news media. Do you pay . . . (MAY EXPLAIN "NEWS MEDIA" MEANS RADIO, TV, NEWSPAPERS, MAGAZINES).

    A lot of attention...4
    Some attention 3
    A little attention 2
    No attention..1
    NO RESPONSE..9

24. How much effort would you say you generally put into getting information from the news media about the presidential election?

    A lot of effort...4
    Some effort...3
    A little effort 2
    No effort...1
    NO RESPONSE..9

Now I'm going to read a list of sources some people use to get political information. Please tell me how much you <u>use</u> each of them for political information about the presidential election.

	a lot	some	not much	don't use	NR
25. First, talk radio programs. Do you use them............................	4	3	2	1	9
26. National Public Radio	4	3	2	1	9
27. Christian radio stations..................	4	3	2	1	9
28. National television news, that is ABC, CBS, or NBC	4	3	2	1	9
29. Local television	4	3	2	1	9
30. Cable newscasts such as CNN or C-SPAN.......................	4	3	2	1	9
31. News magazines such as *Time, Newsweek,* or *U.S. News and World Report*	4	3	2	1	9
32. Daily newspapers...........................	4	3	2	1	9

*IF RESPONDED 2, 3, OR 4 TO Q#32, ASK:

Which newspapers do you read? (RECORD UP TO THREE. IF NO RESPONSE ENTER 999).

_____   _____   _____

33. The Internet	4	3	2	1	9
34. Religious publications	4	3	2	1	9

*IF RESPONDED 2, 3, OR 4 TO Q#34, ASK:

Which religious publications do you read? (RECORD UP TO THREE. IF NO RESPONSE ENTER 999).

_____ _____ _____

35. Religious television	4	3	2	1	9

*IF RESPONDED 2, 3, OR 4 TO Q#35, ASK:

Which religious television shows do you watch? (RECORD UP TO THREE. IF NO RESPONSE ENTER 999).

_____ _____ _____

Please tell me to what extent you agree or disagree with each of the following statements.

	strongly agree	agree	neither agree nor disagree	disagree	strongly disagree	NR
36. I use the news media to see what's going on in government and politics. Do you	5	4	3	2	1	9
37. I use the news media to have something to talk about. Do you	5	4	3	2	1	9
38. I use the news media to be entertained	5	4	3	2	1	9
39. I use the news media to learn about political issues	5	4	3	2	1	9

Next I'm going to ask you some questions about religion.

40. Do you consider yourself . . .
    - Christian   4
    - Jewish   3
    - Muslim   2
    - Or other   1
    - NO RESPONSE   9
    - IF JEWISH, MUSLIM, OTHER, NO RESPONSE GO TO #54
    - IF CHRISTIAN GO TO #41

(CODE THE FOLLOWING AS CHRISTIAN:
METHODIST, DISCIPLES OF CHRIST, ASSEMBLY OF GOD, EPISCOPAL, UNITED CHURCH OF CHRIST, PENTACOSTAL/CHARISMATIC, PRESBYTERIAN, CHURCH OF THE BRETHRAN, MORMON/LATTER DAY SAINTS, LUTHERAN, EVANGELICAL, QUAKERS/FRIENDS CHURCH, BAPTIST, NAZARENE, UNITARIAN-UNIVERSALIST, CATHOLIC, CHURCH OF GOD).

(IF RELIGION NOT ON LIST, SAY: I'm not familiar with that religion. Would that be Christian, Jewish, Muslim, or something else?)

41. On religious matters would you describe yourself as . . .

    Strongly conservative..1
    Conservative...2
    Moderate 3
    Liberal...4
    Strongly liberal 5
    NO RESPONSE/DON'T KNOW..9

42. Do you consider yourself a born-again Christian?

    YES...4
    NO...5
    NO RESPONSE/DON'T KNOW..9

43. Is being charismatic or speaking in tongues an important part of your religion?

    YES...4
    NO...5
    NO RESPONSE/DON'T KNOW..9

44. How often to you attend worship services?

    More than once a week..8
    Every week 7
    Almost every week 6
    Several times a month..5
    About once a month...4
    A few times a year...3
    About once a year..2
    Never..1
    NO RESPONSE/DON'T KNOW..9

45. How would you say you fit into your church? Are you . . .

    A person who contributes to church decisions.........................6
    A leader, but not one of the decision makers 5
    Active, but not a leader..4
    Just an ordinary church member..3
    Not really part of the church at all 2
    Never attend...1
    NO RESPONSE/DON'T KNOW..9

46. What proportion of your friends attend the same church you do?

   Almost all 5
   More that half 4
   About half 3
   Less than half 2
   None 1
   NO RESPONSE/DON'T KNOW 9

Now I'm going to read a list of things some people say about their religion. Please tell me whether you strongly agree, agree, neither agree nor disagree, disagree, or strongly disagree with each statement as it describes you and your beliefs.

	strongly agree	agree	neither agree nor disagree	disagree	strongly disagree	NR
47. My whole approach to life is based on my religion	5	4	3	2	1	9
48. I go to church mainly because I enjoy seeing people I know	5	4	3	2	1	9
49. Jesus is the Son of God	5	4	3	2	1	9
50. Every word of the *Bible* is true	5	4	3	2	1	9
51. Jesus Christ is coming again soon	5	4	3	2	1	9
52. Even today God performs miracles	5	4	3	2	1	9

53. How much does your pastor talk about politics or give an opinion about what government leaders should or should not be doing? Would you say . . .

   A lot 4
   Some 3
   A little 2
   Not at all 1
   NO RESPONSE/DON'T KNOW 9

OK. Just a few questions about you and then we will be finished.

54. What is your marital status? Are you . . .

   Married 5
   Widowed 4
   Separated 3
   Divorced 2
   Single, never married 1
   NO RESPONSE/DON'T KNOW 9

55. What was the last grade you completed in school?
_____(LAST GRADE COMPLETED)

56. And what is your occupation? (Or, what sort of work do you do?) (IF NECESSARY, PROBE: What is your job title? What is your job called? What do you actually do in that job?)

_____
(RESPONDENT'S OCCUPATION/JOB TITLE)

57. And what is your age, please?

_____
(RESPONDENT'S ACTUAL AGE)

58. I'm going to read you a series of income categories. Please stop me when I get to the category that includes your total household income from all sources last year.

(IF RESPONDENT GIVES EXACT INCOME, CHOOSE FIRST CATEGORY THAT INCLUDES THAT AMOUNT).

Less than $5,000 1
$5,000–$10,000 2
$10,000–$20,000 3
$20,000–$30,000 4
$30,000–$40,000 5
$40,000–$50,000 6
$50,000–$60,000 7
$60,000–$70,000 8
$70,000–$80,000 9
$80,000–$90,000 10
$90,000–$100,000 11
$100,000 or more 12
NO RESPONSE/DON'T KNOW 99

59. And finally, what is your race or ethnic identity?
WHITE 1
BLACK 2
OTHER (WHAT? _____) 3
NO RESPONSE/DON'T KNOW 9

Well, that finished the interview. Thank you very much for your time and cooperation!

----

60. RESPONDENT'S SEX (RECORD FROM VOICE – DO NOT ASK)

    MALE   (1)   FEMALE   (2)

61. DATE OF INTERVIEW: _____, 2000

D

Code Book and Code Sheets for Content Analysis

THIS EXAMPLE IS from a longitudinal quantitative and qualitative content analysis of television network news coverage of religion. Stories for this study were identified from abstracts obtained from the Television News Archives at Vanderbilt University. The sample consists of videotapes of all stories aired on days during four constructed weeks whose abstracts mentioned a religion, religious person, or institution or included religious words or images. The sample also includes stories whose abstracts indicated they might plausibly contain reference to religion if examination of the tape indicated there actually was some reference to a religion, religious person, or institution or use of religious language or symbols in the narrative or visuals.

Code Book

This code book contains directions for recognizing religion news stories, coding stories for quantitative analysis, and preparing stories for qualitative analysis.

Recognizing Religion News Stories

Examine the videotape of each story to make sure the story really qualifies as religion news. To qualify as religion news, the story text must mention, refer to, or include quotes or other references to a religious person or institution; use language commonly associated with religion; or be accompanied by visuals showing religious persons, institutions, places, or symbols.

Examples. Stories about people and institutions commonly known to be religious should be included even if they are not specifically identified as religious. If in doubt, consult an encyclopedia or other reference to determine whether institutions are religious or religion-affiliated or whether the person is a member of the clergy or holds a leadership role in a religious organization. However, stories about some religious people or institutions may be eliminated if the context suggests an average person would not expect any information about religion. If in doubt, check with an auditor.

Most stories about the Rev. Jesse Jackson should be included. Stories about Jimmy Carter acting in his capacity as president should generally be excluded; in general, only those in which he speaks from a religious perspective or is shown in a religious setting should be included. Stories about local clergy, generally unknown leaders or members of religious organizations and laypersons should be included only if they are identified as religious in the text, if they are shown in a religious setting or wearing religious symbols such as a cross or yarmulke, or if stories have previously identified them as religious.

Religious settings include places of worship, even if the place is used as the setting for an essentially secular event such as a town meeting. Examples include the Temple Mount and other locations in the Middle East commonly associated with Judaism, Christianity, or Islam. However, exclude otherwise secular stories from a country such as Israel or a city such as Mecca or Jerusalem.

Stories referring to hospitals and schools with religious ties such as Adventist, St. John's, or Notre Dame should be included only if the story includes a religious, moral, or ethical dimension. Stories reporting on this kind of institution as the scene for an essentially secular event such as a medical procedure or a sporting event should be excluded unless the story develops some tie to religion, ethics, or moral values. Similarly, stories referring to cemeteries may generally be excluded; however, those with cemeteries as the setting for events such as a burial that are associated with religious rites such as a funeral should be included.

Stories using religious words such as "holy" or "sacred" should be included even if there is no other religious dimension.

Religious symbols include things such as crosses, the Star of David, and the pentagram, whether they are worn, shown as part of a video from a reporting location, or used as graphics behind an anchor or desk reporter. Also include any stories showing people wearing clergy collars, yarmulkes, or the garb worn by Catholic sisters, Buddhist monks or members of the Hare Krishna religion. Include stories showing women with the bindi on their foreheads. Exclude scenes of people in Middle Eastern dress unless there is something in the story to mark the people or the story as religious.

Coding Stories for Quantitative Analysis

Code the stories from material contained in the videotapes. Use the abstracts only to determine day, date, and network and to check story length, names, and locations. Use information from the code book to select the proper code for all other variables. Refer problem cases to a supervisor or auditor.

Each time you begin work, recode up to ten previously coded stories and calculate Scott's pi. If Scott's pi is satisfactory for all variables, continue coding. If Scott's pi is below .7, resolve the problem before resuming coding. At the end of all coding, recode ten randomly selected stories from each network to make sure reliability is within acceptable limits.

Use the specified codes for each of the following variables:

Identification Number

Give each story a unique identification number. Place that number on the abstract and the appropriate line on the code sheet.

Month
Record the month when the story aired. 1 = January, 2 = February ... 12 = December.

Date
Record the date (e.g., 3 for March 3) within the month when the story aired.

Year
Record the year the story aired.

Day of Week
Record the day of the week when the story aired. 1 = Monday, 2 = Tuesday ... 7 = Sunday.

Network
Record the network airing the story.
   1 = ABC
   2 = CBS
   3 = NBC

Anchor
Create an open-ended list. Give each individual anchor a number.

Code the anchor who actually reports the story or contributes significant information to it.

If the story is reported from the field, record the code for the anchor who introduces the story.

If there are co-anchors, record the code for the one who reports the story or who seems to be in charge of it.

Reporter #1, #2
Create an open-ended list. Give each identifiable reporter his/her own number.

Record up to two reporters. List them in the order they appear in the story.

Story Type
1=News
Use for hard news, current events, and issues. Although these will not be coded in the News Value variables, stories coded here should have some element of timeliness and importance.

2=Feature
Use for personality profiles, holiday stories, and other soft news. News values of timeliness and/or importance will probably be lacking. News values of human interest or novelty will often be present.

3=Comment
Use for analyses and other stories devoted primarily to an anchor/reporter's opinion or interpretation.

Reporting Locations #1, #2, #3
These are the geographic locations from which the story is actually reported. These may or may not be the same as the Actual Location where the event, issue, or other news really occurred.

Create an open-ended list for these locations. Be as specific as possible, but do not try to code separate locations within a city or town. Use the city or town as the most specific location.

Record up to three in the order of reporting.

1=New York/studio

Use this code only if the anchor or an in-studio reporter contributes information to the story. Do not code this as the first reporting location if the anchor's only function is to introduce the story before turning it over to a field reporter. Use it as a second or third location only if the anchor/in-studio reporter adds information following a field report.

999 = No real location or none can be determined.

Actual Location #1, #2, #3

Code the actual place(s) where events, issues, or "news" occurs.
Use the same codes as for Reporting Locations, adding other codes as necessary.
Record up to three, in the order in which they occur in the story.

Religion #1, #2, #3

Create an open-ended list of religions identified in the story.
Record up to three codes in the order mentioned.
Develop separate codes for generic references (Christian, Protestant, Baptist) and for more specific religions such as American Baptist or Sunni Muslim.

Use the code for the most specific religion that can be identified from the text or visuals.

You may infer and code a religion from language or visuals that are specific enough to a generic or more specific religion only if an average person would readily connect the words or visuals to a general or specific religion. For example, code as "Roman Catholic" stories mentioning or showing the Pope or the Vatican even if Roman Catholicism is never mentioned. Code stories about Jesse Jackson as "Christian" if he is referred to as "the Reverend" but as "Protestant" if he is shown preaching or officiating in an obviously Protestant setting. Do not code the story as "Baptist" unless his denominational affiliation is mentioned in the story.

998 = Stories that use religious language or include mentions or visuals that could apply to more than one religion if nothing in the text or visuals connects the reference to particular religions. Examples would include stories using generic religious language such as "holy" or that mention prayer, faith, or spirituality in a general way. However, images of churches may be coded as Christian if the architecture or presence of a cross indicates a Christian church. The image should not be coded more specifically unless there is something in the story to identify it more precisely.

999 = Stories that do not identify a religion. This includes stories about institutions or people that qualify as religion news but do not identify the institution's or person's religion even if that information was included in a previous story. For example, use this code for stories about Jesse Jackson that qualify as religion news if the story does not refer to him by his religious title, mention his religion, or show him in an identifiably religious setting.

Religious Person/Institution #1, #2, #3

Create an open-ended list of organizations, specific institutions, and people who serve as the subject for a story or who are mentioned or shown in it.

Code up to three in the order mentioned.

Use specific codes for well-known or national persons and institutions and for lesser-known ones who may be the subject of more than one story on any one network.

Use generic codes for local congregations, other institutions, and people if the people or organizations are unlikely to appear in several stories on any one network.

999 = No person or institution. An example would be stories that simply use religious language or that refer in a general way to prayer, faith, or spirituality.

Religious News Source #1, #2, #3

These are the institutions or people mentioned as the source for information or who are quoted in a story.

Code up to three in the order mentioned.

Use the same codes as for Religious Person/Institution, adding other categories to the list as necessary. However, use these codes only if it is clear they are the source for the information. Otherwise, use 999.

999 = no religious institution or person mentioned as an information source or quoted in story. Note: This code will be used quite often. The source for many stories will be secular ones.

Subject #1, #2

Use these codes to indicate, in a general way, what the story is about.

Record the main subject as Subject #1. Code a Subject #2 only if there is a second subject.

1=Religion only

Use only if the story is solely about religion. Use other codes for stories that connect religion to other subjects or that simply mention or show religion while reporting on other topics.

2=Government/Politics

Use for stories showing religion as participating in government or the political process or as affecting or being affected by them unless the story fits better in some other category.

3=Law—Church/State

Use for stories of court cases and/or legislation defining church-state relations or religious freedom/repression.

4=Military/War/Terrorism

Use for stories showing ongoing hostilities involving religion and a nation or other organized or named group.

5=Law-Crime

Use for stories showing religion as perpetrator or victim of illegal and/or unethical activities except for those that qualify as "war" or "terrorism."

6=Business/Economy/Taxes

Use for stories showing financial dealings, including financial/entrepreneurial activities and charitable contributions of religious people or institutions. Also use for stories about the interplay between secular business or financial organizations and ones about the economy and/or economic, business, or tax policy.

7=Social Service
Use for stories reporting on welfare and other service activities except medical and education ones and ones concerning projects promoting food distribution or access to resources such as clean water.

8=Medicine
Use for stories about medical procedures and health-care services and institutions.

9=Science
Use for stories about scientific research, developments, processes, etc. unless the story fits better in #8 or #10.

10=Environment/Resources
Use for stories that deal with use of resources such as water, land, energy, and/or the food supply and to ones reporting on activities or policies concerning them.

11=Media/Culture
Use for stories of religious and secular entertainment, press, music, art, etc.

12=Education
Use for stories about religious and secular schools at all levels, miscellaneous education programs, education policy, and issues.

13=Lifestyle
Use for stories about people's way of living and the culture in general. Also include personality profiles unless some other subject dominates.

14=Sports
Use for stories about recreational, amateur, and professional sporting events, issues, and sports figures.

Topic/Theme
Create an open-ended list of codes for events such as holidays, meetings, hearings, crimes, and isolated acts of violence; ongoing projects such as social work; issues and conflicts such as abortion, school prayer, and war/peace efforts in the Middle East.

Category codes should be specific enough to tell what a story is really about but general enough to allow using the same code for related stories.

Create and use generic codes for similar but unrelated stories if the event/project/issue is unlikely to be covered more than once on any one network.

Religion Content #1, #2
Use these variables to record the main kind of information about religion.
Use #1 for the main content. Use #2 only if a story contains a second kind of content.

1=Theology/Beliefs
Use for stories mentioning or showing what religions teach or what religious people actually believe as a matter of faith.

2=Rites/Practices
Use for stories mentioning or showing ceremonies or prescribed behaviors such as prayer, dietary restrictions, and so on. Include worship services unless the focus is on beliefs (use #1) or a particular service as a special, one-time happening (use #3).

3=Events
Use for stories about short-term occurrences such as a worship service, particular funeral, meeting, or battle.

4=Programs
Use for stories about missions, educational programs, health care, and social service activities of an ongoing nature.

5=Issues
Use for stories about religion's interest in moral, ethical, social, or political concerns that involve differences in opinion. Also use for conflict situations such as wars; however, code specific events within ongoing hostilities as an event (#3).

6=History
Use for stories that explain the origin or development of religions, religious practices, or institutions and ones that provide background information on events, programs, situations, or issues.

7=Sociology
Use for stories devoted primarily to information about trends, religious demographics, and/or findings from social science research.

8=Casual Mention
Use for stories that simply mention a religion, use religious language, or show a religious symbol but do not provide any information to explain the mention/use.

News Value #1, #2
These codes are based on journalistic understanding. However, timeliness is assumed; importance will be inferred from the presence of other values.
Code the most important news value as #1. Use #2 only if there is a secondary value.

1=Change
Use for stories reporting any deviation from the status quo, regardless of direction. However, if the change is presented as an aberration or oddity, code as new/novelty, #2.

2=New/Novelty
Use for stories that present something as new, unusual, strange, odd, or different from what one might expect.

3=Conflict
Use for stories that focus on nonviolent disagreements, disputes, or differences between or among parties.

4=Violence

Use for stories that focus on violent disagreements or disputes; war, terrorism, or guerilla activities; or violent crime.

5=Consensus/Conflict Resolution

This is the reverse of Conflict and/or Violence. Use it for stories that report cooperation or agreement among parties or the resolution of hostilities, disputes, or disagreements among them.

6=Human Interest

Use for soft news stories, those showing people overcoming hardships, doing good works, etc.

7=Other

Use for simple announcements and stories that don't fit into categories 1 through 6.

Running Time

This is the length of the entire story,

Use a stopwatch to measure length from the first word/visual through the last word or image.

Check this length against the time listed on the abstract. If the times are different, measure again. If they are still different in spite of careful measurement, use the length measured from the videotape.

Record in seconds, to the nearest five seconds.

Actual Time

This is the time within a story actually devoted to reporting on and showing religion. Actual Time may be the same as or less than the Running Time.

Use a stopwatch to determine this length.

Count and record only the time during which an anchor or reporter mentions or describes religion or during which religious images are used in visuals. These segments may or may not be contiguous. If they are not, Actual Time is the sum of the times for all religion segments.

Be careful not to include any nonreligion content introducing or separating religion content.

Record the combined length of all religion news content in seconds to the nearest five seconds.

Preparing Stories for Qualitative Analysis

Fill out a separate code sheet for each story by placing identifying information for each story in the appropriate spaces at the top of a coding sheet. Be sure the story number, date, day, and network match information on the abstract and the Quantitative Analysis Code Sheet.

For each story:

1. **Transcribe the text** for the story exactly as it was delivered on air. Use the wide, left-hand column for this.
   - Double-space the text.

- Use only the front (the side with the identifying information) of the qualitative coding sheets. If you need more space, use additional sheets. Fill in the identifying information on these sheets. When you finish preparing a story, number the sheets consecutively and staple them together.

  Insert the name(s) of anchors and/or reporters and their reporting location in brackets at the appropriate locations to indicate who is speaking and where that person is located.

  Example: [John Doe, New York studio]; [Mary Roe, somewhere in Israel]

2. **Describe the visuals.** Use the middle column for this.
   - It is not necessary to describe everything in moving images, but you must give the general nature of the scene and indicate any religious elements in it. Also be sure to note/describe any changes of scene in live video and any shifts between graphics and live video.
   - Begin the description for each visual (a scene or graphic) on the line that corresponds to the words spoken as the visual begins.

     Indicate in parentheses the geographic location for each visual and whether it is a graphic or live video.

     Example: (graphic - New York studio); (video - Boston)

3. **Record times in seconds.** Use the narrow, right-hand column for this.
   - Place :00, on the line adjacent to the line where the text for the narrative and the description for the accompanying visuals begin. This is the start time.
   - Use a stop watch to keep track of elapsed times.
   - Next to the appropriate text/description of a visual, insert elapsed times, calculated in seconds from the beginning of the story, to indicate:

     the beginning and end of segments actually devoted to religion
     changes in the person presenting information
     changes in reporting location
     changes from one visual to another
     the end of the story

   Note: The time entered for the end of a story (e.g., :70, should match, to the nearest five seconds, the Running Time on the abstract and Quantitative Analysis Code Sheet. The sum of the times for religion segments should match, to the nearest five seconds, the Actual Time on the Quantitative Analysis Code Sheet.
   - Underline any problematic features in the text or visuals. These include apparent inaccuracies, use of loaded language, unanswered questions, etc.
   - Use the back of the coding sheet to describe the problem or to list things to check further.

Code Sheet for Quantitative Analysis

_____ ID Number	_____ Reporter #2
_____ Month	_____ Story Type
_____ Date	_____ Reporting Location #1
_____ Year	_____ Reporting Location #2
_____ Day	_____ Reporting Location #3
_____ Year	_____ Actual Location #1
_____ Anchor	_____ Actual Location #2
_____ Reporter #1	_____ Actual Location #3
_____ Religion #1	_____ Subject #1
_____ Religion #2	_____ Subject #2
_____ Religion #3	_____ Topic/Theme
_____ Religious Person/Institution #1	_____ Religion Content #1
_____ Religious Person/Institution #2	_____ Religion Content #2
_____ Religious Person/Institution #3	_____ News Value #1
_____ Religious Source #1	_____ News Value #2
_____ Religious Source #2	_____ Running Time
_____ Religious Source #3	_____ Actual Time

Code Sheet for Qualitative Analysis

Identification Number _____ Network _____

Date (Month/Day/Year) _____ Day of Week _____

Text	Visuals	Time

E

Writing Research Reports

THE REQUIREMENT THAT science must be public makes it imperative that you share your findings with others. Academic researchers do this by publishing their work in research journals. Applied researchers typically prepare business reports for their sponsor or client, but they may also publish their work in scholarly journals. Although there are differences between journal style and business style and between the style you will use for formal business reports and informal ones, there are some principles that always apply.

General Principles

Because of the amount of information you undoubtedly accumulated while doing your research, you will have to think carefully about your goal in writing up your findings. As part of this, you will also need to consider who your audience will be, what they will expect, want, or need from your report, and what help they may need to understand your work.

Like all kinds of writing, journal articles and business reports should be accurate, clear, concise, and well organized. In this section we highlight just those considerations that raise special challenges when you write up your research.

- **Follow the rules.** Each journal has its own guidelines for authors; schools have rules for how you should prepare your dissertation or thesis. Sponsors and clients may also have requirements or strong preferences about how you should prepare reports for them. Find out what these guidelines are and follow them explicitly. Some journals automatically reject manuscripts that don't conform to their style. All journals and many sponsors and clients will make you redo reports that don't follow the rules.

    These rules usually cover things such as the length your report should be, how you should organize it, and the style you should use for section headings, visuals, and references. They may also set requirements for the kind of paper, the margins, type style and size you should use, where page numbers should be placed and whether you should single- or double-space your manuscript.

- **Give credit where credit is due.** This has three parts.

    First, if you are writing for a journal, mention the source of any funds you received to do the work. If you are writing for a sponsor or client, the recipient of your report will know who funded your work. But you should still acknowledge any special help you received such as office space or staff support. Acknowledgments usually appear at the bottom of the first page or as a note at the end of a journal article. In formal business reports, they may be part of the front matter or included as an appendix. In informal ones, they may be incorporated into the text.

Second, you should acknowledge those who helped you collect or analyze data or who helped write up the findings. List as coauthors any true collaborators—people who made substantial, individual contributions to the design and implementation of the project. Name others who made some individual contributions to the project in your acknowledgments. You should also recognize in some way the contribution of groups such as students in a class or employees of your or your client's company although it is rarely necessary to name group members individually.

Finally, avoid plagiarism. Science builds on the work of others. Therefore, it is your scholarly duty to acknowledge the contributions of those who have gone before you. Whenever an idea, method, or finding can be traced to another researcher or scholar, say so. Whether you really express another person's work in your own way or just paraphrase from previously published material, acknowledge the borrowing in your text and include a reference. If you borrow another's actual words, give the source in your text, put the borrowed words inside quotation marks, and include a page number in your citation.

- **Respect your data.** Above all, don't make up data or alter it. And don't overstate your case. It can be tempting to overstate the strength or generalizability of your findings or to present only those that conform to your expectations. Dishonesty in reporting findings, drawing conclusions, or making recommendations can cost clients and sponsors dearly if they rely on misleading or incomplete information. It causes problems for researchers who may try to use your work to plan their own. Dishonesty can destroy careers. Few studies are perfect. Acknowledge limitations in your data, and point out what effect those limitations may have on your findings. But remember: Strength and significance are not necessarily synonymous with importance. Unexpected findings may be the most useful.

Journal Style

This is the style for research journal articles, theses, and dissertations. You may also use it if you choose to present your research in book form.

Writing style is fairly formal. Most journals expect or require you to write in the third person—"Findings show . . ."—instead of first person—"We found. . . ." First person is most common and most acceptable for studies using observational research as the method.

Used sparingly, passive constructions are acceptable—"A survey was administered to 397 subjects who were selected randomly." But use active voice—"A randomly selected sample of 397 subjects answered survey questions"—whenever possible. With active constructions, sentences tend to be shorter and somewhat easier to understand.

Most of your readers will have at least some college education and some training in or experience with social science research, so you can use common research terminology without definition. Still, many of your readers will not be experts in your field.

Few, if any, will know as much about your particular work as you do. Therefore, to help your readers understand your work, you should provide definitions or use examples to illustrate concepts with which they may not be familiar.

When you write up your findings for a journal, you can include tables, graphs, and simple diagrams. But most journals will not accept pictures, and they will not let you use color.

Because people will be reading your report for different reasons, journal articles follow a standard organizational pattern. This pattern helps readers find the kind of information they are looking for quickly and easily.

With some exceptions and minor variations, six sections are standard. They are the abstract, introduction, methodology, results, conclusion, and references/notes.

Abstract

Although the abstract is the first section in most journal articles and appears on one of the first pages in a thesis or dissertation, it will most likely be the last section you write. Because the abstract is a brief summary of the work, you will need to write the entire paper before you can distill your material into this minireport.

There are two types of abstracts: informative and descriptive. The **informative abstract** tells what you found and what you conclude or recommend as a result of your research. A **descriptive abstract** tells what is in the longer work. You can think of it as a table of contents written in paragraph-sentence form. Because a descriptive abstract gives no real information about your findings or conclusions, you should avoid using it unless your journal requires this style.

An informative abstract is preferable because some people will not have easy access to your full report. They will have only your abstract, which they found in a compilation such as *Dissertation Abstracts* or in a database like those maintained by some journals and by organizations such as Educational Resources in Communication (ERIC) or the Association for Education in Journalism and Mass Communication.

Because many more people will read your abstract than will read your report, you want to include as much information as possible. But abstracts are very short, ranging from as few as 25 words for some journals and convention programs to 350 words for dissertations. Therefore, they must be very concise.

When you write an abstract, you will probably need to combine information about the study's purpose and methodology into just one or two sentences. That way, you can devote the rest of the available space to your findings. For longer abstracts, you may reserve space for a sentence or two about your conclusions or recommendations. However, you should never include in the abstract any information that is not included in the full report. Neither should you include citations or visuals such as tables or graphs.

Abstracts are always written in paragraph-sentence style. Presenting information in lists is usually unacceptable.

Introduction

The introduction sets up your report. In longer works, this introduction may extend across several chapters. Some journals also divide the introduction into sections, each with its own heading. Others do not. But regardless of format, the typical introduction is made up of a problem statement, a literature review, and a delimited purpose statement.

Problem Statement. This statement introduces your project and begins the task of justifying your work. In the first paragraph or two of your article, you mention the problem or concern giving rise to your work and suggest why it is important. The fact that you are interested in the topic isn't enough. You must justify your work by pointing to a problem that needs a solution or by developing an argument that your work will contribute to the body of scientific knowledge.

Literature Review. This section continues the work of justifying your research by fitting it into a broader context. Because scientific knowledge is cumulative, you use this section to show how your study draws on, checks, or extends previous work.

If you are working in an area where not much work has been done, you may have to choose studies by reasoning through analogy—Researcher k in field x, using method y, found z; "x" is somewhat like my situation. Therefore I will include the study in my literature review because I may also find z if I use method y.

But in most cases, you will find more literature than you can possibly cite. In this situation you should include foundational or major studies, but only the ones that are really relevant to your work. To do otherwise is to invite having your paper rejected because you don't appear to "know the literature." But beyond that, you may narrow the list of works you review in any defensible way. You might, for example, include only the most recent ones or only those that use a particular method or kind of measure.

Your task isn't to include every possible article or to tell everything that is in each article. Direct quotes are used only in cases where a researcher's language is crucial for understanding a unique definition or measure. Paraphrases are also rare. You can usually say everything you need to say about a publication in just a sentence or two. Sometimes you can even summarize several in a single sentence.

Your goal in this section is to distill information from the literature, synthesize it, and then present it in a way that accomplishes your purpose. That purpose may be to

- show a need exists to replicate previous work
- justify hypotheses
- argue for extending previous work to see whether their findings hold true in a different setting or with a different population
- justify a methodology
- show there are gaps or contradictions in the research literature or problems with a method or theory that need to be filled in or corrected

Purpose Statement. Most problems are huge. You can rarely do everything needed to solve them. You probably can't even tell in a single article everything you did to investigate them. Therefore, this purpose statement narrows your problem down to the specific part you are tackling in this report. This statement is rarely longer than a paragraph or two. Typically it ends with your specific research questions or hypotheses. These questions or hypotheses should flow naturally from and be supported by the literature review.

Methodology

This section tells readers how you did the work.

Because you are writing for an educated audience, you won't need to explain how to do standard procedures such as stratified random sampling. However, you do need to give people enough information so they can judge the reliability and validity of your work. If there is any doubt as to whether you should mention or explain something, err on the side of exhaustiveness. Some experts insist that, after reading this section, another researcher should be able to replicate your study without contacting you for more information. Although others aren't that strict, some requirements are fairly standard.

Your readers always need to know:

- what method (experiment, survey, content analysis, focus group, or observation) you used.
- whether anyone besides the author collected or analyzed the data and, if it isn't self-evident, their experience or training.
- where and when you collected your data.
- the population you studied and, if applicable, your sampling frame and sampling procedure.
- your variables.
- conceptual and/or operational definitions for key, nonstandard measures.
- how you analyzed your data.
- whether you checked your work for reliability and validity and, if so, the result of those checks.

Other information is method specific.

- For experiments, tell whether you collected data in the lab or in the field. Also name or describe the experimental design you used and fully describe how you manipulated your independent variable.
- For surveys and other questionnaire research, be sure to indicate whether any findings are based on a subsample. Describe any weighting techniques you used. Also include the exact wording and response options for key questions in this section or in explanatory notes at the end of your article. Usually it is also acceptable to incorporate at least some information on question wording into the section on findings.
- For a content analysis, you won't need to include information on where or when you did the work. But you do need to include dates for the documents as part of your description of the population you studied. Also be sure to include information on intra- and/or intercoder reliability.
- For focus groups, be sure to tell how you recorded the sessions and whether you relied on the recordings or on transcriptions during data analysis. Also describe any supplementary data-collection techniques you used.
- For observational research, tell whether you or your coauthors have any ties to the setting, group, or subjects you are studying. Explain your role and stance during the observations. Also include information on supplementary data collection techniques.

Results

The purpose of this section is to provide the information that answers your research questions or tests your hypotheses. Typically there are three parts to this information: an overview of your data, the findings themselves, and some discussion of them.

Overview. This is a brief introductory section that provides basic demographic information to let your readers know how well your subjects match a population. For a quantitative study and for some qualitative ones, the first paragraph or two may also provide some frequencies or other information on key concepts or variables to help your readers get a general sense of your findings. In observational studies, the overview is also the place for a thorough description of the setting and site.

Findings. The findings are the statistical tests or other evidence from your data that answer the research questions or provide the tests for the hypotheses. You should arrange this section so that you present information in the same order as you listed the questions or hypotheses in your introduction.

Present only the key findings in the text.

For quantitative studies, you can round off, simplify, or combine numbers or percentages in your text. If your readers need more precise figures, you can put them in tables. You can also present supporting or supplementary information in graphs. Tables and graphs help people visualize your findings and make comparisons among them. But if you use them, the text should always refer readers to the appropriate visual. Both text and visuals should include some mention of the statistic you used, its value, and whether findings are statistically significant.

The results section for qualitative studies may include visuals such as maps or diagrams and, on occasion, graphs or tables. However, in qualitative studies the emphasis is on presenting evidence in the form of quotes or paraphrases from the text of documents, interviews, or conversations. For observational research, presenting evidence for findings also requires giving clear descriptions of things the researcher saw or that transpired during observation periods.

Discussion. The discussion puts your findings in context. Some journals treat this as a separate section; others allow you to discuss your findings as you present them. In either place, the discussion compares your findings to the results from the work you cited in the literature review.

As part of the discussion, you should point out similarities between your study and previously cited work. If your results are not what you expected or if the answers to your questions are ambiguous, you should mention any limitations in your study that could have affected your findings. You should also point out any methodological differences between your subjects and those you cited in the literature review even if those differences are not really limitations or flaws in your work.

Conclusion

This section wraps up your presentation. Although it begins with an accurate, objective summary of your key findings, it is the place where you finally get to give an opinion about what your work really means. This section often ends with suggestions for future research. But if you have acknowledged limitations in your work and do not overgeneralize beyond what your data allow, you can also make recommendations to practitioners or offer your thoughts about the importance of your findings, their contribution to the advancement of science, or a possible solution of a practical problem.

References

A list of references at the end of your article lets readers find the information they will need in order to locate the studies you included in the literature review. Therefore, each entry on the list must include the author(s), book, or journal title and the title of articles or chapters in them and the date of publication. For books it also must include the publisher and place of publication. You will also need page numbers for journal articles and for any direct quotations used in your text.

Each journal has its own style that you must follow for citing studies within your text and for preparing your list of references. Most will use an author/date citation style such as that of the American Psychological Association, but *Journalism and Mass Communication Quarterly* and a few others use traditional footnotes.

Traditional footnotes let you include brief explanatory notes within the list of references; the author/date style requires you to create a separate list of notes if you need to provide background information, explanations, or methodological details that some readers may need but that do not fit gracefully into your text.

Formal Business Reports

Like journal articles, these reports are fairly formal. However, they are more likely than journal articles to include photos and drawings and make use of color. Because they are intended to create an archival record of a project as well as meet the needs of the business executives or clients who are their primary audience, they are also much longer.

Typically these reports are written in the third person. But because members of their primary audience are usually less well versed in research methodology than the audience for journal articles, there is less use of technical terms. Where technical terms cannot be avoided, they must be explained more thoroughly and more simply.

Parts of a typical formal report are the front matter; an executive summary; sections for background, methods, results, recommendations, and the conclusion; and appendices. There is no formal literature review. References to other research are simply tucked into the text where they are needed to explain methodology or put results in perspective.

Front Matter

As book-length documents, formal reports usually have a title page, a table of contents, and one or more listings for tables, figures, or other visual displays.

Information on the title page includes the title of the study, the date for the report, and the name and affiliation of its authors.

The most formal reports may also have a letter of transmittal. This cover letter, addressed to the person who commissioned the research, briefly describes the purpose of the project and tells what the report encompasses.

Executive Summary

This is the business equivalent of the abstract.

Because business reports are usually prepared for decision makers, the executive summary emphasizes recommendations and conclusions instead of findings per se.

Because these decision makers are often quite busy, this information may be presented as a list instead of in paragraph/sentence form.

Information on the purpose and methodology for the study is included in very abbreviated form. Methodological information may be mentioned along with the purpose for the study as a way of introducing the recommendations, or it may be relegated to a single concluding sentence.

Background

This section provides justification for your research by explaining why the research was done and/or why research findings are needed. It also sets up the rest of your report.

To provide justification, you use archival data or quotes to document the existence of a problem and show its history and severity. You then use that information to establish the purpose of your research and the appropriateness of your research questions or hypotheses.

Methodology

Formal business reports require even more methodological information than would be required for a journal article. However, this section may provide just the basics. Detailed information on procedures for data collection or data analysis may be put into appendices. Most formal reports also include copies of questionnaires, protocols, or code sheets and code books as appendices.

Results

The content, organization, and writing style for this section are much the same as for the journal article. However, graphs are used more frequently than they are in journal articles because the audience for business reports usually finds them easier to understand than the tables that are more common in journal articles. Whenever possible, the emphasis in the discussion section will be on comparing current findings to findings from research previously conducted for the commissioning organization rather than on comparisons to findings published in the research literature.

Recommendations

Because readers of business reports are primarily interested in finding out what the research means to their organization, business reports usually devote an entire section to the kind of recommendations that are placed in the conclusion of journal articles. On occasion, however, these recommendations may be incorporated into the results section by first presenting a finding and then telling what that finding means to or suggests for the client organization.

Conclusion

This is the wrap-up. It usually consists of a brief summary of the major findings and recommendations. It may also include a statement about suggested future research.

References

These are the same as for a journal article. The list should provide all the information a reader would need to find any studies, including proprietary ones conducted on behalf of the client organization, that are mentioned in the text.

Appendices

This is the place for supplementary information. Questionnaires, protocols, and other detailed methodological information belong in appendices. Information on the firm conducting the research or copies of individual researchers' vitas may also be included as an appendix. Some reports will also have a glossary of technical terms.

Informal Business Reports

Informal business reports may be prepared as a written report or an oral one. Although we present this kind of report last, applied researchers prepare informal reports more often than they prepare journal articles or formal business reports.

These reports are designed for decision makers and others who are usually less well versed in research methodology than the audience for other kinds of reports. Therefore, the writing style is simpler and more personal.

To help make your work understandable to nonspecialists, you can write or speak directly to your audience by using first- and second-person pronouns as we have done in this sentence. You should also use graphs instead of tables if you can.

In many cases it will also be appropriate to substitute common words for technical terms even if the common words are less precise. When you can't avoid using technical language, you will have to explain the terms very carefully. But those occasions should be rare.

Most of the people who will read or hear your report have no real need for or interest in learning the precise details of your work. Therefore, informal reports do not contain a literature review or methodological information much beyond a simple mention of the method and how you chose your subjects. This makes them relatively short. It also allows you to place your emphasis on what your audience really wants to know.

Because your audience just wants to know what you found and what your findings mean for them, the emphasis is on recommendations and/or conclusions. However, the organizational pattern for these reports varies according to presentation style, which in turn influences report length and content.

A report presented as a memo might simply say "Results of the focus groups we conducted in July indicate that the XYZ corporation should . . .".

A somewhat longer written report or an oral one might begin with a combined purpose/method statement followed by a longer section weaving together findings and recommendations: "Results from the survey we conducted in September show that we are doing a good job. Most respondents said X; however, some said Y. This suggests we should . . .".

For longer informal reports, you may want to use a simplified approximation of the pattern for a formal business report such as executive summary, background/problem, recommendations, findings/support, and method.

You probably won't need to include any visuals or a list of references in very short memo reports, but you may want to include them in longer reports. For longer ones you may also insert a methodology section, add it as an attachment, or prepare a separate technical report for your client's files.

For oral reports you can create transparencies or use presentation software to project your main points, graphs, or tables for people to look at as you talk. You can also distribute an executive summary, tables, and graphs or a list of references as handouts.

Glossary

academic research: see basic research.

alternate forms reliability: a type of measurement validity established by examining the correlation between two forms of the same test or measure (Chapter 5).

alternative hypothesis: a hypothesis paired with a null hypothesis that states that the independent variable has an effect on the dependent variable (Chapter 3).

analysis of variance (ANOVA): a statistical technique commonly used to determine whether there are significant differences on a dependent variable across two or more groups. This procedure is commonly used with experimental designs (Chapter 6).

anonymity: an ethical safeguard to protect the subjects of a study. In this situation the subject's right to privacy is protected because the subject's identity is not known to the researcher and is thus protected from disclosure (Chapter 2).

antecedent variable: a variable that occurs before others in a causal or temporal sequence (Chapter 3).

applied research: research conducted to solve a concrete, real-world problem or address a policy question. Research with practical applications (Chapter 1).

audit: in public relations, research conducted to examine an organization's internal or external communication. Also, research conducted to establish an organization's situation, clarify issue positions, or determine its image (Chapter 3).

audit procedure: a method for verifying the reliability of a measure by having an outsider or a panel of judges assess the fit between a measure or an indicator and the concept it represents (Chapter 5). This procedure can also help increase the reliability of the findings, interpretations, and conclusions in a research study (Chapters 8 and 10).

authority: a way of knowing that relies on the opinions of people who have greater knowledge as a result of their education, experience, or position in society. Truth is what those in positions of power or influence say it is (Chapter 1).

basic research: research conducted to create, test, or improve theory and thus contribute to the advancement of knowledge. Also called theoretical or academic research (Chapter 1).

bivariate statistics: statistics used to test for a relationship between two variables. Some commonly used bivariate statistics are chi-square, phi, Cramer's V, gamma, Kendall's tau, Spearman's rho, and Pearson's correlation coefficient (Chapter 7).

blocking: adding a nuisance variable or confound to a research design as a way of controlling for it (Chapter 6).

causal relationship: a relationship between two variables in which a change in one variable (independent) produces or "causes" a change in another variable (dependent) (Chapters 3 and 6).

causal validity: an aspect of internal validity that deals directly with whether one can rule out alternative explanations in order to conclude that the independent variable does indeed cause or have an effect on the dependent variable (Chapter 2).

census: a study in which data are collected from every element in the target population (Chapter 4).

central limit theorem: a theorem that states that the mean values for repeated samples drawn from a population will cluster around the true mean value of the population for any given characteristic. Plotting these sample averages or means will produce a normal or bell-shaped curve around the population parameter (Chapter 4).

check question: a question that asks the respondent the same thing as an earlier question but does so in a different way. This type of question helps identify individuals who are providing inconsistent responses (Chapter 7).

chi-square: a measure of association used to determine whether there is a relationship between two variables that may be measured at the nominal level (Chapter 7).

classic experimental design: a true experimental design in which subjects are assigned to experimental and control groups using randomization. Measurements for the dependent variable for both groups are taken before (pretest) and after (posttest) the experimental group is exposed to the stimulus, or independent variable, which is measured through manipulation (Chapter 6).

close-ended questions: see fixed-response question.

Cloze procedure: a procedure to help determine the readability of a message by having subjects read passages that have every k^{th} word in the text replaced by a blank. Reading ease is based on the proportion of subjects who fill in the blanks correctly (Chapter 8).

cluster sampling: a probability sampling procedure in which the population of interest is broken down into groups called "clusters" so that a probability sample of clusters can be selected. Cluster sampling may involve several steps. See multistage cluster sampling (Chapter 4).

code book: a document that provides operational procedures for recognizing concepts and assigning them to coding categories. May also include directions for recognizing members of a target population. Used in content analysis and with open-ended survey questions and observational measures (Chapters 7 and 8).

code of ethics: a list of principles and guidelines developed by professional organizations to guide research practice and to clarify which behaviors are considered appropriate (Chapter 2).

code sheet: a form used in content analysis to record data for quantitative analysis. A code sheet includes a list of all relevant concept categories to be coded and either a list of coding categories for each concept category or a place to record numbers assigned during the coding process (Chapter 8).

coding: the process of assigning numbers to categories or amounts of a concept operationalized through observation (Chapters 5, 7, and 8).

coding categories: the response options associated with concept categories in content analysis (Chapter 8).

cohort analysis: a technique for determining how cohorts, or groups of subjects of a similar age, change over time. A type of longitudinal research. Often conducted in conjunction with a trend study (Chapter 7).

composite measure: a measure created by combining two or more measures into a single measure. Indexes and scales are examples of composite measures (Chapter 5).

computer-assisted telephone interviewing (CATI): The questionnaire is "programmed" into the computer along with relevant skip patterns. A CATI system combines the tasks of interviewing, data entry, and some data cleaning into a single system (Chapter 7).

concept: an abstract idea formed by generalizing from particulars and summarizing related observations (Chapters 3 and 5).

concept categories: the categories of the concepts of interest into which units are coded in content analysis (Chapter 8).

conceptual definition: a statement that specifies or describes what the researcher means by a particular concept or term (Chapters 3 and 5).

conceptual fit: the degree of fit or agreement between the specified meaning of a concept (the conceptual definition) and the indicator actually used to measure the concept (Chapter 5).

concurrent reliability: a type of measurement reliability established by checking the strength and statistical significance of the correlation between the measure in question and some other measure (Chapter 5).

concurrent validity: a type of measurement validity established by finding a statistically significant, positive correlation between the measure or test in question and another measure or test for which validity has already been established (Chapter 5).

confidence interval: a range (or interval) within which the true population value on a measure is estimated to lie at a specific level of confidence. Used to qualify estimates based on a probability sample (Chapter 4).

confidence level: the degree of certainty used in deciding whether data support a hypothesis. The probability findings from a sample would hold true if one were to collect data from all members of a population. Commonly used confidence levels are .01 and .05 (Chapters 4, 5, 6, and 7).

confidentiality: an ethical safeguard to protect the subjects of a study. In this situation the researcher knows, or can figure out, who the participants are but will separate names and identifying information from the data and not reveal this information to anyone (Chapter 2).

confound: an extraneous factor in a research study that may affect scores on a dependent variable. Confounds may be controlled for through statistical techniques, blocking, or holding constant (Chapter 6).

construct: a complex concept. It subsumes several concepts or distinct dimensions. An abstract entity that cannot be observed directly (Chapter 3).

construct validity: a type of measurement validity established by showing that a measure is related to other measures as specified by a theory (Chapter 5).

content analysis: the systematic study of the content, and sometimes the meaning, of a document. One of the five research methods (Chapters 3 and 8).

content validity: a type of measurement validity that rests on logic alone to determine whether a measure reflects the content of a particular concept (Chapter 5).

contingency question: a type of question used in a survey questionnaire in which a subject's answer to this question determines which questions the subject will be asked to answer next (Chapter 7).

continuous variable (or measure): a variable that can take on any value over a range of values and can meaningfully be broken into smaller parts or fractions. An example is height, measured in feet and/or inches (Chapter 5).

control group: the subjects in an experiment who do not receive the stimulus (Chapter 6).

control variable: a variable whose influence on a relationship the researcher wishes to eliminate or hold constant (Chapter 3).

convenience sampling: a form of nonprobability sampling in which the researcher selects subjects or cases that are readily available, "convenient," such as college students in a classroom. The goal is only to find a number of research subjects quickly and easily (Chapter 4).

counterbalanced experimental design: a quasi-experimental design appropriate for factorial studies in which there are several levels or categories of one or more independent variables. In this design, all subjects are exposed to multiple treatments, but the order of exposure to treatments is rotated. Also called a Latin square or a repeated measures design (Chapter 6).

counterbalancing: a technique used in experiments with several stimuli in which subjects are exposed to the various stimuli in an attempt to hold treatment effects constant (Chapter 6).

covert observation: a method of data collection in which the observed individuals are unaware of the researcher's presence (Chapter 10).

Cramer's V: a measure of association that helps determine the strength of the relationship between two variables measured at the nominal level. Scores on this statistic range from 0 to 1 with 1 indicating a perfect relationship and 0 indicating no relationship (Chapter 7).

criterion validity: a type of measurement validity established by showing that a measure relates as expected to some outcome termed the "criterion" (Chapter 5).

critical studies: a type of textual analysis that emphasizes "reading between the lines" to uncover or deconstruct the hidden or real meaning of messages and to evaluate or criticize a messenger's motives and/or the effect of these messages. Postmodernism, Marxist, and feminist scholarship are variants of this way of studying messages (Chapter 8).

Cronbach's alpha: an inter-item correlation summary statistic that is used to check the reliability of a composite measure or index (Chapter 5).

cross-tabulation: a procedure for displaying bivariate data to show the number and/or percentage of subjects with a certain score on one variable who have each of the possible scores on the second variable. Commonly used by survey researchers in conjunction with measures of association (Chapter 7).

cross-cultural survey design: a single survey conducted in several locations or with several populations or several surveys conducted simultaneously, each in a different area or with a different population. The research design resembles a quasi-experiment (Chapter 7).

debriefing: a procedure used in experiments when the subjects cannot be told in advance of the true nature of a study or all of its possible effects. Subjects are informed at the end of the study of the purpose of the study to clear up any deceptions used and to obtain feedback from the participants (Chapter 6).

deduction: reasoning from the general to the particular (Chapter 1).

deductive research: the classic research model based on deduction. The researcher starts with a theory, develops hypotheses, gathers data, tests the hypotheses, and then draws a conclusion about the theory (Chapter 1).

Delphi technique: a focus group variant used to help set priorities or reach consensus. A staff group generates questionnaires and circulates them to a group of respondents who answer the questions, add comments, and make suggestions about the questionnaire and then return it to the staff group, who summarize the responses and generate a new set of questions. The back-and-forth process continues until closure is reached (Chapter 9).

dependent variable: the variable that the researcher is trying to explain. Its value is presumed to depend on or vary with the value on the independent variable(s) (Chapter 3).

descriptive research: a study conducted to obtain a snapshot of reality or to collect facts about a specified population or sample (Chapter 3).

descriptive statistics: statistics designed to describe or summarize the distribution on one variable or the relationship among variables (Chapter 3).

discourse analysis: a type of textual analysis that focuses on the structure and function of messages and/or the ways in which people use language or interact with each other to make sense of their world (Chapter 8).

disk-by-mail survey: a self-administered survey in which a survey questionnaire on computer disk is sent to respondents who answer the questions and then return the disk to the researcher (Chapter 7).

dispersion: the amount of variability in a set of scores or values on a variable (Chapter 4).

double-barreled question: a single question that in reality requires two different responses from the subjects (Chapter 7).

double-bind question: a question in which the response, because of its wording, implies an affirmative answer to an unasked question (Chapter 7).

dummy variable: a categorical variable that has been recoded to assign the values of 0 and 1 to each category, in which 1 indicates the presence of the category and 0 represents the absence of the category. Creating a dummy variable makes it possible to use statistics designed for higher levels of measurement with variables measured at the nominal level (Chapter 5 and 7).

element: the unit of a population. A unit can be a person, a group, or a thing (Chapter 4).

equivalent group pretest-posttest design: a quasi-experimental design in which very similar subjects or in-tact groups are assigned to experimental and control groups using matching techniques (Chapter 6).

ethnography: a type of observational research in which researchers totally immerse themselves in a cultural context. Most common in anthropology and sociology (Chapter 10).

evaluation research: a research study that is conducted to assess the impact of policies, programs, or procedures (Chapter 6).

exhaustive: a set of measurement categories constructed so that every case (or subject) can be assigned to one and only one category or attribute of a variable (Chapter 5).

experiment: data-collection method in which the researcher divides subjects into a test group that receives a stimulus or treatment and a control group that is not exposed to the stimulus and observes the effect on the dependent variable. This method is the best for showing cause-and-effect relationships. One of the five research methods. (Chapters 3 and 6).

experimental design: the total plan for assigning subjects to groups and then collecting data. These designs are commonly classified as true experiments, quasi-experiments, or pre-experiments depending on how well they control the common threats to validity (Chapter 6).

experimental group: the group in an experiment that receives the stimulus or manipulation of the independent variable (Chapter 6).

explanatory research: studies that investigate the reasons things are the way they are or work the way they do by establishing explanations or causes for observed effects (Chapter 3).

exploratory research: studies that are intended to provide a beginning familiarity with a topic about which very little is known. The goal is to find out what is out there (Chapter 3).

external validity: the degree to which the findings from a sample of subjects can be generalized to an entire population or to other subjects in other locations and/or at other times (Chapter 2).

face validity: the simplest form of content validity. The researcher claims that obviously, on its face, the measure is a reasonable one. That is, the measure appears to be measuring what it is intended to measure (Chapter 5).

face-to-face interview: a researcher-administered survey in which subjects are interviewed in person (Chapter 7).

factor analysis: a multivariate statistical technique used to assess whether a large number of variables can be reduced to a smaller number of factors or sets of variables. This method is typically used to find out whether items can be combined to create a composite measure or to confirm that a composite measure is unidimensional (Chapter 5 and 9).

factorial experimental design: an experimental design in which the effect of two or more independent variables or factors on a dependent variable are analyzed simultaneously (Chapter 6).

feeling thermometer: a scaling technique in which subjects are shown a picture of a thermometer with temperatures from zero to 100° and asked to use those temperatures to indicate how warmly they feel toward some referent (Chapter 5).

field experiment: an experiment that is conducted in a natural setting rather than in a laboratory or other place controlled by the researcher (Chapter 6).

field notes: an observational researcher's record of observations. Written notes may be supplemented by using cameras or recording equipment (Chapter 10).

filter question: a question used in a survey questionnaire to separate respondents who are eligible to answer a question or set of questions from those who should not answer (Chapter 7).

fixed-response question: a question that provides respondents with a limited and predetermined number of responses from which to choose (Chapter 7).

Flesch's formula: see reading-ease formula.

focus group: data-collection method making use of small group discussions facilitated by a moderator trained to encourage participation and keep the conversation on track without leading it in a particular direction. One of five research methods (Chapters 3 and 9).

fog index: a readability measure based on sentence length and the number of syllables per word. Developed by Robert Gunning (Chapter 8).

formative research: in public relations, research undertaken to help decide what to do. Also research conducted to monitor, evaluate, and improve campaigns, programs, or projects that are in progress (Chapter 3).

gamma: a measure of association suitable for use with ordinal measures. Scores range from −1 to 1 with −1 indicating a perfect negative association, 1 indicating a perfect positive relationship, and 0 indicating no relationship (Chapter 7).

"going native": in observational research, a situation in which the researcher becomes overly involved, thereby losing objectivity (Chapter 10).

group-administered survey: a method of survey data collection that combines features of a researcher-administered survey and a self-administered survey. The researcher provides general instructions and distributes questionnaires to respondents who are assembled in one location; the subjects then complete the questionnaire (Chapter 7).

guinea pig effect: a reactivity threat to internal validity. Subjects adjust their answers to conform to what they believe a researcher wants or what they believe is expected of them as research subjects. A researcher-demand effect (Chapter 6).

Guttman scale: a scaling technique that measures intensity or commitment by presenting subjects with a series of statements on a topic, which can be arranged in a hierarchical fashion. Also called scalogram analysis (Chapter 5).

Hawthorne effect: reactivity threat to internal validity caused by subjects changing their behavior in response to a change in their environment (Chapter 6).

history: a threat to internal validity in which what is happening in the world apart from the research setting may affect scores on the dependent variable (Chapters 2 and 6).

holding constant: a control technique, often involving the use of statistics, that involves limiting the range of scores on a nuisance variable or confound (Chapter 6).

hypothesis: a statement about the expected relationship between two or more phenomena that can be tested by empirical means. It is a statement about the relationship that ought to be observed in the real world if the theory is correct (Chapter 1).

independent variable: the variable that influences the dependent variable. The presumed cause (Chapter 3).

index: a composite measure based on summing, averaging, or otherwise combining the responses to multiple questions that are intended to measure the same concept (Chapter 5).

induction: reasoning from the particular, that is, from individual instances, to the general (Chapter 1).

inductive research: research based on induction. The researcher first makes observations, then develops and combines propositions to create a theory (Chapter 1).

inferential statistics: statistics that are used to estimate how likely it is that findings based on data from a probability sample will hold true for the population from which the sample was drawn (Chapter 3).

informant: in observational research, a person who helps a researcher gain access and also provides insight into the people and setting being studied (Chapter 10).

institutional review board (IRB): a committee mandated by the National Research Act that reviews research proposals to make sure that subjects are not harmed, have been informed of the risks and benefits of the study, and have given informed consent (Chapter 2).

instrumentation: any measurement technique or procedure. Especially the manipulation used to measure the independent variable in experimental research (Chapter 6).

intercoder reliability: assessment of the reliability of a measure by calculating the degree of agreement between the coding done by two or more independent coders (Chapter 5 and 8).

internal validity: refers to whether the study is really measuring or observing what the researcher thinks it is. Includes measurement validity and causal validity. Establishing internal validity requires ruling out the possibility that extraneous factors are affecting the results (Chapters 2).

Internet survey: a self-administered survey in which a researcher posts the survey questionnaire on the Internet for respondents to fill out (Chapter 7).

interval level of measurement: differences between measurement categories are fixed and equal, but there is no absolute zero (Chapter 5).

intervening variable: a variable that comes between or influences the relationship between an independent and dependent variable (Chapter 3).

intracoder reliability: assessment of the reliability of a measure by calculating the degree of correspondence or agreement between two different codings done by the same coder (Chapter 5 and 8).

intuition: a way of knowing that relies on the assumption that something is true because it is "obvious" or just "common sense." The criterion for truth is the reasonableness of the argument (Chapter 1).

isomorphism: similarity of form or structure (Chapter 5).

jury validation: technique for assessing internal validity in which a panel of experts or a group of subjects similar to those who will be used in the actual study judge whether measures are reasonable (Chapter 5).

K-R 20: a technique used to assess the reliability of a measure by examining the correlation of a subject's score on individual items with the score the subject received for the total set of all items (Chapter 5).

Kendall's tau: a measure of association used to determine the strength of the association between two ordinal level variables. Scores range from –1 to 1 with –1 indicating a perfect negative association, 1 indicating a perfect positive relationship, and 0 indicating no relationship (Chapter 7).

Kuder-Richardson formula 20: see K-R 20.

laboratory experiment: an experiment that is conducted in a laboratory or a controlled setting (Chapter 6).

latent content: below-the-surface meaning of a message (Chapter 8).

Latin square: see counterbalanced experimental design.

law: a proposition that has been repeatedly verified and is widely accepted as true (Chapter 1).

leading question: a question worded to suggest or "lead" a respondent to a certain response (Chapter 7).

level of analysis: the major entity of social life on which the research question focuses. Common levels of analysis are the individual, interpersonal, group, organizational, and geographic levels (Chapter 3).

level of measurement: classification of measures according to information content. The four levels of measurement are nominal, ordinal, interval, and ratio (Chapter 5).

Likert scale: scaling technique that measures the strength and direction of people's attitudes or opinions by having subjects use a five-point scale to indicate how strongly they agree or disagree with a series of statements (Chapter 5).

longitudinal survey design: a survey design in which questionnaires are administered over an extended period of time in order to examine changes over time or draw causal inferences. Trend and panel studies are two types (Chapter 7).

mail survey: a self-administered survey in which a researcher mails a questionnaire to respondents who write their answers on the form and return it to the researcher (Chapter 7).

mall intercept: a type of convenience sample in which researchers interview every kth person entering a store or shopping mall (Chapter 4).

manifest content: obvious, surface, or more literal meaning of messages (Chapter 8).

manipulation: artificially creating one or more versions of a stimulus in order to measure an independent variable. One of three measurement methods. Most often used in experiments (Chapters 5 and 6).

manipulation check: a procedure used to determine whether the manipulation of the independent variable in an experiment actually had its intended effect (Chapter 6).

matching: a procedure used in experiments to assign subjects to treatment groups when randomization isn't possible or is impractical. Individual subjects or in-tact groups that are very similar are paired or matched on important characteristics so that one member of each pair can be assigned to the experimental condition and the other to the control group (Chapter 6).

maturation: a threat to internal validity in which physiological or social changes occurring in subjects during the course of a study may explain scores on a dependent variable (Chapter 2 and 6).

mean: the arithmetic average computed by adding up the values of all cases and dividing by the total number of cases. A measure of central tendency (Chapter 3).

measurement: the process of assigning numerals or numbers to phenomena according to certain rules (Chapter 5).

measures of association: statistics that summarize the significance, strength and/or direction of the relationship between two variables. Common measures of association are chi-square, Cramer's V, phi, gamma, Kendall's tau, Spearman's rho, and Pearson's r (Chapters 3 and 7).

measures of central tendency: statistics that describe how individual values on a single variable cluster together. The "typical value" for a variable. Common measures of association are the mean, the median, and the mode (Chapters 3 and 7).

measures of dispersion: statistics that show how subjects' scores on a single measure are spread or dispersed across the set of all possible values. Common measure of dispersion are the range, standard deviation, and variance (Chapter 3).

median: the point that divides a distribution in half such that 50 percent of the cases fall above the value and 50 percent of cases fall below the value. A measure of central tendency (Chapter 3).

method: a particular technique for doing a study or a particular way of collecting or analyzing data. There are five research methods: experiments, surveys, content analysis, focus groups, and observational research (Chapter 3).

methodology: the overall plan for conducting research or gathering or analyzing data. The collection of all methods used in a study (Chapter 3).

mode: the most frequent value in a distribution. A measure of central tendency (Chapter 3).

moderator's guide: a document used in focus groups that lists the procedures for conducting sessions, the topics to be covered, and/or the questions to be asked of participants (Chapter 9).

mortality: a threat to internal validity caused by differential loss of subjects between the beginning and end of a study (Chapter 2 and 6).

multistage cluster sampling: choosing subjects by sampling in two or more stages, at least one of which includes choosing clusters or groups of subjects (Chapter 4).

multistage sampling: choosing subjects by sampling in two or more stages (Chapter 4).

multivariate statistics: statistical procedures used to examine the relationship among three or more variables at the same time. Common ones include analysis of variance and regression analysis (Chapter 7).

mutually exclusive: measurement categories constructed so that there is one and only one category where a subject fits (Chapter 5).

network sampling: see snowball sampling.

nominal group: a small-group method for reaching consensus or setting priorities through orderly, guided discussion. Variant of focus groups (Chapter 9).

nominal level of measurement: measures that assign subjects to one of two or more categories of an indicator for which there is no mathematical interpretation. The categories vary in kind or quality but not in amount (Chapter 5).

nondirectional hypothesis: see two-tailed hypothesis.

nonparametric statistics: statistics designed to be used with measures at the nominal or ordinal level (Chapter 5).

nonprobability sampling: any sampling technique that does not ensure that subjects are representative of a larger group (Chapter 4).

normal curve: a symmetric, bell-shaped curve that possesses specific mathematical characteristics (Chapter 4).

null hypothesis: a hypothesis stating no relationship exists between the variables of interest (Chapter 3).

observation: the procedure of measuring by directly or indirectly watching or observing subjects in order to place them in measurement categories. One of three methods of operationalizing a concept (Chapter 5).

observational research: researchers gather firsthand data by watching subjects to find out how they behave in their natural setting. One of five research methods (Chapters 3 and 10).

one-group pretest-posttest design: a pre-experimental design in which a single group is given a pretest, exposed to the independent variable, and then given a posttest (Chapter 6).

one-shot case study: a pre-experimental design in which a treatment is administered to a group and then the dependent variable is measured. There is no control group or pretest in this design (Chapter 6).

one-tailed hypothesis: a hypothesis that states the direction of the expected relationship between two variables. Also called a unidirectional hypothesis (Chapter 3).

open-ended question: a question that allows subjects to answer in their own words (Chapters 5 and 7).

operational definition: a detailed description that specifies the operations or procedures necessary to recognize and measure a particular concept (Chapter 3).

ordinal level of measurement: categories of an indicator specify only the rank or order of subjects along some dimension, but the distance between adjacent categories is unspecified and most likely not the same (Chapter 5).

overregistration: a situation in which a sample frame includes elements that are not part of the defined population (Chapter 4).

overt observation: a method of data collection in which the individuals being observed are aware of the researcher's presence (Chapter 10).

paired-choice ranking scale: a scale created from subjects' ranking of items in each pair of items in a set of all possible pairs (Chapter 5).

panel design: a longitudinal survey in which the same sample of individuals is interviewed at several different points of time in order to study change at the individual level (Chapter 7).

parameter: a characteristic (e.g., number, mean, or proportion) of a population or universe. The true value (Chapter 4).

parametric statistics: statistics designed to be used with measures at the interval or ratio level (Chapter 5).

partial correlation: a statistic used to determine the size, direction, and statistical significance of a relationship between two interval or ratio level variables while controlling for the effect of one or more other variables (Chapter 7).

Pearson's correlation coefficient: a measure of association that summarizes the type and strength of a statistical relationship between two interval or ratio level variables. Scores range from –1 to 1 with –1 indicating a perfect negative association, 1 indicating a perfect positive relationship, and 0 indicating no relationship. Also called Pearson's r (Chapter 7).

periodicity: refers to naturally occurring temporal or seasonal variations in document content. This variation leads to bias in systematic random samples because every k^{th} document may possess characteristics other documents do not. A common problem in content analysis (Chapter 8).

phi: a measure of association suitable for nominal measures, at least one of which is a dichotomous variable. Values range from 1 for a perfect association to 0 for no relationship (Chapter 7).

population: the set of all elements from or about whom data are needed. The population may be people, organizations, documents, or other entities. Sometimes called a universe (Chapter 4).

positive response bias: the tendency of subjects to give a positive or "yes" answer, especially when they are guessing or not really paying attention. Researchers work around this problem by wording some questions so that people who really feel favorably toward some referent must give an apparently negative answer (Chapter 5).

posttest-only control group design: a true experimental design in which there is no pretest; subjects are randomly assigned to treatment and control groups and given a posttest (Chapter 6).

predictive research: studies designed to find out about or make predictions or projections about what will happen in the future or in other settings (Chapter 3).

predictive validity: a technique for assessing the validity of a measure by establishing that a measure or test can accurately forecast or predict scores on a criterion variable (Chapter 5).

pre-experimental design: a design that lacks one or more features of a true experimental design such as a control group or random assignment. These designs are relatively simplistic and do not effectively control for most threats to internal validity (Chapter 6).

primary source: literature in which authors describe their own work. Useful primary sources include books, dissertations, theses, and articles published in academic journals (Chapter 3).

probability sampling: any technique for choosing subjects from a sample frame so that the sample will be representative of the target population as a whole. This procedure is based on the process of random selection. Also called random sampling or random selection (Chapter 4).

probe: an unscripted question used by interviewers to get respondents to answer a particular question or answer the question more fully (Chapter 7).

proposition: an if-then statement that links two or more concepts together (Chapter 3).

protocol: the set of directions and procedures for conducting a research project. In survey research, the directions, list of topics, and questions given to interviewers to aid them in being consistent in conducting in-depth interviews (Chapter 7).

psychographic research: survey research conducted to obtain data on people's lifestyles, attitudes, and values (Chapter 7).

purposive sampling: a form of nonprobability sampling in which subjects are chosen because they can be expected to provide useful information (Chapter 4).

push poll: an unethical survey in which subjects are first asked a series of questions, most often about their issue positions or feelings about a candidate, and then "pushed" toward a desired position through follow-up questions designed to plant doubts in their minds about their original positions (Chapter 7).

Q-methodology: a data-gathering technique in which subjects individually rank or sort a large number of statements according to some criterion (Chapter 9).

Q-sort: the act of arranging or sorting statements in a study using Q-methodology (Chapter 9).

qualitative research: research in which observations are not reduced to numbers; findings are presented in narrative form. Research is usually inductive in nature (Chapter 1).

quantitative research: research in which numbers are attached to observations in such a way that statistics can be used to analyze the results and present findings. Research is usually deductive in nature (Chapter 1).

quasi-experimental design: an experimental design that lacks at least one feature of a true experiment, most often randomization, but still provides some control over threats to internal validity (Chapter 6).

questionnaire: a fixed set of questions intended to be asked in the same way of all subjects. Used primarily in survey research (Chapter 7).

quota sampling: a form of nonprobability sampling that involves nonrandom selection of subjects so that there will be a specified proportion or quota of subjects with each demographic or other characteristic of interest (Chapter 4).

random-digit dialing: an automated method of selecting telephone numbers so that all numbers have an equal chance of being selected for a sample (Chapter 4).

random error: a type of error over which the researcher has no control. This type of error is unlikely to distort findings in favor of or against any particular result (Chapter 2).

randomization: in experiments, the use of probability techniques to assign subjects to treatment conditions so that each subject has an equal and known chance of receiving the stimulus or being in the control group. As the result of randomization, the treatment and control groups should be equivalent on all possible confounds before the experiment begins (Chapter 6).

random sampling: see probability sampling.

random selection: see probability sampling.

ranking scale: any measure that provides information on how subjects prioritize each item in a set of items (Chapter 5).

ratio level of measurement: differences between measurement categories are fixed and equal and there is an absolute zero. With a true zero it is possible to multiply and divide terms (Chapter 5).

readability studies: a type of content analysis designed to draw inferences from writing style about whether people can understand a message. Most often used in the field of education (Chapter 8).

reading-ease formula: readability formula based on the number of words per sentence and the number of syllables per word. Developed by Rudolf Flesch (Chapter 8).

regression analysis: a statistical procedure used to predict or explain scores on a dependent variable measured at the interval or ratio level on the basis of subjects' scores on one or more independent variables (Chapter 7).

regression to the mean: a threat to internal validity resulting from selecting subjects for a study on the basis of extreme scores on the dependent variable. Subjects who score at the extremes have a tendency to behave less atypically the next time (Chapter 2 and 6).

reliability: the internal consistency or stability of a measure; consistency and reproducibility of findings from a research study (Chapters 2 and 5).

repeated-measures design: see counterbalanced experimental design.

research demand: see guinea pig effect.

research ethics: moral principles and recognized rules of conduct that govern the activities of researchers (Chapter 2).

research journal: a periodical publication that prints researchers' reports of their own scholarly research. A kind of primary literature (Chapter 3).

researcher-administered survey: a method of survey data collection in which the researcher or research assistants conduct face-to-face or telephone interviews with subjects (Chapter 7).

response rate: the percentage of individuals selected for a sample who actually participate in a given study (Chapter 7).

reverse directory: a phone book in which the telephone numbers are organized alphabetically and then numerically by address rather than alphabetically by a business name or personal surname (Chapter 4).

rhetorical criticism: a type of textual analysis that uses standards of excellence to interpret and evaluate messages (Chapter 8).

sample: a subset of elements, or cases, selected from a population or universe (Chapter 4).

sample frame: an actual list identifying all the individuals, households, documents, or other elements specified by the definition of the target population. If no list exists, the sample frame indicates a procedure, such as random-digit dialing, for locating relevant elements (Chapter 4).

sample weighting: a statistical procedure that adjusts the distribution of characteristics of subjects in a sample to more nearly reflect their distribution in the target population (Chapter 4).

sampling distribution: a probability distribution of all possible values of a statistic that would occur if all possible samples of a specific size were drawn from a given population (Chapter 5).

sampling error: the likely difference between an actual population value or parameter (e.g., the mean) and the population value or statistic as estimated by the sample (Chapter 4).

sampling interval: the ratio of the number of elements in the sampling frame to the desired sample size. Used to select every k^{th} case in systematic random sampling (Chapter 4).

sampling unit: the element or set of elements selected for study in content analysis (Chapter 8).

scale: any composite measure that taps strength, amount, direction, salience, importance, or some combination of these. Examples include Likert scales, semantic differential scales, Guttman scales, and Thurston scales (Chapter 5).

scalogram analysis: see Guttman scale.

science: a way of knowing based on the systematic gathering and testing of evidence through observation (Chapter 1).

scientific research: the systematic, objective collection and analysis of information to uncover facts, patterns, and relationships in order to create knowledge that enhances understanding of the world or provide answers to practical questions (Chapter 1).

Scott's pi: a statistic used to assess the intra- or intercoder reliability of a measure (Chapters 5 and 8).

screen: a question or set of questions in a survey questionnaire designed to select only appropriate respondents for a sample. A type of filter question (Chapters 4 and 7).

secondary analysis: analysis of pre-existing data (Chapter 3).

secondary source: literature in which an author reports on someone else's research. Secondary sources include general circulation magazines and newspapers as well as many books aimed at a general audience (Chapter 3).

segmentation study: a study that divides subjects into categories in order to examine their behavior or opinions in a way that helps organizations target their products, services, or messages to the appropriate public (Chapter 7).

selection: a threat to internal validity resulting from use of nonprobability techniques to choose subjects for a study or to assign them to treatment and control groups for an experiment (Chapter 2, 4 and 6).

self-administered survey: a method of survey data collection in which respondents fill out a questionnaire on their own, outside the presence of a researcher. Common types are mail, disk-by-mail, and Internet surveys (Chapter 7).

self-report: answers to questions or materials from documents generated by research subjects. One of three ways of operationalizing a concept (Chapter 5).

self-selection sampling: a form of nonprobability sampling in which subjects choose for themselves, without first being directly and personally contacted by a researcher, whether to be a part of a study (Chapter 4).

semantic differential scale: a scaling technique that provides information on direction and intensity by having subjects rate a specified phenomenon on a number of seven-point scales bounded at each end by words that are polar opposites (Chapter 5).

separate-sample pretest-posttest experimental design: a quasi-experimental design in which separately selected random samples of subjects are used as test and control groups. Often used with survey research to create a design approximating an experimental design (Chapter 6).

simple random sampling: a probability sampling procedure in which each unit in the sample frame has an equal and known chance of being included in the sample (Chapter 4).

situational analysis: in public relations, research conducted to monitor the environment or gather background information from the media or from publics that can be used to help decide what to do in order to achieve organizational goals. A type of audit (Chapter 3).

sleeper question: a trick question inserted into a survey questionnaire to help identify subjects who are merely guessing in response to factual questions or who are providing opinions on subjects about which they have no real knowledge (Chapter 7).

snowball sampling: a nonprobability sampling procedure in which individuals are asked to identify additional members of the target population who, in turn, are asked to name others. Also called network sampling (Chapter 4).

social science: the use of scientific methods to investigate individuals, societies, and social processes (Chapter 1).

Solomon four-group design: a true experimental design that combines the classic experimental design and the posttest-only control group design by randomly assigning subjects to one of four possible conditions (Chapter 6).

split-half reliability: a technique for assessing reliability by randomly dividing subjects into two groups and then comparing scores of the two groups on some measure (Chapter 5).

standard deviation: a measure of variability or dispersion that indicates the average spread of observations around the mean. The square root of the variance (Chapter 4).

Stapel scale: a scaling technique that uses just a single word along with five positive and five negative numbers plus a midpoint indicated as zero (Chapter 5).

Starch test: a posttest recall study conducted by the Starch INRA Hooper firm to evaluate advertising effectiveness. Involves face-to-face interviews with subjects who are shown periodicals and then questioned at length to see whether they saw certain ads and, if so, what they remember about each one (Chapter 7).

static group comparison design: a pre-experimental design in which there are two groups and a posttest, but no random assignment to the groups (Chapter 6).

statistic: a characteristic of a sample, such as the number or proportion. Statistics are used to estimate population parameters (Chapter 4).

statistical control: use of statistics to remove confounds or hold them constant (Chapter 6).

strategic research: in public relations, research conducted to help determine objectives and plan campaigns (Chapter 3).

stratified random sampling: a probability sampling technique in which the sample frame is divided into groups or strata based on some population characteristic so that independent random samples can be selected from each stratum (Chapter 4).

straw poll: a quasi-survey of a convenience sample of subjects. Often used by journalists as a way of finding sources to comment on some topic (Chapter 4).

subject-researcher interaction effects: a reactivity threat to internal validity in which subjects' responses to a researcher may affect scores on a dependent variable. A guinea pig effect (Chapter 2, 6 and 7).

summative research: in public relations, research conducted to learn whether campaign goals have been met (Chapter 3).

survey: a research method in which data consist of answers to questions asked in the same way and same order of a large probability sample of subjects. One of the five research methods, this is the best for collecting large amounts of data and for generalizing findings to an entire population (Chapters 3 and 7).

systematic error: a type of error in research that biases the findings in one direction or another (Chapter 2).

systematic random sampling: a form of probability sampling in which, after selecting a random start point, every k^{th} element is selected from the sample frame (Chapter 4).

target population: population to which a researcher would like to be able to generalize results (Chapter 4).

telephone survey: researcher-administered survey conducted over the telephone. Probably the most widely used type of survey in the United States. (Chapter 7).

tenacity: a way of knowing based on evidence from the past. The assumption is that what has worked before will still work today (Chapter 1).

test-item analysis: a procedure for assessing the reliability of a measure by calculating the correlation of subject's scores on individual items, which can be scored using dichotomous categories such as right/wrong, to the subject's score for the total score of all test items. K-R 20 is an example (Chapter 5).

test-retest reliability: a procedure for assessing reliability by means of showing that scores on a measure or test administered at two points in time are highly correlated (Chapter 5).

textual analysis: a form of content analysis in which message content and characteristics are examined using qualitative methods (Chapter 8).

theoretical research: see basic research.

theory: a logically interrelated set of propositions about empirical reality (Chapter 1).

Thurston scale: a scaling technique that measures direction and intensity or salience. Distances between intervals on this kind of scale are fixed (Chapter 5).

time series design: a quasi-experimental design that involves taking measures on the dependent variable at several points in time before and after administering the stimulus or independent variable (Chapter 6).

tracking study: a longitudinal study conducted to monitor performance. Often used in advertising to learn how well a campaign is working (Chapter 3).

treatment group: see experimental group.

trend study: a longitudinal study designed to detect change over time that involves conducting multiple surveys over a period of months or years using a new sample from the same population each time (Chapter 7).

triangulation: a method of improving internal validity by using multiple methods and/or multiple measures to operationalize a concept or gather data (Chapters 5 and 10).

true experiment: a design that includes assignment of subjects to a test and control group through probability techniques, manipulation of the independent variable, posttest measurement, and often also pretest measurement of the dependent variable. These designs provide the most control over possible threats to internal validity (Chapter 6).

t-test: a statistical procedure used to examine differences between two groups' scores on a dependent variable measured at the interval or ratio level. Often used to analyze the results from an experiment (Chapter 6).

two-tailed hypothesis: a hypothesis that states that a relationship exists between two variables without specifying the direction of that relationship. Also called a nondirectional hypothesis (Chapter 3).

underregistration: a situation in which the sample frame does not include many relevant elements in a target population (Chapter 4).

unidirectional hypothesis: see one-tailed hypothesis.

unit of analysis: in a content analysis study, the communication component that is actually coded for analysis (Chapter 8).

univariate statistics: statistics used to analyze one variable (Chapter 7).

universe: see population.

useability study: research, often using a pre-experimental design, conducted to find out whether subjects can follow a set of directions or understand a manual or other training materials or whether training programs or other educational materials are effective. Commonly conducted by educators and technical writers (Chapter 3 and 6).

validity: the degree to which a study or indicator measures what it is intended to measure (Chapters 2 and 5).

variable: any concept operationalized so that it can take on two or more values (Chapter 5).

Index

Abstract
  descriptive, 295
  informative, 295
Access for observational research, 246-247
Accuracy, reliability and, 30-31, 106
Acknowledgments in reports, 294
Alternate forms reliability, 107
Alternate hypothesis, 49
American Association for Public Opinion Research (AAPOR), 151
American Sociological Association (ASA) Code of Ethics, 263-270
Analysis of covariance (ANCOVA), 43, 138
Analysis of variance (ANOVA), 43, 122, 138, 140, 141-142, 208
Analyzing data. *See* Data analysis
Anonymity, 26, 152, 244, 265
Antecedent variable, 49-50
Appendices, business report, 301
Applied research, 14
Arbitron, 150
Association, measures of, 43, 168, 170-173
  chi-square, 73, 79, 170-171, 173, 208
  Cramer's V, 43, 171, 173
  gamma, 172, 178
  interpreting, 178-179
  Kendall's tau, 43, 171-172, 178-179
  Pearson's r, 43, 79, 94, 106, 172-173, 178, 208
  phi, 43, 171
  sample size requirements, 79
  Spearman's rho, 43, 172, 178
Audiotaping, 251-252
  focus group sessions, 226
Audit, 41, 110, 133, 206
Authority, 5-6
Authorship credit, 270

Background section, business report, 300
Basic research, 14
Bell-shaped curve, 65
Beta, 181, 183-184
Bias
  accidental, 81
  positive-response bias, 96
  sample, 81-82
Bivariate statistics, 43, 94, 168, 173-175, 208
Business reports
  formal, 299-301
    appendices, 301
    background, 300
    conclusion, 300
    executive summary, 299-300
    front matter, 299
    methodology, 300
    recommendations, 300
    references, 301
    results, 300
  informal, 301-302

Call-in polls, 76, 80
Campbell, Donald, 31, 125
Category, concept, 201-202
Causal studies, 44
Causal validity, 31
Causation, showing, 120-121
Cause-and-effect relationship, 44, 120-121
Census, 63
Central limit theorem, 65
Central tendency, measures of, 43, 168
Chall, Jeanne, 196
Chat room, 230
Check question, 160
Children, informed consent with, 267-268

Chi-square, 43, 79, 170-171, 173, 208
Cloze procedure, 196
Cluster sampling, 77-78, 80, 81-82
Code book, 202-203, 204, 283-291
Code sheet, 202-203, 204, 292
Coding materials for content analysis, 201-206, 283-292
Coefficient alpha, 107
Cohort analysis, 154
Composite measure, 100-104
Computer use for content analysis, 203-204
Concept, 55
    operationalization of, 88-91
    unidimensional versus multidimensional, 100
    variables and, 91
Concept category, 201-202
Conceptual definition, creation of, 87-88
Conceptual fit, 88
Concerns, fundamental, 23-35
    costs, 23-24
    ethics, 24-27
    main points, 35
    quality control, 27-34
Conclusion section
    business report, 300
    research report, 298
Concurrent reliability, 106-107
Concurrent validity, 110, 207
Conference telephone calls, 230
Confidence interval, 66-67, 81
Confidentiality, 26, 263-266
    anonymity, 26, 152, 244, 265
    discussing, 264-265
    electronic transmission of information and, 265
    focus group, 219
    limits of, 264
    maintaining, 264
    preservation of confidential information, 266
    surveys and, 152
Confounds, 120
    blocking, 121
    elimination, 121
    holding constant, 121-122
Consent
    children and, 267-268
    focus group, 218-219
    observational research, 243-244
    process, 267
    for recording device use, 252
    scope of, 266-267
    of students and subordinates, 267
    surveys and, 152
Constant measure, confounding variables as, 121-122
Constraints, identification of, 44
Construct, 55
Constructed week, 200
Construct validity, 110
Contamination problems, 134
Contempt charges, 245
Content analysis, 53, 193-211
    choosing documents, 197-201
        databases, 198-199
        indexes, 198-199
        periodicity and, 199-201
    code book, 202-203, 204, 283-291
    code sheets, 202-203, 204, 292
    communications applications, 194
    computer use for, 203-204
    data analysis, 201-202, 208-209
        qualitative, 201, 204, 208-209
        quantitative, 201-202, 204, 208
    ethics of, 195
    main points, 209-211
    measurement, 201-202
        category construction, 201-202
        enumeration, 202
        units of analysis, 201
    overview of, 193-194
    preparing documents for, 204-205
    quality control, 205-207
        external validity, 207
        internal validity, 206-207
        missing materials and, 207
        reliability, 205-206
    reporting on, 297
    searching materials, 198-199, 203-204, 209
    strengths and weaknesses of, 193-194
    types of, 195-197
        readability studies, 195-196
        textual analysis, 196-197
Content validity, 110
Contingency question, 160, 163
Contingency table, 173
Continuous measure, 95
Control, experiment
    confounding variables and, 121-122
    counterbalancing, 122
    matching, 122
    randomization, 122
    statistical control, 122
Control group, 119

Control variable, 50
Convenience sampling, 75, 80, 81, 82, 154
Cook, Thomas, 31
Correlation, 101. *See also* Association, measures of
    partial, 43, 179, 180
Cost
    limitations of, 23-24
    opportunity, 24
    overhead, 23
Counterbalanced designs, 131
Counterbalancing, 122
Covert observation, 248-249
Cramer's V, 43, 171, 173
Criterion validity, 110
Critical studies approach, 197
Cronbach's alpha, 102-103, 107, 108
Cross-cultural survey designs, 153
Cross-tabulation table, 173-174, 175-178

Dale, Edgar, 196
Data analysis
    content analysis, 201-202, 208-209
    ethics, 27
    experiment, 135-140, 141-142
    focus group, 231-232, 232-234
    level of analysis, 41, 42
    observational research, 255-256
    quantitative versus qualitative, 14-15
    survey, 167-185
Databases, 198-199
Data sets, online, 51
Debriefing, 26, 124
Deception, 26, 243, 268
Deductive reasoning, 12-14
Delphi technique, 228
Demographics, survey questions on, 159, 163
Dependent variable, 49-50, 119. *See also* Experiments
Descriptive abstract, 295
Descriptive studies, 42
Diary, research, 251
Difference, tests of, 43, 136, 168
Discourse analysis, 197
Discussion section, research report, 298
Dispersion, 43, 65
Disproportionate likelihood sampling, 73
Document analysis. *See* Content analysis
Double-barreled question, 162
Double-blind question, 162
Dummy variable, 94, 181, 182, 185, 208
Duplicate coverage, 207

Eigenvalue, 104
Electronic transmission of confidential information, 265
Elements, population, 63-64
E-mail, 230
Equivalence in experiments, 134
Equivalent group pretest-posttest design, 129
Error
    random, 28-29
    in reasoning, 8
    sampling, 65-68, 78-80
    systematic, 28-29
Ethics
    American Sociological Association (ASA) Code of Ethics, 263-270
    authorship credit, 270
    codes of, 25, 263-270
    confidentiality, 263-266
        discussing, 264-265
        electronic transmission of information, 265
        limits of, 264
        maintaining, 264
        preservation of confidential information, 266
    consent, informed
        with children, 267-268
        process, 267
        scope of, 266-267
        of students and subordinates, 267
    content analysis, 195
    data analysis, 27
    deception, 243, 268
    experiment, 124
    falsifying data, 27
    focus group, 218-219
    inducements, offering, 269
    law and, 24-25
    observational research, 243-245
    plagiarism, 270
    in planning and implementation, 268-269
    privacy, 26
    recordings, use of, 268
    reports and, 27, 269-270, 294
    sponsors, working with, 26-27
    standards, 263
    surveys and, 151-152
    treatment of subjects, 25-26
Ethnography, 241
Evaluation research, 120
Executive summary, business report, 299-300

Experiments, 53, 119-143
  causation, showing, 120-121
  communication applications, 119-120
  confounds, 120, 121-122
  control, 121-122
  data analysis, 135-140, 141-142
  design, 124-132
    Campbell and Stanley notation system, 125
    preexperiments, 131-132
    quasi-experiments, 128-131
      counterbalanced designs, 131
      equivalent group pretest-posttest design, 129
      nonequivalent control group design, 129-130
      separate-sample pretest-posttest design, 130
      time series, 130-131
    true experiments, 125-128
      classic, 125-126
      posttest-only control group, 126
      Solomon four-group design, 126
      variants, 126, 128
  ethics, 124
  main points, 140-143
  manipulation, control of, 121
  quality control, 132-135
    external validity, 135
    internal validity, 133-135
    reliability, 132-133
  recording variables, 139-140
  reporting on, 297
  sample size, 80
  setting for, 123-124
Explanatory studies, 42
Exploratory studies, 42
External validity, 31, 34
  content analysis, 207
  experiment, 135
  focus group, 231
  observational research, 254-255
  sample size and, 79
  surveys, 166-167
  threats to, 32-33

Face-to-face interview, 156, 166, 167
Face validity, 110, 207
Factor analysis, 43, 101, 104-105
Factorial design, 128, 131
Factor loadings, 104, 105
Factors, 128
Falsifying data, 27
Fico, Frederick, 200

Field experiment, 123-124
Field notes, 89, 250, 251, 252
Field research. *See* Observational research
Filter question, 160, 163
Findings, research report of, 273-274, 298
Fiske, Marjorie, 218
Fixed-response question, 149, 163
Flesch, Rudolph, 195, 196
Focus groups, 53, 217-236
  alternatives to, 226-230
    conference telephone calls, 230
    Delphi technique, 228
    e-mail, 230
    Internet chat room, 230
    nominal group technique, 227-228
    Q-methodology, 228-230, 232
    videoconference, 230
  composition of, 221-222
  cost, 217, 221
  data analysis, 231-232
    qualitative analysis, 232-234
    quantitative analysis, 232
  ethical concerns, 218-219
  main points, 234-236
  number of, 221-222
  origin of, 218
  planning, 219-226
    groups, 221-222
    location, 226
    materials, 224-225
    moderator, 222-223
    participants, 219-221
    recording provisions, 226
    scheduling, 225
  quality control, 230-231
    external validity, 231
    internal validity, 231
    reliability, 230-231
  questionnaires, 224-225
  reporting on, 297
  size, 80, 221
  uses of, 218
  weakness of, 217
Fog Index, 195, 196
Forced sort, 229-230
Forecasting, 131
Formative research, 41
Frequency distribution, 168, 169
Front matter, business report, 299
Funding, source of, 271

Gamma, 172, 178
General Social Survey, 51
Glaser, Barry, 13

Group-administered surveys, 157
Guinea pig effect, 123, 133
Gunning, Robert, 195, 196
Guttman, Louis, 98
Guttman scale, 98-99

Hawthorne effect, 123, 133
History as threat to internal validity, 32, 127, 134-135
Hypothesis, 12
  alternate, 49
  evaluating research report, 272
  null, 49
  one-tailed (unidirectional), 49
  in planning research projects, 48-50
  research questions compared, 48-49
  testing and quantitative research, 15
  two-tailed (nondirectional), 48
  wording, common forms of, 50

Independent variable, 49-50, 119. *See also* Experiments
  in factorial design experiments, 128
  in surveys, 152-153
Indexes, searching, 198-199
Index measure, 101, 102-103
Inducements for research participants, 269
Inductive reasoning, 13-14
Informant, 247, 254
Information, confidential. *See also* Confidentiality
  anonymity of sources, 265
  anticipation of uses of, 265
  electronic transmission of, 265
  preservation of, 266
Informative abstract, 295
Informed consent. *See* Consent
Instructions, survey, 163-164
Instrumentation
  internal validity, effects on, 32
  reliability and, 132
Interaction effects on validity, 33
Interference
  pretest-posttest, 133
  test-retest, 126
Internal validity, 31-34
  causal validity, 31
  content analysis, 206-207
  experiment, 133-135
  focus group, 231
  measurement validity, 31
  observational research, 253-254
  surveys, 166
  threats to, 32, 127

Internet
  chat room, 230
  surveys, 157
Inter-University Consortium for Political and Social Research, 51
Interval level measurement, 93, 94
Intervening variable, 50
Interviewer, survey
  reliability and, 165
  validity and, 166, 167
Interview surveys, 155
Intra- and intercoder reliability, 108, 206
Introduction section, report, 295-296
  literature review, 296
  problem statement, 296
  purpose statement, 296
Intuition, 5-6

Journals
  communication research, list of, 46-47
  writing research reports for, 294-299
Jury validity, 110

Kendall, Patricia L., 218
Kendall's tau, 43, 171-172, 178-179
Keywords, 203, 204-205, 209
Knowing, ways of
  errors in reasoning, 8
  science as, 7-10
  traditional, 5-6
Kuder-Richardson formula 20 (K-R 20), 108, 109

Laboratory experiment, 123
Lacy, Stephen, 200
Latent meaning, 196
Latin square, 131
Law
  ethics and, 24-25
  libel, 245
  recording device use and, 252
  scientific, 10
Leading question, 161
Least significant difference (LSD), 140
Level of analysis, 41, 42
Lexis/Nexis, 199
Libel, 245
Likert, Rensis, 96
Likert scale, 96
Limiting factors
  cost, 23-24
  ethics, 24-27
Literature
  content evaluation (*See* Content analysis)

Literature *(continued)*
  journals, list of, 46-47
  primary and secondary sources, 45, 47
  research reports, evaluating, 271-274
  reviewing, 45-47
  review section, research report, 296
Load, factor, 104, 105
Logistics
  focus group, 225-226
  planning, 56
Longitudinal designs, 153-154

Magnitude of effects, calculating, 127
Mail survey, 157-158, 159, 167
Mall intercept, 75, 156, 157
Manifest meaning, 196
Manipulation
  checks, 133
  control of, 121
  measurement by, 88-89, 91
Matching, 122, 129
Maturation as internal validity threat, 134-135
Mean, 43, 135-136, 168
Measurement validity, 31
Measures and measurement, 87-111
  composite measures, 100-104
  conceptualization, 87-88
  content analysis, 201-202
  factor analysis, 101, 104-105
  levels of, 91-95
    implications of, 94-95
    interval, 93
    nominal, 92
    ordinal, 92-93
    ratio, 93
    statistic choice and, 94-95
  main points, 111
  operationalization, 88-91
    examples, 91
    manipulation, 88-89
    observation, 89-90
    self-reports, 90-91
  quality control, 104-110
    reliability, 106-109
    validity, 109-110
  scales, 95-100
    Guttman, 98-99
    Likert, 96
    ranking, 99-100
    semantic differential, 97-98
    thermometer, 99
    Thurstone, 98
  sources for existing measures, 101

Mediamark Research, Inc. (MRI), 151
Median, 43, 168
Merton, Robert K., 218
Message structure programs, 203
Methodology
  business report section, 300
  development of, 54-55
    measures, creating, 55
    subject choice, 54-55
  evaluating research reports for, 272-273, 274
  research report section, 296-297
Metric, 93
Mode, 43, 168
Moderator, focus group, 222-223
  guide, 224
  number of, 222-223
  qualitative analysis by, 233
  skills required, 222
  training, 222
Mortality
  external validity, threat to, 33, 167
  internal validity, threat to, 32, 134-135
Multidimensional concepts, 100
Multiple analysis of covariance (MANCOVA), 43, 138
Multiple analysis of variance (MANOVA), 43, 138
Multiple treatment interference, 33
Multistage cluster sampling, 77-78, 81-82
Multistage sampling, 76-78
Multivariate analysis, 94
Multivariate statistics, 43, 175, 179, 181

Naive science, 7
National Election Studies, 51
Naturalistic inquiry. *See* Observational research
Network sampling, 74-75
Nielson, 150
Noise, 28
Nominal group technique, 227-228
Nominal measurements/data, 92, 94, 170, 171
Nondirectional hypothesis, 49
Nonequivalent control group design, 129-130
Nonparametric statistics, 94, 95
Nonprobability sampling
  in multistage sampling, 76
  sample size, 80-81
  surveys, 154-155
  types of, 73-76
    call-in polls, 75

convenience sampling, 75
mall intercept, 75
purposive sampling, 74
quota sampling, 73-74
self-selection sampling, 75-76
snowball (network) sampling, 74-75
straw poll, 75
Normal curve, 65
Note taking, 250-251
Null hypothesis, 49
Numerals, 92

Observational research, 53, 241-258
collecting and recording information, 249-253
managing data, 252-253
note taking, 250-251
recording device use, 251-252
techniques for, 250-252
types of information, 249-250
what to collect, 250
communications applications, 242-243
data analysis, 255-256
qualitative, 255-256
quantitative, 255
ethics, 243-245
main points, 257-258
overt versus covert observation, 248-249
overview of, 241-242
participation versus nonparticipation by researcher, 247-248
planning, 245-249
choosing settings and sites, 245-246
gaining access, 246-247
role, 247-248
stance, 247, 248-249
quality control, 253-255
external validity, 254-255
internal validity, 253-254
reliability, 253
reporting on, 297
weaknesses of, 241, 242
Observation as operational procedure, 89-90, 91
Observation notes, 250-251, 252
One-group pretest-posttest, 131
One-shot case study, 131
One-tailed (unidirectional) hypothesis, 49
Open-ended question, 149, 163
Operational definition, creation of, 88
Operationalization
defined, 88
manipulation, 88-89

observation, 89-90
self-reports, 90-91
Opportunity cost, 24
Oral reports, 302
Ordinal measurements/data, 92-93, 94, 171, 172
Overhead, 23
Overt observation, 248-249

Paired choice, 100
Panel study, 153-154, 166
Parameter, population, 64
Parametric statistics, 94, 95
Partial correlation, 43, 179-180
Participant observation. *See* Observational research
Participants, focus group, 219-221
difficult, dealing with, 223
ensuring participation, 220-221
locating, 220
payment, 221
recruiting, 220
Pearson's r, 43, 79, 94, 106, 172-173, 178, 208
Percentile, 43
Periodicity in content analysis studies, 199-201
Phi, 43, 171
Plagiarism, 270, 294
Planning research projects, 39-58
experimental design, 124-132
focus groups, 219-226
logistics, 56
main points, 57-58
method, choosing, 53-54
methodology, development of, 54-55
observational research, 245-249
preplanning, 39-52
constraint identification, 41-44
deciding how to proceed, 51-52
level of analysis, 41, 42
literature review, 45-47
problem, definition of, 39-41
purpose, determination of, 41-44
refinement, 47-50
setting parameters, 39-44
Politics, Religion, and Media Use Survey 2000, 275-282
Polls. *See also* Surveys
push, 151
Population
census, 63
in content analysis, 197-198
defining appropriate, 63

Population *(continued)*
    sampling (*See* Sampling populations)
    target, 63
Positive-response bias, 96
Posttest-only control group design, 126
Precision, reliability and, 30-31, 106
Predictive validity, 110
Pre-experiments, 131-132
Primary sources, 45, 47, 272
Principle Component Analysis, 104
Privacy, 90. *See also* Anonymity;
    Confidentiality
    minimizing intrusions on, 265
    observational research and, 243-244
Probability sampling
    in multistage sampling, 76
    overview of, 64
    size of sample, 78-80
    theory, 65-68
    types of, 68-73
        cluster sampling, 77-78
        simple random, 68-70
        stratified random, 72-73
        systematic random, 70-71
Probe question, 160
Problem, defining research, 39-41
Problem statement, research report, 296
Procedural instructions, 163
Protocol, survey, 158-159
Pseudonyms, use of, 26, 244
Psychographic research, 150
Public Opinion Quarterly, 68
Purpose of study, determining, 41-44
Purpose statement, research report, 296
Purposive sampling, 74, 82
Push polls, 151

Q-methodology, 228-230, 232
Qualitative analysis
    content analysis, 201, 204, 208-209, 290-291
    focus group, 232-234
    observational research, 255-256
    quantitative compared, 14-15
    reporting findings of, 298
    survey, 181-185
Qualitative research. *See* Observational research
Quality control
    content analysis, 205-207
    experiment, 132-135
    focus groups, 230-231
    of measures, 104-110

observational research, 253-255
random error, 28-29
reliability, 29-31
surveys, 165-167
systematic error, 28-29
validity, 31-34
Quantitative analysis
    content analysis, 201-202, 204, 208, 284-290
    experimental data, 135-142
    focus group, 232
    observational research, 255
    qualitative compared, 14-15
    reporting findings of, 298
    surveys, 168-181
Question. *See also* Questionnaire; Surveys
    check, 160
    contingency, 160, 163
    double-barreled, 162
    double-blind, 162
    filter, 160, 163
    leading, 161
    open-ended versus fixed-response, 149, 163
    probe, 160
    research, 48-49
    screening, 77, 160
    sleeper, 160
    wording of, 160-162
Questionnaire. *See also* Surveys
    in Delphi technique, 228
    focus groups, 224-225
    preparation, 158-165
        helps, 163-164
        instructions, 163-164
        layout, 164-165
        length, 159
        order of questions, 162-163
        types of questions, 159-160
        wording of questions, 160-162
    pretesting, 166, 167
    reporting on research, 297
    in self-administered surveys, 157-158
Quota sampling, 73-74, 80, 81, 82

Random-digit dialing, 68-69, 155
Random error, 28-29
Randomization and experiment control, 122
Random-number table, 69, 70
Random sampling
    simple, 68-70
    stratified, 72-73
    systematic, 70-71

Range, 43
Ranking scales, 99-100
Ratio level measurement, 93, 94, 95
Reactivity effects, 254
Readability studies, 195-196
Reader's Guide to Periodic Literature, 198-199
Reading Ease Formula, 195, 196
Reasoning
 deductive, 12-14
 errors in, 8
 inductive, 13-14
Recommendations, business report, 300
Recordings, use of, 268
Recording variables, 139-140
References
 in business report, 301
 in research report, 299
Regression analysis, 43, 94, 181, 182-184, 208
Regression to the mean, 134
Reliability, 29-31
 coefficient, 206
 content analysis, 205-206
 experiment, 132-133
 focus group, 230-231
 of measures, 106-109
  alternate forms reliability, 107
  concurrent reliability, 106-107
  Cronbach's alpha, 107, 108
  intra- and intercoder reliability, 108, 206
  K-R 20, 108, 109
  Scott's pi, 108
  split-half reliability, 107
  test-item analysis, 109
  test-retest reliability, 107
 observational research, 253
 survey, 165
Repeated measure design, 131
Reports, research
 ethics, 27, 269-270
 evaluating, 271-274
  background check, 271-272
  of findings, 273-274
  for methodology, 272-273, 274
  usefulness, 274
 writing, 293-302
  abstract, 295
  business reports, formal, 299-301
  business reports, informal, 301-302
  conclusion, 298
  general principles, 293-294

 introduction section, 295-296
 journal style, 294-299
 literature review, 296
 methodology section, 296-297
 problem statement, 296
 purpose statement, 296
 references, 299
 results section, 297-298
Research arrangements, reactive effects of, 33
Research demand, 123
Research questions, 48-49, 272
Response rate, survey, 149, 156, 157-158, 166-167
Results section
 business report, 300
 research report, 297-298
Reverse directory, 78
Rewards for participation in surveys, 167
Rhetorical criticism, 197
Riffe, Daniel, 200
Role in observational research, 247-248
Rotation method, 104

Sample bias, 81-82
Sample size
 factors affecting, 78
 nonprobability samples, 80-81
 probability samples, 78-80
Sample weighting, 73
Sampling error, 65-68
 calculating for mean scores, 66
 sample size and, 67, 78-80
Sampling frame, 63-64
Sampling interval, 70
Sampling populations
 disproportionate likelihood sampling, 73
 multistage, 76-78
 nonprobability, 64, 73-76, 80-81
  sample size, 80-81
  types of, 73-76
   call-in polls, 75
   convenience sampling, 75
   mall intercept, 75
   purposive sampling, 74
   quota sampling, 73-74
   self-selection sampling, 75-76
   snowball (network) sampling, 74-75
   straw poll, 75
 probability, 64-73, 78-80
  overview of, 64
  size of sample, 78-80
  theory, 65-68

Sampling populations *(continued)*
   types of, 68-73
     cluster sampling, 77-78
     simple random, 68-70
     stratified random, 72-73
     systematic random, 70-71
   sampling frame, 63-64
   size of sample, 68, 78-81
   validation of sample, 81-82
Sampling screens, 77
Sampling units, 197, 201
Scales, 95-100
  Guttman, 98-99
  Likert, 96
  ranking, 99-100
  semantic differential, 97-98
  Stapel, 97
  thermometer, 99
  Thurstone, 98
Scalogram analysis, 98
Scanning documents, 204
Scheffé test, 140
Science
  naive versus true, 7
  as a way of knowing, 7-10
Scientific research, 7-10. *See also* Social science research
  issues in, 12-15
Scott's pi, 108, 206
Screener question, 160
Screens, 77
Searching materials in content analysis, 198-199, 203-204, 209
Secondary analysis of data sets, 51
Secondary sources, 45, 47, 272
Segmentation studies, 150-151
Selection
  external validity, threat to, 32
  internal validity, threat to, 32, 127
Self-reports, 149. *See also* Surveys
  as operational procedure, 90-91
Self-selection sampling, 75-76, 80, 82
Semantic differential scale, 97-98
Separate-sample pretest-posttest designs, 130, 153
Setting, observational research
  choosing, 245-246
  evaluating, 246
  gaining access to, 246-247
Significance, statistical, 168, 173
Simmons Market Research Bureau (SMRB), 151
Simple random sampling, 68-70, 79

Site for observational research, 245-247
Situational analysis, 41
Skip interval, 70-71
Sleeper question, 160
Snowball sampling, 74-75, 81, 82, 154
Social science research, 3-16
  basic versus applied, 14
  data analysis, 14-15
  deductive versus inductive reasoning, 12-14
  main points, 15-16
  nature of, 10-12
  need for, 3-5
  ways of knowing
    errors in reasoning, 8
    science as, 7-10
    traditional, 5-6
Solomon four-group design, 126, 127, 133, 134, 138
Sorting
  forced, 229-230
  in Q-methodology, 229-230, 232
  unforced, 229-230
Spearman's rho, 43, 172, 178
Split-half reliability, 107
Sponsors
  access from, 247
  ethics and, 26-27
Stance in observational research, 247, 248-249
Standard deviation, 43, 65-66
Stanley, Julian, 31, 125
Stapel scale, 97
Starch test, 150
Static group comparison, 132
Statistic, population, 64
Statistical control, 122
Statistical regression as internal validity threat, 32
Statistics. *See also specific statistics*
  categories of, 43
  level of analysis and choice of, 94
  for reliability assessment, 108
  sample size requirements, 79
Strategic research, 41
Stratified random sampling, 72-73, 80
Straus, Anselm, 13
Straw poll, 75
Subjects
  ethical treatment of, 25-26
  focus groups, 219-221
Subjects, choosing, 54-55, 63-83
  main points, 82-83

sampling populations
   multistage, 76-78
   nonprobability, 64, 73-76, 80-81
   probability, 64-73, 78-80
   sampling frame, 63-64
   size of sample, 68, 78-81
   validation of sample, 81-82
Summative research, 41
Surveys, 53, 149-187
   communications applications, 150-151
   data analysis, 167-185
      qualitative, 181-185
      quantitative, 168-181
         association, measures of, 168, 170-173, 178-179
         bivariate statistics, 168, 173-175
         cross-tabulation table, 173-174, 175-178
         multivariate statistics, 175, 179, 181
         regression analysis, 181, 182-184
         univariate statistics, 168
   data collection techniques, 155-158
      face-to-face, 156-157
      self-administered, 157-158
      telephone, 155-156
   design, true survey, 152-154
      cross-cultural designs, 153
      longitudinal designs, 153-154
   ethics, 151-152
   introductory statement, 151-152
   limitations of, 149
   main points, 186-187
   panel study, 153-154
   Politics, Religion, and Media Use Survey 2000, 275-282
   quality control, 165-167
      external validity, 166-167
      internal validity, 166
      reliability, 165
   quasi-surveys, 154-155
      interview surveys, 155
      nonprobability sample surveys, 154-155
   questionnaire preparation, 158-165
      helps, 163-164
      instructions, 163-164
      layout, 164-165
      length, 159
      order of questions, 162-163
      types of questions, 159-160
      wording of questions, 160-162
   reporting on, 297
   response rate, 149, 156, 157-158, 166-167
   trend study, 153

Systematic error, 28-29
Systematic random sampling, 70-71, 199

Target population, 63
Telemarketing, 151
Telephone call-in polls, 76, 80
Telephone survey, 155-156, 159, 165, 167
   random-digit dialing, 68-69
Tenacity, 5-6
Testing
   external validity, effects on, 33
   internal validity, effects on, 32, 127
   reactive effects of, 33
Test-retest reliability, 107
Textual analysis, 196-197
Theory, role in research, 12-14
Thermometer scale, 99
Third variable problem, 44
Thurstone, L. L., 98
Thurstone scale, 98
Time series, 130-131, 138
Tracking studies, 42, 131
Transitional instructions, 164
Trend studies, 153
Triangulation, 255
Truth, 5-6
T-test, 43, 136-138, 208
Tukey test, 140
Two-tailed (nondirectional) hypothesis, 48
Types of studies, 41-42, 44

Unforced sort, 229-230
Unidimensional concepts, 100
Unidirectional hypothesis, 49
Univariate statistics, 43, 94, 168, 208
Universe, 63
University of Michigan Survey Research Center, 99
U.S. Bureau of the Census, 51
Usability studies, 42, 120, 132

Validity
   causal, 31
   concurrent, 110, 207
   content analysis, 206-207
   experiment, 133-135
   external, 31, 32-33, 34
   face, 110, 207
   factors affecting, 32-33
   focus group, 231
   internal, 31-34
   jury, 110
   measurement, 31

Validity *(continued)*
  of measures, 109-110
    construct validity, 110
    content validity, 110
    criterion validity, 110
  observational research, 253-254
  predictive, 110
  survey, 153, 166-167
Variable
  antecedent, 49-50
  confounding, 120, 121-122
  control, 50
  dependent, 49-50
  dummy, 94, 181, 182, 185, 208
  independent, 49-50
  intervening, 50
  recording, 139-140
Varimax, 104
Videoconference, 230
Videotaping, 251-252
  focus group sessions, 226

Writing research reports,
  293-302

Z-Score, 43